Fluidization and Multiphase Flow 2018

Topical at the 8th World Congress on Particle Technology

Orlando, Florida, USA
22 - 26 April 2018

ISBN: 978-1-5108-6979-0

Fluidization and Multiphase Flow 2018

Topical at the 8th World Congress on Particle Technology

Orlando, Florida, USA
22 - 26 April 2018

AIChE
The Global Home of Chemical Engineers

Printed from e-media with permission by:

Curran Associates, Inc.
57 Morehouse Lane
Red Hook, NY 12571

Some format issues inherent in the e-media version may also appear in this print version.

Printed by Curran Associates, Inc. (2018)

AIChE
120 Wall Street, FL 23
New York, NY 10005-4020

Phone: (800) 242-4363
Fax: (203) 775-5177

www.aiche.org

Additional copies of this publication are available from:

Curran Associates, Inc.
57 Morehouse Lane
Red Hook, NY 12571 USA
Phone: 845-758-0400
Fax: 845-758-2633
Email: curran@proceedings.com
Web: www.proceedings.com

TABLE OF CONTENTS

Some abstracts in this book refer to images that were not made available by the publisher

507304 A Hydrodynamic Study of Subway Grating and Disk and Donut Trays in a 0.6-m Diameter Fluidized Bed Stripper

Abstract:

Fluidized bed strippers are flowing fluidized beds where an up-flowing gas removes (strips out) a gaseous product entrained in a down-flowing stream of the emulsion phase. The stripper has to facilitate the transfer of the gas in the interstices of the emulsion phase and that adsorbed on the particles into the bubbles of the stripping gas. Major industrial applications of fluidized bed strippers are in fluid catalytic cracking (FCC) and fluid coking processes.

The counter-current contacting between the down-flowing emulsion phase stream and the up-flowing bubbles stream in fluidized bed strippers is enhanced by the use of an assembly of horizontal or inclined baffles. The baffles are important because in un-baffled fluidized beds, the movement of the bubbles is unrestricted. So, as they rise, interact, coalesce and grow in size they tend to move laterally toward the center or, in some instances, the bubbles can consolidate in a jet stream that bypasses significant portions of the bed in what has been termed gas bypassing. Therefore, open beds would give too low of a mass transfer to be effective strippers. Installing baffles enhances gas-solids interaction by controlling the bubble size and improving bubble distribution over the stripper cross section. As is typical for two-phase countercurrent flow unit operations, fluidized bed strippers can also suffer from flooding at relatively high solids mass fluxes.

Literature shows that disk and donut trays give excellent stripping efficiencies. They are among the most widely-used fluidized bed stripper internals in the fluid catalytic cracking (FCC) process. Disk and donut internals are, however, expensive and prone to flooding. This study compares the hydrodynamics of simple and cheaper subway grating trays with disk and donut trays in a 0.6-m-diameter fluidized-bed stripper. The stripper column was 7.6 m tall and the test solids were FCC catalyst particles containing 8% fines, less than 44 microns. Solids flowing out of the stripper were transferred by a 0.25-m-diameter, 3.4-m long vertical standpipe to a 15-m tall, 0.3-m-diameter riser that conveyed the solids back to the stripper via primary and secondary cyclones. The measurements performed in the study included differential pressure and local bubble properties.

Results showed that the subway grating stripper built a higher-pressure head, operated more smoothly without flooding over a wider range of operating gas velocities and solids fluxes compared to the disk and donut stripper. Literature studies showed, however, that the stripping efficiency of subway grating tray strippers is significantly lower than that of disk and donut strippers. Therefore, while hydrodynamics suggests a superior subway grating stripper, the future challenge will be to find ways of raising its stripping efficiency. Installing more subway grating trays in the same vertical height will most likely improve stripping.

506869 Exceptional Intruder Sphere Sinking Due to Local Fluidization of Apparently Fixed Powder Bed

Abstract:

It is well known that a powder bed can be fluidized when the velocity of an airflow through the bottom of the bed is higher than the minimum fluidization velocity u_{mf}, and that the powder bed is a fixed bed when the air velocity is smaller than u_{mf}. The fluidized bed has liquid like properties such as density and viscosity. When an intruder sphere is put into the fluidized bed, the sphere floats or sinks based on the difference between the fluidized bed density and the sphere density. The float-sink can be utilized for density separation; the continuous separators are commercialized for waste treatments in Japan. Fundamentals of the fluidized bed, such as properties of air bubbles and convective flow of particles, have been widely investigated by many researchers. Generally, of course, the air velocity $u_0 > u_{mf}$ is employed for the fluidized bed research. We have been focusing on the spheres' float-sink in the fluidized bed. During the float-sink experiments, we accidentally found that spheres can sink even if u_0 is smaller than u_{mf}. Therefore, we investigated the sphere sinking in the powder bed for $u_0 < u_{mf}$. We varied the air velocity ratio u_0/u_{mf} and the density ration ρ_{sphere}/ρ_{bed} (sphere density/powder bed density). The spheres sinking was observed in the range $0.83 < u_0/u_{mf} < 1.0$, indicating that the powder bed behaves as a solid for $u_0/u_{mf} \leq 0.83$ and as an apparently fixed bed for $0.83 < u_0/u_{mf} < 1.0$. The final depth of sphere sinking increases with increasing the sphere density in the range $1.3 < \rho_{sphere}/\rho_{bed} < 2.6$. However, surprisingly, smaller density spheres sink "deeper" than larger density spheres in the range $1.0 < \rho_{sphere}/\rho_{bed} < 1.3$. We recorded the state of the top of the power bed by a video camera. When $\rho_{sphere}/\rho_{bed} < 1.3$, air bubbles are observed to percolate from the powder bed surface at the location of the sinking sphere, whereas for $\rho_{sphere}/\rho_{bed} \geq 1.3$, such percolation bubbles are not present. The rising air stream can cause *local fluidization* around the sphere when the sphere density is close to the powder bed density. As a result, air voids can then form underneath the sphere and migrate to the powder bed surface where they are observed as emerging percolation bubbles. As the sphere density ρ_{sphere}/ρ_{bed} decreases from 1.3 to 1.0, the decreasing weight of the sphere allows larger regions of local fluidization to develop. This creates larger voids that allow the sphere to sink deeper in spite of the decreasing sphere density.

508575 Structured Flow of Bubbles in Pulsed Fluidized Beds: Pattern Stabilization and Propagation

Abstract:

Gas-solid fluidized bed reactors are widely applied in different industrial processes, such as chemical, environmental and pharmaceutical industries, where efficient contacts between solid and gas phases are critical. The solids mixing and transport process are highly dependent on the bubble dynamics. Nevertheless, the nonlinear collective behavior arising from the dissipative interparticle collisions, and the seemingly chaotic coalescence and the breakup of bubbles give rise to complex and unsteady bubbly flows which are hard to predict, challenging every aspect of the optimization, control, design, and scale-up of gas-solid fluidized bed reactors [1].

Over the past years, lots of efforts have been made to manipulate the bed hydrodynamics via introducing additional degrees of freedom. Some intrusive methods, such as fractal injectors and internal baffles, are imposed to suppress the chaos of bubble dynamics and channeling effects [2,3]. Among those, supplying a periodically pulsed gas flow has been demonstrated as an efficient non-intrusive method to structure gas-solid fluidized beds. When subjected to a periodic pulsed flow, gas-solid quasi-2D fluidized beds can emerge a structured flow of bubbles, where gas bubbles rise in a staggered fashion with a characteristic wavelength, forming regular hexagonal patterns. Under certain operating conditions, changing the inlet pulsed flow profile could simply customize the bubble size and its spatial distribution while maintaining the ordered structure [3,4].

Although pattern formation provides great potentials of engineering fluidized bed reactors, its fundamental aspects remain largely unknown. A few computational attempts have been conducted to model this phenomenon, but none of them have yet captured the regular pattern of bubbles convincingly [5,6]. Only recently, we reported the first successful simulation reproducing the dynamics of pattern formation with a CFD-DEM (computational fluid dynamics-discrete element method) modeling approach [7]. These numerical results provide valuable physical insights into the fundamentals of fluidized bed dynamics.

This contribution investigates the experimental observations on the formation and the stability of regular patterns in quasi-2D pulsed fluidized beds, subjected to various conditions. We discuss the response of pulsed bed dynamics to different pulsed flows, particles shape, static bed heights, and we also report a phase diagram demonstrating the favorable operating conditions for pattern formation. By comparing the experimental results with CFD-DEM computational modeling, we also study how the patterns propagate and being disturbed with bubbles rising towards the top surface. These results are valuable from a fundamental point and provide great insights into the scale-up applications as well.

[1] S. Pannala, M. Syamlal, T. J. O'Brien, Computational Gas-Solids Flows and Reacting Systems: Theory, Methods and Practice, IGI Global, Hershey, 2011.

[2] J.R. Van Ommen, J. Nijenhuis, M.-O. Coppens,Reshaping the structure of fluidized beds, Chem. Eng. Prog. (2009) 49-57.

[3] M.-O. Coppens, J.R. Van Ommen, Structuring chaotic fluidized beds, Chem. Eng. J. 96 (2003) 117-124.

[4] M.-O. Coppens, M.A. Regelink, C.M. van den Bleek, Pulsation induced transition from chaos to periodically ordered patterns in fluidised beds, Proceedings of the 4th World Congress on Particle Technology, Sydney, 2002.

[5] K. Wu, L. de Martín, L. Mazzei, M.-O. Coppens, Pattern formation in fluidized beds as a tool for model validation: A two-fluid model based study, Powder Technol. 295 (2016) 35-42.

[6] X. Wang, M. Rhodes, Pulsed fluidization—a DEM study of a fascinating phenomenon, Powder Technol. 159 (2005) 142-149.

[7] K. Wu, L. de Martín, M.-O. Coppens, Pattern formation in pulsed gas-solid fluidized beds – the role of granular solid mechanics, Chem. Eng. J. (2017) (in press).

509583 Intermediate Regime in Particle-Fluid Flow

Abstract:

Simulations of the multi-phase flow in fluidised beds are presented. Some of the results rely on a model describing the solid stresses in particle laden flow. The model is shown to be valid for slow and intermediate flows and is developed following the theory developed by Tardos et al. (2003). The study extends previous work by Makkawi et al. (2006), where a steady one-dimensional particle flow model was used to demonstrate the smooth merge of quasi-static flow with â €œliquid-likeâ€ behaviour. The work aims at studying the solid stress in conjunction with a number of numerical codes. Validation is then undertaken using electrical capacitance tomography (ECT) measurements.

The time series data of solid fraction fluctuation, averaged within 14.5-18.5 cm above the distributor level, at 3 different fluidisation velocities are studied. In addition, comparisons are undertaken using different simulation software, namely OpenFOAM, Fluent and MFix.

The comparisons undertaken are used to detect which modelling framework and which software are more suitable to represent the behaviour of particle laden flow. The merge between flow regimes and the characteristics of the intermediate regime bring to the conclusion that a more complex constitutive law for the solid stresses is needed.

References

[1] G.I. Tardos, S. McNamara, I. Talu, Powder Technology. 2003;131(1):23-39.

[2] Y.T. Makkawi, P.C. Wright, R. Ocone, Powder Technology. 2006;163(1):69-79.

508334 Modelling of Chemical Reactions in Metallurgical Processes

Abstract:

Developing ways for the direct reduction of iron ores has attracted much research interest in the last three decades, since it can be considered as a core process in steel industry [1]. New advances in the iron making technologies offer up potential savings in energy and emissions. One of the most advantageous direct reduction processes are fluidized bed reactors. However, due to the harsh conditions inside these reactors, most investigations are carried out through computational tools. One such tool is the CFD-DEM method, in which the gas phase reactions and governing equations are calculated in the Eulerian (CFD) side, whereas the particle reactions and equations of motion are calculated in the Lagrangian (DEM) side.

In this work, two of the most common types of models to represent the reaction between solid particles submerged in fluids are implemented into the CFD-DEM library. These models are the Shrinking Particle Model (SPM), where the solid particle reacts with the fluid thereby changing its size, and the Unreacted Shrinking Core Model (USCM), where after the particle reacts with the fluid a product layer is formed around the particle that impedes the reaction rate [2].

The communication framework between the CFD and DEM sides are initially verified with the use of SPM. A case where a single particle reacts with the incoming fluid is carried out and the species mass balances are verified. In the metallurgical process of iron-ore reduction the oxygen is extracted from the iron-oxide through intermediary steps, therefore the SPM becomes insufficient. In this case, the different reaction steps of iron-ore reduction needs to be considered which leads to using the three-layered USCM representation. The layers that need to be considered are the Fe (Iron-Layer), FeO (Wüstite), Fe_3O_4 (Magnetite) and Fe_2O_3 (Hematite), which is the innermost core of the particle. The gaseous reactant diffuses through the porous layers of iron shell, wüstite and magnetite layers, while simultaneously a portion of the diffused gas reacts at the layer interfaces. Finally, the reactant gas reaches the hematite core and reacts with the core surface producing magnetite.

For the investigation of metallurgical processes, both of the mentioned models is implemented into the CFD-DEM library in such a manner that only the required data is communicated between the two phases with an adaptable communication interval. A similar case as with the SPM is also carried out with the implemented USCM for an iron-ore. The overall reduction rates are validated by comparing them to available literature. After the initial validation cases with only a single particle, the model capabilities are demonstrated with a lab-scale fluidized bed where various reactant gases such as hydrogen, carbon monoxide and a mixture gas of H_2 and CO are utilized. The simulation results are in good agreement with the fractional reduction obtained from the experiment. These results give us further insights about the general behavior of the reducing and product gases, the gas ratio, and temperatures as well as the effects that particle properties such as the particle porosity, diameter and pore size has on the rate of reduction.

507932 Modeling Mass Transfer in Hydraulic Conveying of Salt Particles Using CFD-DEM Numerical Simulation

Abstract:

Hydraulic conveying of particulate solids is a common transportation method widely used in various fields and industries, such as the energy, pharmaceutical and agricultural. In most of hydraulic conveying systems there is neither mass transfer interaction nor dissolution of the particulate solids into the conveying fluid phase, mainly due to design considerations and the fact that the solids production processes are commonly separated from the transportation procedures. Nonetheless, there are processes which involve dissolution of the solid into the continuous phase, and result in a significant change in the two phase flow dynamics. One example which is being investigated in this research is the conveyance of salt particles via an unsaturated brine mixture.

The studied phenomenon in this research is the dissolution of the solid particles into the unsaturated brine which occurs during hydraulic conveying of salt particles. The observed multiphase system is constantly prone to create a thermodynamic balance, which derives a spontaneous dissolution of the solid salt into the mixture. The aim of the current study is to investigate this phenomenon via combined CFD-DEM numerical simulations. The mass transfer between the phases and its respective influence on the flow characteristics throughout these types of hydraulic conveying flows are the focus of our investigation.

Our multiphase flow system was modeled using an Euler-Lagrange model. The conveying brine takes place as a multi-species Euler continuum which its physical and chemical properties are described by various models taken from past works. The particulate NaCl matter is represented by a lagrangian discrete phase where each particle momentum, heat and mass transfer equations are considered. As for the computational implementation of the model, ANSYS Fluent was used for the computation of the Eulerian phase with an addition of User defined functions for all properties, such as viscosity, thermal conductivity and density. As for the discrete phase, our in-house BGU-DEM code [1], [2] integrated to Fluent computational scheme also by a set of user defined functions, and a full coupling via source terms of mass, momentum and energy are considered.

Three mass transfer models where investigated, and one model proposed by [3], was chosen for the hydraulic flow. The model was validated using experimental data collected by [4]; the simulations showed very good agreement with the experiments and approved the multiphase model as proper-based tool to describe dynamic flows which are characterized by dissolution of particles.

The main benefit from the Lagrangian representation comes with the attendance of a discrete approach which enable us to track each particle in the flow individually, and by that to extract additional dynamic and kinetic data about the physical problem we choose to simulate. A lot of research in the field of mass transfer in a solid-liquid interface was conducted and exhibited in past works. Also, many works have presented an investigation and characterization of particulate flows with an Euler-Lagrange approach. But very few works involved the dissolution of particles in a flow, and its study via a CFD-DEM simulation as done in our work. The conduction of this study unveils an innovative approach which presents a full coupling and the individual influence of each particle being dissolved over the flow characteristics and vice versa.

Additionally, we obtained discrete dynamic data on the particles longitudinal and radial concentration, diameter distribution and variation, linear and angular velocity etc. which are all under the direct and indirect influence of the present mass transfer.

Furthermore, the numerical model was compared with experimental data collected in lab, of various static dissolution experiments of small NaCl particles (2-5 mm). The model shows a substantial over â€“estimation of the dissolution. One main explanation is the dynamical terms which the mass transfer models used in the simulations were derived from, that constrained our model to dynamical dissolution problems only, being investigated and presented in future research.
Reference

[1] T. Brosh and A. Levy, "Modeling of Heat Transfer in Pneumatic Conveyer Using a Combined DEM-CFD Numerical Code," *Drying Technology,* vol. 28, no. 2, pp. 155-164, 2010.

[2] T. Brosh, H. Kalman, A. Levy, I. Peyron and F. Ricard, "DEMâ€“CFD simulation of particle comminution in jet-mill," *Powder Technology,* vol. 257, pp. 104-112, 2014.

[3] G. A. Aksel'rud, A. E. Boiko and a. A. E. Kashcheev, "Kinetics of the solution of mineral salts suspended in a liquid flow," *Journal of engineering physics,* vol. 61, no. 1, p. 885â€“889, 1991.

[4] L. Calcada, L. Martins, C. Scheid, S. Magalhaes and A.L.Martins, "Mathematical model of dissolution of particles of NaCl in well drilling: Determination of mass transfer convective coefficient," *Journal of Petroleum Science and Engineering,* vol. 126, pp. 97-104, 2015.

509633 Bubbling and Slugging of Geldart Group a Particles in Small Diameter Columns: Experiments and 3D Numerical Simulations

Abstract:

In order to increase Concentrated Solar Power (CSP) plants efficiency, current research aims at increasing operating temperatures of the heat transfer fluid, and consequently increasing the efficiency of the power cycle. In addition, heat transfer fluids classically used for CSP have several limitations: restricted range of operating temperature, safety issues, corrosion and maintenance costs. Since solid particles are not subject to such limitations, provide a good thermal capacity, and a low-cost heat transfer and storage; an innovative alternative is to use an air-fluidized dense particle suspension as the heat transfer fluid, also called Upflow Bubbling Fluidized Bed (UBFB) [1][2]. The Next-CSP (http://next-csp.eu) European project aims at improving the reliability and performance of CSP plants with this new technology, by demonstrating the concept feasibility with a $4MW_{th}$ working pilot, and upscale it to the industrial scale by designing a $150MW_{el}$plant.

The key to the proposed process is the effective heat transfer from the solar heated surfaces to the heat transfer fluid, which is directly linked to the gas-solid hydrodynamics. Moreover, the use of a UBFB receiver limits the operable particles down to Geldart group A or near-B particles because of their low minimum fluidization velocity: since the fluidization air will exhaust the UBFB at the bed temperature, sensible heat loss increases significantly with increasing air velocity [3].

Heat Transfer Coefficient (HTC) was measured up to 400-1100 W/m^2K during on-sun experiments with small 1m height receiver tubes [4] while industrial scale receiver tubes will reach up to 10m height. In addition, hydrodynamics in the UBFB are expected to be strongly dependent on the tube height. Indeed, it has been shown that common unrestrained bubbling can be transformed into slugging when the bubble size approaches the size of the tube [5], which is known to strongly reduce the HTC.

In order to better understand the process and to optimize the design of the solar receiver it is of paramount to know how particles behave inside the bundle of small tubes within it and consequently being able to limit the formation of slugs.

Experimental measurements of fluidization of Geldart group A powder in small internal diameter tubes were conducted [5]. Then three-dimensional numerical simulations were carried out using the unstructured parallelized multiphase flow code, NEPTUNE_CFD, based on a Eulerian n-fluid modelling approach, but first simulations had trouble to replicate the experimentally measured bed height due to an inaccurate inclusion of the Particle Size Distribution (PSD) span and particles non-sphericity into calculation model [6].

In this paper, a new methodology was developed to include the impact of the PSD span into the expression of the drag force. In addition, the impact of particles irregular non-spherical shape [7] was also taken into account into the drag force term calculation. Both these corrections were based onto experimental measurement of morphology and laser diffraction particles size analysis. Also, the impact of non-sphericity onto particles-particles collisions and wall-particles collisions was evaluated.

Those three-dimensional numerical simulations were compared with experimental measurements of bed expansion, slugs size and frequency, and with high speed imaging analysis of the particles local motion. Fluidization velocities ranged from $7xU_{mf}$ to $56xU_{mf}$, corresponding to the velocity range reached in the industrial scale receiver. Therefore numerical models were found in good agreement toward experiments concerning their ability to represent wall effects and slugging.

Acknowledgements

This project has received funding from the European Union's Horizon 2020 research and innovation program under grant agreement No 727762, Next-CSP project. This work was granted access to the HPC resources of CALMIP under the allocation P1132 and CINES under the allocation gct6938 made by GENCI.

REFERENCES

[1] Benoit, H., Spreafico, L., Gauthier, D. & Flamant, G., 2016. Review of heat transfer fluids in tube receivers used in concentrating solar thermal systems: Properties and heat transfer coefficients. Renew. Sustain. Energy Rev. 55, 298–315.

[2] Zhang, H., Benoit, H., Gauthier, D., Degrève, J., Baeyens, J., Pérez, López.I., Hemati, M. & Flamant, G., 2016. Particle circulation loops in solar energy capture and storage; gas-solid flow and heat transfer considerations. Appl. Energy 161, 206–224.

[3] Boissière, B., 2015. Étude hydrodynamique et thermique d'un nouveau concept de récepteur solaire à suspensions denses gaz-particules (Hydrodynamic and thermal study of a new concept of solar receiver using dense gas-particle suspensions). PhD thesis, Institut National Polytechnique de Toulouse (INP Toulouse).

[4] Benoit, H., López, P. I., Gauthier, D., Sans, J. L., & Flamant, G., 2015. On-sun demonstration of a 750°C heat transfer fluid for concentrating solar systems: Dense particle suspension in tube. Solar Energy, 118, 622–633.

[5] Kong, W. B., Tan, T. W., Baeyens, J., Flamant, G. & Zhang, H. L., 2017. Bubbling and Slugging of Geldart Group A Powders in Small Diameter Columns. Ind. Eng. Chem. Res. 56, 4136–4144.

[6] Ansart, R., García-Triñanes, P., Boissière, B., Benoit, H., Seville, J. P. K., & Simonin, O., 2017. Dense gas-particle suspension upward flow used as heat transfer fluid in solar receiver: PEPT experiments and 3D numerical simulations. Powder Technology, 307, 25–36.

[7] Bagheri, G. H., & Bonadonna, C., 2016. On the drag of freely falling non-spherical particles. Powder Technology, *301*, 526–544.

EFFECT OF FUEL HEAP SHAPE ON AIR FLOW IN A SIDE-FED BIOMASS GASIFIER BASED ON DEM-CFD SIMULATION

Yuan Tan, Technical University of Munich, Chair for Materials Handling, Material Flow, Logistics, Garching, Germany
Michael Rackl, Technical University of Munich, Chair for Materials Handling, Material Flow, Logistics, Garching, Germany
Johannes Fottner, Technical University of Munich, Chair for Materials Handling, Material Flow, Logistics, Garching, Germany

Abstract

In recent years, because of the rapid reduction of other conventional energy and growing earth population, biomass usage has increased considerably. As a promising renewable energy technology, biomass gasification offers the highest efficiency among all biomass conversion routes. The subjects in this paper are updraft, side-fed and fixed bed gasifiers based on the method of air supply and the position of the feeding system. It is assumed in most research that the fuel bed's upper surface is even and smooth. In this case, simulative investigation of gasification with idealized conditions may differ from real applications. Even so, studies with respect to the effect of slope form on air flow within the fuel heap cannot be found in published research.

To investigate the effect of the fuel heap shape on local gasification conditions, a small-scale fixed-bed updraft gasifier was modeled. The simulation model consists of a cylindrical container as the gasifier, with a default lateral feeding position and air supply from below. Based on DEM/CFD co-simulations in two open source software, LIGGGHTS® and OpenFOAM®, the air flow was examined.

The simulation results were investigated qualitatively and quantitatively in terms of air flow pattern, including the uneven pressure and temperature distribution at/over the slope surface. Besides, the effects of operating parameters of the feeding system have been highlighted with respect to modification of the slope form. The results showed that the combustion efficiency at the side of the feeding system is the lowest due to insufficient air flow. According to these first investigations, real fuel heap shape should be taken into consideration for the simulative study of gasification affected by air flow in the future.

Keywords: Biomass gasification, Side-fed and updraft gasifier, Fuel heap shape, CFD-DEM

1. Introduction

Biomass boilers for heat generation in private households (up to max. 30 kW) are often designed as updraft gasifiers. In contrast to direct combustion, biomass is stacked up as a fixed bed, volatilized by thermal energy and the product gas ignited, as displayed in Fig. 1. While biomass is combusted in the lower end of the furnace, pyrolysis and oxidation are forced to take place as a result of the secondary air supply and set heat (energy) free. Volatilization of biomass leads to higher energetic efficiency than direct combustion. However, in the case of updraft gasifiers with secondary air supply, it is impractical to feed fuel from above [1], since the fuel stream would interfere with the secondary combustion zone above the secondary air inlet. As Fig. 1 depicts, residential-size biomass updraft boilers are typically fed from the side by means of a single stoker screw [2].

In state-of-the-art works, particle-related thermo-chemical processes in biomass boilers are simulated by means of coupled discrete element method (DEM) and computational fluid dynamics (CFD) models [3, 4, 5]. The combined DEM/CFD approach also allows for implementation of thermo-chemical reaction equations and subsequent simulation of flue gas composition, local concentrations of CO_2 and O_2 or reaction temperatures and overall boiler efficiency.

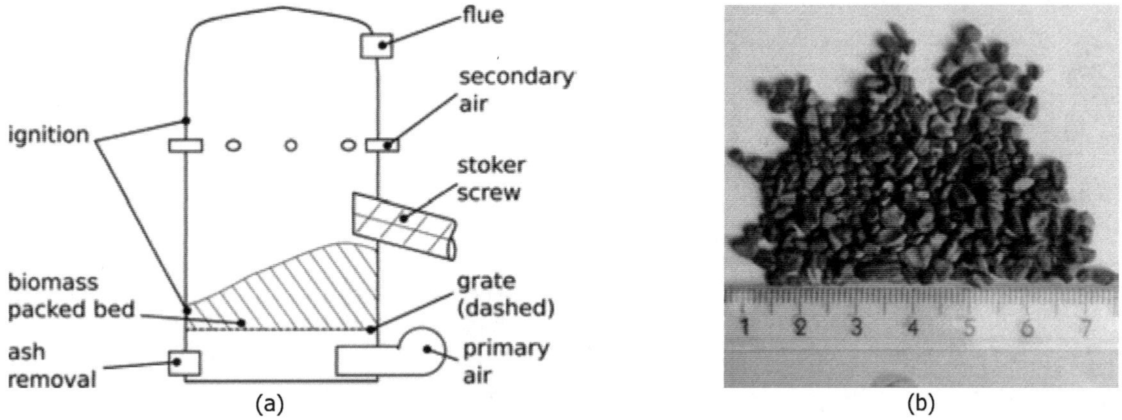

Fig. 1 (a) Sketch of a small-scale biomass boiler and (b) image of olive stones investigated

While simulations and modeling techniques have become more and more complex and today also feature particle shrinking as these are combusted [6, 7], many studies assume that the initial biomass fixed bed's height is constant over the grate. In a recent study, Wiese et al. [6] explicitly included the fixed bed build up caused by a lateral feeding device in their simulation, which resulted in an uneven fixed bed height profile on the grate. The group demonstrated that this has a considerable influence on the overall boiler efficiency and combustion/volatilization process and suggested that the air status above the fixed fuel bed should be investigated in future works.

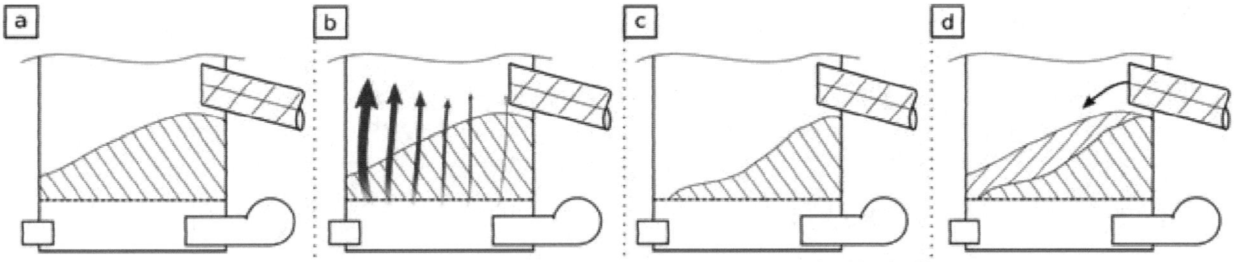

Fig. 2: Prospective burning process under asymmetric fuel bed effect

As demonstrated in [6], an asymmetric fuel bed affects the combustion process such that fuel is not combusted evenly. Fig. 2a-d depicts the lower part of the side-fed gasifier from Fig. 1a. In Fig. 2a the fuel bed has been built up by running the stoker screw feeder, which results in an uneven fuel distribution across the grate. Fig. 2b shows the streamlines of air induced by the primary air fan. The streamlines' line thickness indicates their respective air velocity. This velocity pattern is expected, since the higher bed height on the right corresponds to a larger aerodynamic resistance and thus slows down the stream of air. Given that the combustion/volatilization process strongly depends on oxygen supply, the fuel on the left will be turned over more quickly and a fixed bed shape as shown in Fig. 2c follows. When more fuel is added to the fixed bed (Fig. 2d), it will slide down the remains of the present fuel and form a distribution, similar to the initial shape (Fig. 2a), with partially converted fuel at the bottom right.

The DEM/CFD-based study here deals with the formation of the fixed fuel bed in laterally fed updraft biomass gasifiers and its effect on the air velocity distribution, temperature as well as pressure drop over the grate. Additionally, the inlet air velocity was taken into discussion, which is considered as another decisive parameter during ignition process in the boiler.

Nomenclature

d_p	particle diameter	R_{pf}	fluid-particle interaction force caused by variations in the stress tensor ξ
f_d	drag force	t	time
$f_{\nabla p}$	pressure gradient force		
f_{pf}	fluid-particle interaction force on particle		*Greek symbols*
f_{pp}	particle-particle interaction force on particle	ρ_f	fluid density
f_{pw}	wall-particle interaction force on particle	Re_p	Reynolds number
$f_{\nabla\cdot\tau}$	viscous force	u_f	particle velocity
g	gravitational acceleration	u_p	particle velocity
ΔL	length of granular packed bed	φ	sphericity of part
m_p	particle mass	ε_f	volume fraction of fluid
Nu	Nusselt number	ε_p	volume fraction of particle
p	fluid pressure	ξ	fluid stress tensor
Pr	Prandtl number	τ	shear stress
Re_p	Reynold number		

2. Computational methods

The governing theories for DEM/CFD co-simulation and the following modeling, which were applied in this study, are described in this section.

2.1 DEM/CFD co-simulation setup

DEM/CFD co-simulation belongs to the so-called four-way coupling methods [8], which include intersection between particles and fluid (two-way), particle swarm (local effect on the motion of particles around), and particle collision with solid wall. The primary mathematical expressions were concluded by [9] and [10] and are listed in Equations 2-1 and 2-2 (for gas state) and Equations 2-3 and 2-4 (for solid state).

Tab. 1: Governing mathematical equations for DEM/CFD co-simulation.

Equation	Ref.	
$\frac{\partial \varepsilon_f}{\partial t} + \nabla\cdot(\varepsilon_f u_f) = 0$	[9]	Eq. 2-1
$\frac{\partial(\rho_f \varepsilon_f u_f)}{\partial t} + \nabla\cdot(\rho_f\varepsilon_f u_f u_f) = -\nabla p + \nabla\cdot\tau -\cdot(\frac{R_{pf}}{\varepsilon_f} - \rho_f\varepsilon_p g) + \rho_f\varepsilon_f g$	[9]	Eq. 2-2
$m_p \frac{du_p}{\partial t} = m_p g + f_{pf} + \Sigma_{N_p} f_{pp} + \Sigma_{N_w} f_{pw}$	[10]	Eq. 2-3
$f_{pf} = f_d + f_{\nabla p} + f_{\nabla\cdot\tau}$	[10]	Eq. 2-4

2.2 Fuel data and modeling

The fuel material investigated was olive stones, a renewable fuel, which can be turned into bioethanol and produced in plant. Some researchers have reported that, because of the low cost for transporting and transforming olives stones into biofuel, it should be used as an alternative to petrol or diesel [11]. A photograph of this kind of fuel is shown in Fig. 1b. It can be seen that the size distribution is almost uniform. For DEM/CFD co-simulation, the particles were modeled as spheres with an upscaling diameter of 10 mm, based on measurement of olive stone samples from Spain and corresponding calibration. If particle dimensions would match the real olive stones, computational effort would be too large to obtain results within reasonable amounts of time. The DEM/CFD co-simulation in this study was carried out using the method proposed by LIGGGHTS® (LAMMPS improved for general granular and granular heat transfer simulations) and OpenFOAM® (Open-source field operation and manipulation). The employed simulation parameters, such as key material properties, boundary conditions and convective/conductive heat transfer coefficient, are summarized in Tab. 2. Two of them, i.e. thermal conductivity and thermal capacity, were adopted from previous studies [12, 13]. The others were determined based on measurement or calibration.

Tab. 2: Material properties and parameters used in DEM/CFD co-simulation

Parameter	Unit	Value
Sphere diameter	mm	10
Material density	kg/m^3	540
Bulk density	kg/m^3	681
Young's modulus	-	$2.5x10^7$
Poisson ratio	-	0.25
Coefficient of restitution	-	0.5
Coefficient of static friction between particles	-	0.8
Coefficient of rolling friction between particles	-	0.11
Thermal conductivity	W/mK	0.17[12,13]
Thermal capacity	J/kgK	1.2[12,13]
Initial temperature (Fixed fuel bed)	K	298
Initial temperature (Inlet)	K	283
Pressure above heap surface	Pa	1

The side-fed gasifier is a common design for such boilers and was simplified into a cylindrical container with a lateral inlet. Its inner diameter and thickness are 290 mm and 10 mm, respectively. The geometry including the dimension and location of the inlet and the meshed model are shown in Fig. 3a and Fig. 3b, respectively. The model is divided into two computational domains (see the two dashed boxes in Fig. 3b) to meet the requirement of OpenFOAM®. Furthermore, both domains are meshed with non-uniform cubic elements for CFD computation. Since an unresolved co-simulation is settled in consideration of large particle quantity, the finest computational element is at least the same size, or significantly larger than the dimension of a particle sphere. The meshes are shown in Fig. 3b.

During the simulation, particles with required mass flow were randomly generated in the so-called "particle factory" (as noted in Fig. 3a), which is used to model the stoker screw outlet. The constant mass flow was based on experimental measurements. In this way, the real packing structure, including the particle position, orientation, and porosity distribution, was obtained in the DEM model. Simultaneously, inlet air flow was provided from the bottom of the cylinder and the flow field information, such as velocity and temperature, was numerically analyzed. Clearly, instead of an even and smooth surface, a fuel bed with natural slope was formed and the angle kept changing during the feeding process, as the screenshots presented in Fig. 4 show. The particle generation proceeded until the fuel bed reached the bottom of the stoker screw outlet.

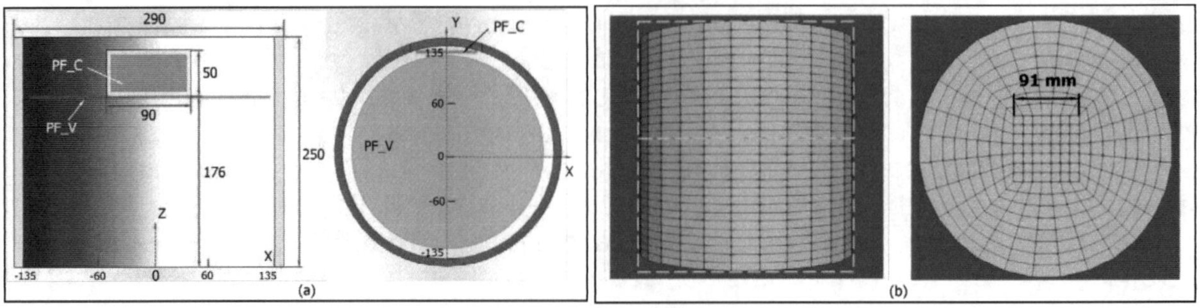

Fig. 3: (a) Boiler geometry (PF_C: Particle factory for case study; PF_V: Particle factory for validation); (b) Meshing for DEM/CFD co-simulation

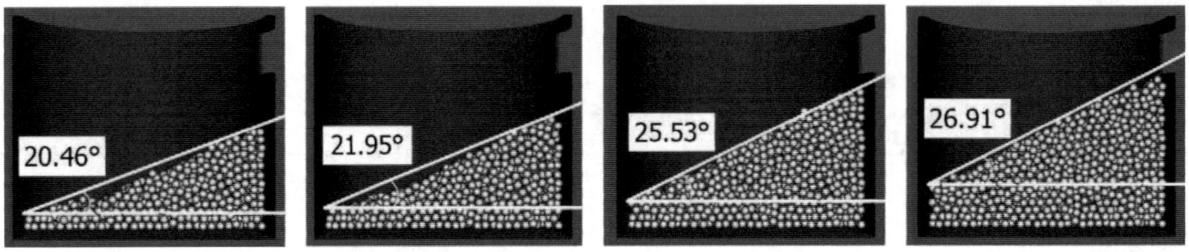

Fig. 4: Slope angles of fixed fuel bed at four characteristic moments

In the current study, two influential factors are investigated: heap shape and inlet air flow. The different values for each factor are listed in Tab. 3. A total of 41 simulations (6 and 35 for model validation and case study, respectively) were conducted, which will be discussed detailed in section 3 and 4. It is important to notice that 0.2 m/s is the mean velocity, which was calculated based on experiment measurement during real gasification procedure. Except from these two parameters, the simulation conditions were kept constant. The simulation durationis determined to 30 s, within which enough fuel particles were fed so that the peak of fuel bed reached the stoker screw outlet and the provided air flow reached outlet of boiler in this condition.

Tab. 3: Variables of heap geometry and inlet air flow

Model validation	Slope angle (°)	0						
	Inlet air velocity (m/s)	0.01	0.05	0.1	0.2	0.5	1	
Case study	Slope angle (°)	20.46	21.95	24.12	25.53	25.95	26.22	26.91
	Inlet air velocity (m/s)	0.01	0.05	0.1	0.2	0.5		

3. Model validation

It is necessary to validate the reasonability and feasibility of the proposed simulation model. A comparison of the pressure drop and energy transfer coefficient between analytical and simulation results was carried out to validate the DEM-CFD model. Considering that the original analytical expressions were developed for fuel beds with even surfaces, the fuel particles for the model validation were generated by rainfall method, within the particle factory "PF_V" in Fig. 3a. It was located parallel to the bottom of boiler and at the same level as the "PF_C". While the pressure drop is calculated based on the Ergun equation (Eq. 3-1 in Tab. 4) [14], the energy transfer coefficient is evaluated by using classical Nusselt equation (Eq. 3-2 to Eq. 3-4 in Tab. 4) provided by Ranz and Frantz et al. [15, 16]. Besides, the pressure field is represented as "gauge pressure", which is equal to difference between absolute pressure and local atmospheric pressure. Moreover, the air flow is assumed as the incompressible flow in OpenFOAM®. Therefore the gauge pressure is normalized by air density (ρ_f) and presented with the unit "m^2/s^2".

Tab. 4: Governing equations of pressure drop and Nusselt number

Parameter		Ref.	
$\Delta p = \Delta L \cdot [150 \cdot \frac{(1-\varepsilon)^2}{\varepsilon^2} \cdot \frac{\mu_f u_f}{(\varphi d_p)^2} + 1.75 \cdot (\frac{1-\varepsilon}{\varepsilon^3}) \cdot \frac{\mu_f u_f}{\varphi d_p}$		[14]	Eq. 3-1
$\mathrm{Nu} = 2 + 0.6\varepsilon_f{}^{\mathrm{n}} \cdot \mathrm{Re}_p{}^{0.5} Pr^{0.33}$	$(Re_p < 200)$	[15]	Eq. 3-2
$\mathrm{Nu} = 2 + 0.5\varepsilon_f{}^{\mathrm{n}} \cdot \mathrm{Re}_p{}^{0.5} Pr^{0.33} + 0.02\varepsilon_f{}^{\mathrm{n}} \cdot \mathrm{Re}_p{}^{0.8} Pr^{0.33}$	$(Re_p < 1500)$	[16]	Eq. 3-3
$\mathrm{Nu} = 2 + 0.000045\varepsilon_f{}^{\mathrm{n}} \cdot \mathrm{Re}_p{}^{1.8}$	$(Re_p > 1500)$	[16]	Eq. 3-4

Tab. 5: Results of Nominate Ergun pressure drop and Nusselt number

Air velocity (m/s)	0.01		0.05		0.1		0.2		0.5		1	
	An	Si	An	Si	An	Si	An	Si	An	Si	An	Si
Nominate Ergun pressure drop (Δp)	0.247	0.269	1.810	1.821	4.787	5.098	8.361	9.638	18.86	20.65	316.8	248.4
Nusselt number (Nu)	2.104	0.157	2.33	0.781	2.328	1.465	2.612	2.005	2.912	2.295	3.309	3.819

The analytical (An) and simulation (Si) results are listed in Tab. 5. As plotted in the Fig. 5, although not all pressure drop values from the simulation show strong consistency with analytical values, relative errors were under the reached maximal value (22%) generated in a tutorial of LIGGGHTS®. On the other hand a noticeable discrepancy of Nusselt number can be observed, except from the one at inlet air velocity of 1 m/s. The reason behind this is that the analytical model was carried out for case under high air flow velocity, which effects on the laminar boundary of particle surface. Besides, the packed bed with high bulk density in this study was largely different from the fluid bed studied in [17, 18], in which the theory and calculations of the two parameters were originally developed. The simulation results with varying air flow velocity from 0.01 m/s to 1 m/s suggested that the faster air velocity is, the more consistent simulation and analytical results are. However, if the particles are blown up under over-speed air flow, the results will not match the analytical results obtained based on equations in [15, 16].

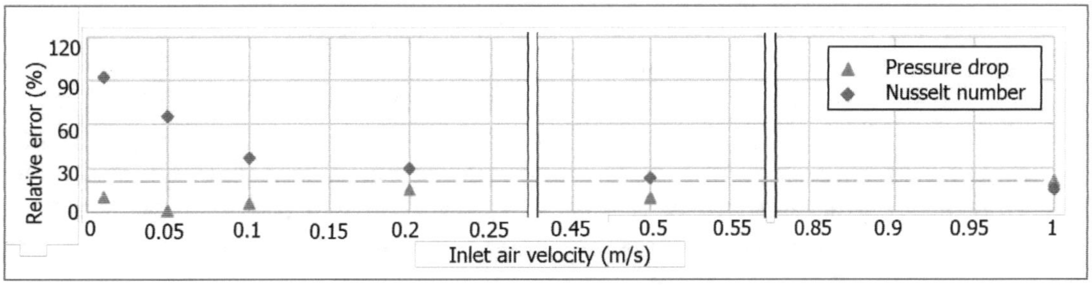

Fig. 5: Relative errors of pressure drop and Nusselt number between simulation results and theoretical values

4. Results and discussion

All 35 combinations of slope angle and air velocity at inlet (Tab. 3) were simulated. Then, the influence of each of the two parameters on the pressure gradient, temperature distribution, and air velocity field over the fixed bed surface were analyzed. Similar to that in section 3, the YZ plane view of each simulation was used for performance evaluation and this selected plane was divided into three fields for discussion: (1) Y = [-135, -60], (2) Y = [-60, 60], and (3) Y = [60, 135] (as the coordinate system in Fig. 3a). Furthermore, the result at a velocity of 0.2 m/s with seven different angles, and that at 25.53° with five different velocities, were discussed in detail.

3.1 Air velocity over heap surface

When the air just leaves the fuel bed (in the frame of a parallelogram Fig. 6), the velocity decreases along the slope from field No. 1 to No. 3, regardless of the slope angles or inlet air velocity. Longer interaction time between the solid (fuel particles) and fluid (air) phases causes more loss of kinetic energy of air flow, as indicated by legend in Fig. 6. Furthermore, it is difficult to describe the velocity field in the area with a certain distance from the heap surface. Basically, the faster the inlet air velocity, the higher the possibility that the velocity vector changes its direction because of surrounding air. Although air velocity is a vector in 3D coordination system, the velocity components in direction of X- and Y-Axis are too small to impact the resultant velocity in this study because of the constant temperature of fuel bed. A quantitative analysis will not be conducted in this paper, which is more meaningful only if the temperature of fuel bed changes like in reality.

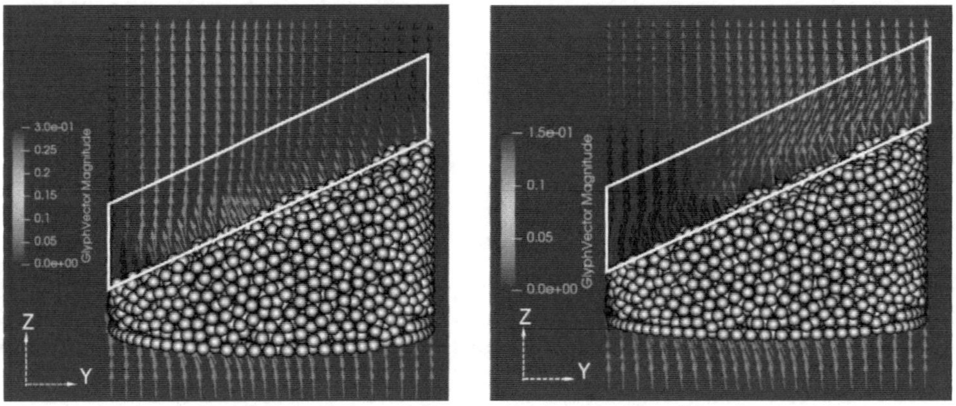

(a) (b)

Fig. 6: Air velocity field: (a) Slope angle: 25.53° and u_f = 0.1 m/s; (b) Slope angle: 25.53° and u_f = 0.2 m/s

3.2 Influence of slope angle

Fig. 7 shows the pressure gradient and temperature distribution at the outlet of the boiler with an inlet air velocity of 0.2 m/s and varying slope angles (20.46° – 26.91°). It can be seen that as the slope angle increases, the gauge pressure decreases rapidly along Y-Axis. Despite of this general tendency, the pressure distribution in field No. 3 is complicated: When the slope angles are smaller than 25.53°, the local pressure increased slightly; on the contrary, the pressure in other simulations kept decreasing along Y-axis towards stoker screw outlet, with an exception of the one with the largest slope angle, which showed only mild pressure fluctuation due to impact of fuel feeder.

Different from pressure gradient, the temperature distributions above the fuel bed are comparable for all simulations. Undoubtedly, the more accumulated fuel the air flow is confronted with, the more heat it will absorb, leading to increased temperature. And until the air flow past through a certain amount of fuels, its temperature will stay stable, as the curve in field No. 3 of Fig. 7b. Thus, the steeper the slope angle is, the smaller the temperature difference is.

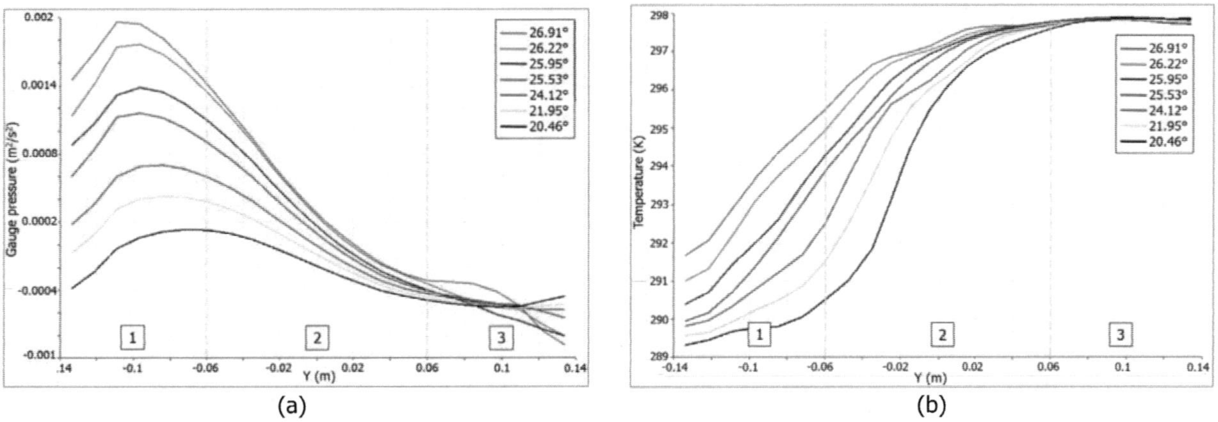

Fig. 7: (a): Pressure gradients at outlet (φ = 25.53°); (b): Temperature distribution at outlet (φ = 25.53°)

The simulation results at other velocities are basically similar to that of above and, therefore not further discussed. One exception is the simulation with a velocity of 0.01 m/s and varying slope angles, as shown in Fig. 8. Apparently, gauge pressure gradient and temperature distribution in Field No. 2 and No. 3 are much different from that in Fig.7. It is worth mentioning that the relative error of the validating case was nearly 100 %, which as explained in section 3. Therefore this entirely different pattern could be due to the own limitation of the DEM/CFD co-simulation.

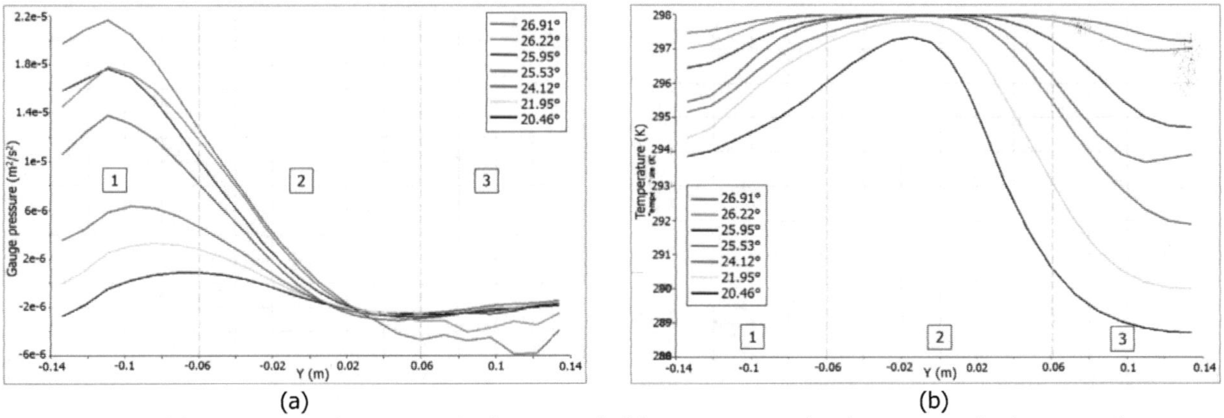

Fig. 8: (a): Pressure gradients at outlet (φ = 25.53°); (b): Temperature distribution at outlet (φ = 25.53°)

3.3 Influence of inlet air velocity

Fig. 9 depicts the pressure gradients and temperature distributions along the heap slope in Y-axis at five different inlet air velocities, with a constant slope angle of 25.53°. It can be seen that at low air velocities (0.01 m/s and 0.05 m/s), the pressure at outlet remains almost unchanged. However, when the velocity increases from 0.1m/s to 0.5m/s, the effect of air velocity on outlet pressure changed significantly. In addition, the pressure gradients are significantly affected by the solid wall. It can be seen from the sections in field No.1 and No.3 of pressure curve, that there are trend reverse in both fields. And the influence of inlet air velocity on gauge pressure becomes much more noticeable as it increases considerably.

As regards temperature distributions, the deviation among the cases at difference air velocities suggests that 0.2 m/s and 0.5 m/s are too fast for heat exchange between particles and air flow.

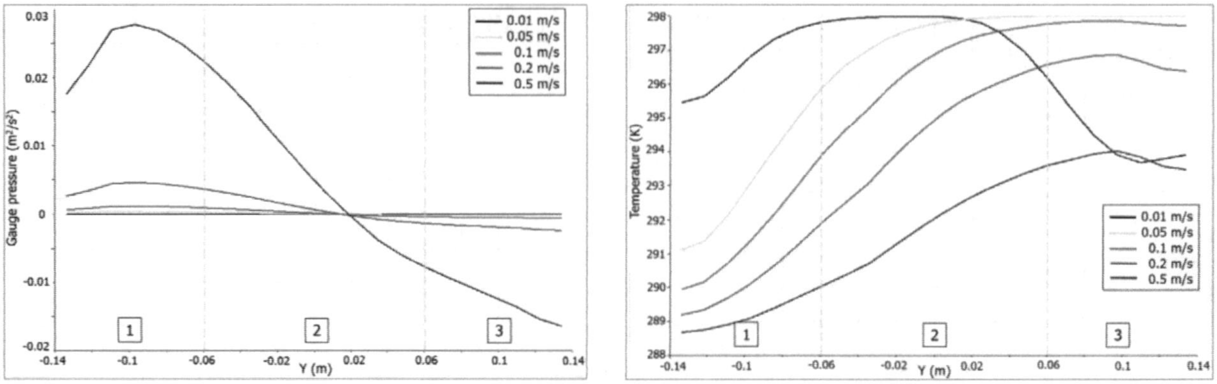

(a) (b)

Fig. 9: Left: Pressure gradients at outlet (u_f = 0.2 m/s); Right: Temperature distribution at outlet (u_f = 0.2 m/s)

From the simulation results, it is noteworthy that a reduction in slope angle and inlet air velocity leads to an increase in heat energy transfer, which is different from that for the gauge pressure gradient. Moreover, the rates of temperature change at different inlet air velocities (with constant slope angle) or at different slope angles (with constant velocity) are comparable. Tab. 7 lists the differences between maximum and minimum values of gauge pressure of each simulation. It can be seen that the influence of slope angle, i.e., heap shape, is stronger than that of inlet air velocity on temperature over fuel bed.

Tab. 6: Differences between maximum and minimum values of temperature (K)

Slope angle (°)		20.46	21.95	24.12	25.53	25.95	26.22	26.91
Inlet air velocity (m/s)	0.05	7.8	7.5	7.1	6.8	6.2	5.4	4.5
	0.1	8.3	8.2	7.9	7.7	7.4	6.7	6.1
	0.2	8.2	7.7	7.5	7.6	7.5	6.7	6.7
	0.5	5.4	5.6	5.1	5.3	5.6	5.2	5.2

5. Conclusion and outlook

This study discussed the influence of fuel heaps in a side-fed biomass boiler on three major process variables, namely: gauge pressure gradient, temperature distribution as well as air velocity field. Comparisons were made among varying simulation conditions. By this understanding of its influence on the first two variables is extended quantitatively as well as on the last one qualitatively. To validate the reliability of numerical models, analytical results were compared with that of preliminary simulations. Afterwards, the outcomes of parameter studies showed that process parameters were strongly affected by changing the fuel heap shape. Combined with another elementary process parameter, the inlet air velocity, the presented analysis was more comprehensive and closer to the real process. Besides it was also found that the solid wall plays a nonnegligible role for proper assessment of the influence of both slope angle and inlet air velocity on pressure gradient.

Due to inhomogeneous distribution of pressure and temperature in the combustion zone, the combustion of individual fuel particles in the boiler is not in sequence anymore and this can even affect the subsequent pyrolysis and oxidation process. This means that undesirable burning of fuel may be the consequence of the uneven heap shape. Therefore the concluded results about the influence of heap shape as well as inlet air velocity are meaningful in view of further attempts to optimize the burning process in biomass boilers. Moreover, it was proved that the current DEM/CFD co-simulation model is not applicable to fixed bed conditions, where the bulk density is relatively high compared to low inlet air velocity , for example the investigated fuel, olive stone, with respect to air velocity of 0.01m/s. It can be expected that parameter modification is required for accurate simulation in light of the popularity of wide range of biomass fuels.

As the first step of the complete analysis of the reality-based model, homogenous and constant fuel bed temperatures and a simplified process were assumed. Indeed, information during this real burning process is difficult or even impossible to measure experimentally with sufficient accuracy, such as the air flow velocity and pressure drop. Hence, the current study provides a useful DEM/CFD co-simulation model, which is also available for further research with modified process parameters to match reality.

6. Acknowledgement

This project has received funding from the European Union's Horizon2020 Program (H2020/2014-2020), under grant agreement no. 654446. The authors would like to thank CIEMAT (Soria, Spain), a Spanish Public Research Institution, for providing us with the data of particle size distribution, moisture content measurements and the bulk densities according to DIN and EN.

7. Literature

1. Pedroso, Daniel Travieso; Machín, Einara Blanco; Silveira, Jose Luz; Nemoto, Yasuyuki (2013): Experimental study of bottom feed updraft gasifier. In: Renewable Energy 57, S. 311–316. DOI: 10.1016/j.renene.2013.01.056.
2. Rackl, Michael; Günthner, Willibald A. (2016): Experimental investigation on the influence of different grades of wood chips on screw feeding performance. In: Biomass and Bioenergy 88, S. 106–115. DOI: 10.1016/j.biombioe.2016.03.011.
3. Simsek, E.; Brosch, B.; Wirtz, S.; Scherer, V.; Krüll, F. (2009): Numerical simulation of grate firing systems using a coupled CFD/discrete element method (DEM). In: Powder Technology 193 (3), S. 266–273. DOI: 10.1016/j.powtec.2009.03.011.
4. Mahmoudi, Amir Houshang; Markovic, Miladin; Peters, Bernhard; Brem, Gerrit (2015): An experimental and numerical study of wood combustion in a fixed bed using Euler–Lagrange approach (XDEM). In: Fuel 150, S. 573–582. DOI: 10.1016/j.fuel.2015.02.008.
5. Mahmoudi, Amir Houshang; Besseron, Xavier; Hoffmann, Florian; Markovic, Miladin; Peters, Bernhard (2016): Modeling of the biomass combustion on a forward acting grate using XDEM. In: Chemical Engineering Science 142, S. 32–41. DOI: 10.1016/j.ces.2015.11.015.

6. Wiese, Jens; Wissing, Frank; Höhner, Dominik; Wirtz, Siegmar; Scherer, Viktor; Ley, Ursula; Behr, Hans Martin (2016): DEM/CFD modeling of the fuel conversion in a pellet stove. In: Fuel Processing Technology 152, S. 223–239. DOI: 10.1016/j.fuproc.2016.06.005.
7. Mohseni, Mohammad; Peters, Bernhard; Baniasadi, Mehdi (2017): Conversion analysis of a cylindrical biomass particle with a DEM-CFD coupling approach. In: Case Studies in Thermal Engineering 10, S. 343–356. DOI: 10.1016/j.csite.2017.08.004.
8. Crowe, C.T., 2005. Multiphase Flow Handbook. Taylor and Francis, Boca Raton, USA. ISBN: 9780849312809
9. Anderson, T.B.; Jackson, R.: A fluid mechanical description of fluidized beds. In: Industrial & Engineering Chemistry Fundamentals, Jg. 6 (1967) Nr. 4, S. 527–539.
10. Hager, A.: CFD-DEM on multiple scales – An extensive investigation of particle-fluid interactions, Christian Doppler Labor für partikuläre Strömungen, Linz, 2014.
11. Wile. (2008) "Powered by olive stones? Turning waste stones into fuel". Retrieved from https://phys.org/news/2008-10-powered-olive-stones-fuel.html. 01.04.2018 Accessed.
12. Arce, M. E.; Míguez, J. L.: Biomass fuel and combustion conditions selection in a fixed bed combustor. In: Energies, Jg. 6 (2013), S. 5973–5989.
13. Guo, W.; Lim, J.: Determination of effective thermal conductivity and specific heat capacity of wood pellets. In: Fuel, Jg. 103 (2013), S. 347–355.
14. Ergun, S.; Orning, A. A.: Fluid flow through randomly packed columns and fluidized beds T. In: Industrial & Engineering Chemistry, Jg. 41 (1949) Nr. 6, S. 1179–1184.
15. Ranz, W.E., Marshall, W.R., 1952. Evaporation from drop. Chem. Eng. Prog. 48 (3), 141–173.
16. J.F. Frantz, Chem. Eng. 69 (1962) 89.
17. Li, J.; Mason, D.J.: A computational investigation of transient heat transfer in pneumatic transport of granular particles. In: Powder Technology, Jg. 112 (2000), S. 273–282.
18. Kloss, C.; Goniva, C.: Models, algorithms and validation for opensource DEM and CFD-DEM. In: Progress in Computational Fluid Dynamics, Jg. 12 (2012), S. 140-152.

507392 Particles Wall Coating Due to Electrostatic Charge Generation in Gas-Solid Fluidized Beds in Turbulent Versus Pre-Turbulent Flow Regimes

Abstract:

Continuous contacts amongst the particles and the particles and the fluidization vessel wall give rise to electrostatic charge generation in gas-solid fluidization processes. This might lead to significant operational challenges including fluidizing particles adhesion to reactor wall, particle agglomeration, and electrostatic discharges. One industry that has significantly suffered from negative effects of fluidized bed electrification is the petrochemical industry, where fluidized bed reactors are used for gas-phase ethylene polymerization to produce polyethylene. In such reactors, electrostatic charge generation results in catalyst and polyethylene particles to adhere to the reactor wall, causing a problem known as â€œsheetingâ€. The formation of particle sheets can require shutdown periods for clean-up which results in significant economic losses. To better understand the particles reactor wall coating due to electrostatics in such process, it is beneficial to conduct studies in relation to polyethylene reactor electrifications under industrially-relevant operating conditions of elevated pressures and turbulent flow regime. In this work, the effect of gas velocity on the degree of particles wall coating was studied with a specific focus on the transition to turbulent flow regime. Experiments were conducted in a pilot-scale fluidized bed operating under industrially-relevant conditions of a pressure of 2600 kPa (abs) and fluidizing gas velocities of 1.5, 3, and 5 times of U_{mf} (pre-turbulent regimes) and 7.5 times of U_{mf} (turbulent regime). Polyethylene particles directly received from a commercial reactor were fluidized in this work. Increasing the gas velocity and transitioning to the turbulent flow regime improved particle-wall contacts; and thus, augmented the extent of wall fouling, which indicates the increase in bed electrostatic charge generation. The amount of particles wall coating was approximately five times larger in turbulent flow regime (7.5 U_{mf}) in comparison to that for the lowest gas velocity examined in bubbling flow regime (1.5 U_{mf}). The particles coating on the column wall consisted of a thick bottom layer which extended to the expanded bed height, and a thin top layer which extended to the top of the column near the outlet. The particles net specific charge in the top and bottom layers did not vary with the increase in gas velocity. However, the net charge of these particles increased. The fine particles entrained from the bed had a net negative charge resulting in a net positive charge to be left behind in the bed contributing to the increase in the magnitude of wall fouling at higher gas velocities, especially in turbulent flow regime.

507583 Process Intensification By Miniaturization of Fluidized Beds

Abstract:

Chemical micro-process technologies are relatively new concepts that offer much promise in terms of achieving process intensification by taking advantage of excellent mass and heat transfer characteristics of micro-structured devices. Recently there is a trend of miniaturization of fluidized bed systems to utilize further enhancement in mixing, heat and mass transfer provided by fluidized particles. However, there are some fundamental differences with classical macro-scale counterparts beds which need to be taken into account when designing these kinds of devices. The two major differences are the unavoidable wall effects due to small bed sizes and more importantly surface forces which start to be comparable or even dominate over gravitational force. These can be used to define the boundary between micro and macro-scale fluidization at 1 cm with a stricter limit at 1 mm in line with general microfluidics.
In terms of applications, the micro-fluidized bed holds promise in the already mentioned micro-process and micro-fluidic context. Besides fundamental studies on mixing and mass transfer enhancement, the current applications are mainly in high-throughput screening and testing of catalysts. The miniaturized fluidized bed also enables fast and reliable measurements of solid catalyzed reaction kinetics. Some more specific applications are in a low-cost device for detection of infectious bacteria, granulation screening of pharmaceutical tablet fillers, photocatalytic reactors and compact carbon dioxide HVAC units.

Acknowledgment

The authors acknowledge the support of the European Commission under the HORIZON 2020 Programme, for EU Project 680565 â€" â€˜Intensified by Designâ€™ (IbD).

507730 Scaling Analysis of Rectangular Spouted Bed Dynamics

Abstract:

2017 AIChE 8th World Congress on Particle Technology

Topic/Group: Fluidization & Multiphase Flow

Tentative Session: Fundamentals of Fluidization

Title: Scaling Analysis of Rectangular Spouted Bed Dynamics

Authors:

Steven L. Rowan: Steven.Rowan@netl.doe.gov

Jingsi Yang: Jingsi.Yang@netl.doe.gov

Ronald W. Breault: Ronald.Breault@netl.doe.gov

Key Words: Spouted Beds, Scaling, Cold Flow, Energy

Abstract:

Spouted bed reactors are used extensively in the chemical and food industries for the processing and drying of large, coarse particles that are not suitable for use in other multiphase flow systems such as fluidized beds. A thorough understanding of the dynamic performance of these units is critical when scaling up from small bench-scale units to larger production scale systems. An experimental study has been conducted to examine the hydrodynamic characteristics of rectangular spouted beds of geometric scales.

For this study, experimental data has been obtained using a small scale 4-inch x 1-inch rectangular spouted bed, as well as a larger 12-inch x 3-inch bed, using a variety of bed material, including 707 μm glass beads, 860 μm high density polyethylene, and 750 μm hollow alumina spheres. Hydrodynamic similarity between the two spouted bed reactors is explored via a statistical (skewness and kurtosis), frequency (power spectral density), and autocorrelation analysis of differential pressure data. In addition to analysis of differential pressure measurements, high speed video was recorded and used to determine the bed expansion ratio and the fountain height as a function of gas velocity. From the data obtained, correlations for various spouting regime transition velocities are proposed.

509344 On the Mechanism of Anomalous Sphere Sinking in Apparently Fixed Particle Beds (Discrete Particle Simulation and Ultra-Fast MRI Measurement)

Abstract:

It is well known that a particle bed behaves as a fluid in the case it is aerated with gas velocity higher than the minimum fluidization velocity Umf. This is called as a fluidized bed and widely used in a large number of industrial processes. Under Umf, on the contrary, gravity force working on a particle overcomes fluid force and a particle bed behaves as a solid. Oshitani et al. (Phys. Rev. Lett., 116, 068001, 2016) reported anomalous sinking of intruder spheres in granular beds aerated under but close to Umf. They observed deep and slow sinking of spheres in the case density of sphere is close to the bulk density of the bed. They attributed this phenomena to local fluidization around the spheres while its detailed mechanism was not clear due to opaque nature of granular beds. In the present study, anomalous sinking and stopping mechanism of intruder spheres in an apparently fixed bed aerated just below Umf were explored and revealed in detail by means of discrete particle simulation and ultra-fast MRI measurement. Our numerical simulation and MRI measurement enable us to observe detailed flow dynamics and forces working on the spheres with sufficient spatial and temporal resolutions. Fluid force, contact force from surrounding particles and the gravity force working of the sphere are almost balanced in the case sphere density matches with the bulk density of a bed and small deviations from the equilibrium make slow sinking. The deviation is due to the detachment of small bubbles from the sphere surface. Bubble detachment from the sphere surface loosens local gas pressure gradient that supports the sphere weight and the sphere falls at the time.

Flow Structure Study on Expanded Bed in Gas Fluidization of Fine Ellipsoids

Jieqing Gan[*], *Zongyan Zhou, Aibing Yu*

Laboratory for Simulation and Modeling of Particulate Systems, Department of Chemical Engineering, Monash University, VIC 3800, Australia

Abstract

Combined approach of computational fluid dynamics for gas phase and discrete element method for particles (CFD-DEM) is used to exhibit the formation of expanded bed in gas fluidisation of fine ellipsoids. The results illustrate that in the start-up stage, the bulk disconnects into two major bulks with a lowest contact force on the axial direction, especially for flat and elongated particles. The positions of the lowest point of contact force slightly change with time for oblate and prolate particles, but it moves smoothly from bottom to top for spheres. A slower gas introduction process has a mildest change in bed height (porosity). The fractures between major bulks gradually disappear when the cohesive forces become smaller. In the transition stage, when aspect ratio deviates from 1.0, the pressure drop fluctuation significantly increases. In the stable stage, when the aspect ratio decreases or increases from 1.0, the contact forces and van der Waals forces increases, and the particle-fluid forces become less uniformly distribution in the beds. Ellipsoidal particles undergo an orientation rearrangement during the expansion process.

Key words: CFD-DEM, expanded bed, fine ellipsoid, interparticle force.

Introduction

Geldart A (1973) particles, referred as 'aeratable', are characterized with an interval of gas flow beyond the point of minimum fluidization over which the bed expands in an apparently ideal manner, without bubbling. One focus on the flow regime is the stabilizing expanded bed, which has been a topic for many years. Criteria have been obtained for its stabilization in terms of interparticle force or particle-fluid forces [1, 2]. Another focus on expanded bed is the velocity range of the interval. Some correlations have been established to predict its range of the two demarcation point, that is the minimum fluidization velocity U_{mf} [3] and minimum bubbling velocity U_{mb} [4-10]. The presence of interparticle force was found to affect the bed bubbling [11, 12].

However, the formation process of the expanded bed is rarely reported. A few questions are still not clear. For example: what factors affect the formation process? How are the expanded bed formed? What are the microdynamics (e.g. particle velocity, orientation and interparticle forces) inside the bed at different stages of expanded bed formation? An insight into these questions could be useful for better understanding of the macroscopic formation of expanded beds. However, experiments experience difficulties in capture the instant information at particle scale, such as, particle contacts, orientation and forces. CFD-DEM becomes an alternative in the study of particle-fluid flow fundamentals. In this approach, the solid phase is based on the DEM, and gas phase is treated as a continuum phase in a similar way to the two fluid model. It has been used in the study of gas-fluidized beds of fine spheres [1, 13] and ellipsoids [13-15].

Although most particles in reality are non-spherical, only a few efforts had been made to investigate the effect of particle shape on the flow behaviour, and the role of particle forces on the formation of expanded of fine powders under the effect of particle shape is also not fully understand. This paper aims at investigating the formation process of expanded bed of fine ellipsoids using CFD-DEM method. The bed structure changes in three typical formation stages are discussed in terms of flow pattern, pressure drop, porosity, particle velocity, particle orientation and force ratio.

[*] Corresponding author. Email: jieqing.gan@monash.edu; Tel: 61 3 9905 0850.

Model description and simulation conditions

The governing equations for the translational and rotational motion of particle i with radius R_i, mass m_i, and moment of inertia I_i can be written as

$$m_i \frac{d\mathbf{V}_i}{dt} = \mathbf{f}_{f,i} + \sum_{j=1}^{k_c} (\mathbf{f}_{c,ij} + \mathbf{f}_{d,ij} + \mathbf{f}_{v,ij}) + m_i \mathbf{g} \quad (1)$$

$$I_i \frac{d\boldsymbol{\omega}_i}{dt} = \sum_{j=1}^{k_c} (\mathbf{M}_{t,ij} + \mathbf{M}_{r,ij} + \mathbf{M}_{n,ij}) \quad (2)$$

where $\mathbf{v}_i$ and $\boldsymbol{\omega}_i$ are the translational and angular velocities of the particle, respectively, and k_c is the number of particles in interaction with the particle. The forces involved are: particle-fluid interaction force $\mathbf{f}_{f,i}$, the gravitational force $m_i\mathbf{g}$, and interparticle forces between particles, which include elastic force $\mathbf{f}_{c,ij}$, viscous damping force $\mathbf{f}_{d,ij}$ and van der waals force $\mathbf{f}_{v,ij}$. These interparticle forces can be resolved into the normal and tangential components at a contact point. The torque acting on particle i by particle j includes: tangential torque $\mathbf{M}_{t,ij}$, and rolling friction torque $\mathbf{M}_{r,ij}$ and an addition torque $\mathbf{M}_{n,ij}$ because the normal contact force and van der waals force don't necessarily pass through the particle centre.

The governing equations for continuum fluid phase is given as

$$\frac{\partial \varepsilon_f}{\partial t} + \nabla \cdot (\varepsilon_f \mathbf{u}) = 0 \quad (3)$$

$$\frac{\partial (\rho_f \varepsilon_f \mathbf{u})}{\partial t} + \nabla \cdot (\rho_f \varepsilon_f \mathbf{u}\mathbf{u}) = -\nabla p - \mathbf{F}_{fp} + \nabla \cdot \varepsilon_f \boldsymbol{\tau} + \rho_f \varepsilon_f \mathbf{g} \quad (4)$$

where $\mathbf{u}$, ρ_f, p and $\mathbf{F}_{fp} = (\sum_{i=1}^{k_V} \mathbf{f}_{f,i}) / \Delta V$ are the fluid velocity, density, pressure and volumetric fluid-particle interaction force, respectively. $\boldsymbol{\tau}$ and ε_f are the fluid viscous stress tensor and porosity, respectively. The equations used to calculate the particle-particle interaction forces and torques and particle-fluid force can be found elsewhere [13-15].

The simulation begins with the poured packing of ellipsoidal particles in a rectangular box. Then gas is introduced from the bed bottom. The fluidization process is considered to have reached the pseudo-stable state when the bed pressure drop fluctuates around fixed values. Spheroids with aspect ratios varying from 0.25 to 3.5 are used in this work to produce shapes from oblate to prolate spheroids. Spheroids have three principal diameters. Let a be the principal radius in the polar direction, and b and c the principal radii in the equatorial plane. Here, the particle shape factor, namely aspect ratio α is defined as, $\alpha = a/c$. The particle properties used in the simulation are listed in Table 1.

Table 1. Simulation settings for particles and bed.

Parameters	Value
Particle size d_v, μm	50
Particle density ρ_p, kg/m^3	1349.5
Particle shape, [-]	Oblate(α<1), sphere(α=1), prolate (α>1)
Particle aspect ratio, [-]	0.25, 0.5, 0.75, 1.0, 1.5, 2.5, 3.0, 3.5
Number of particles N, [-]	18000
Bed size, [-]	$60d_p \times 480d_p$
CFD Mesh, [-]	20 ×160
Young's modulus E, N·m^2	1.0×10^7
Poisson's ratio ν, [-]	0.30
Friction coefficient μ, [-]	0.3
Rolling friction coefficient μ_r, d_v	0.002
Normal damping coefficient γ_n, [-]	2.0×10^{-5}
Hamaker constant H_a, J	0, 1.05×10^{-21}, 2.1×10^{-21}, 4.2×10^{-21}
Cut off distance h_{min}, nm	1, 5, 10

Results and discussion

When the gas velocity surpasses the U_{mf}, the bed is in expanded bed regime. The expanding process can divided into three stages, namely, start-up (corresponding to the increase in porosity at the beginning), transition (defined as the range of sharp descending in porosity), and stable (where all forces and porosity keep stable with time) [16].

Strat-up stage

From previous studies on spheres [1, 17] and ellipsoids [14], in the ascending process of the start-up stage, particles in the bed will disconnect into two bulks. To find out the reason, the effects of gas introduction method and particle cohesion are studied.

In our previous study [14], the gas is directly introduced with a constant gas velocity of U_g, but in most of the fluidization experiment, gas is gradually introduced to a specified gas velocity U_g. In present work, three gas introduction methods are studied: constant gas velocity introduction (case (a)), fast increasing gas velocity from U_{mf} to U_g within 0.25s (case (b)), and slowly introduce gas from velocity of U_{mf} to U_g within 0.75s (case (c)). The flow pattern, gas velocity and porosity (calculated from bed height) changes are plotted in Fig. 1. From Figs. 1(a), (b) and (c), it can be seen that for the three cases, there are fractures in the particle bulk. As shown in Fig. 1(d), case (a) the firstly to exhibit a highest porosity peak (corresponding to highest bed height), the gas does not have time to permeate the upper bed before they are lifted up by the bottom gas. The case (c) is the latest to show the lowest porosity peak as it reaches U_g in the latest, and the gas has enough time the permeate the upper bed and go through the whole bed. However, the finial stable porosity is quite close for the all three cases. Therefore, the gas introduction mode cannot eliminate the break-up of particle bulk, but a slower gas introduction process has a mildest change in bed height (porosity). In the following discussion, we still use the constant gas velocity introduction mode.

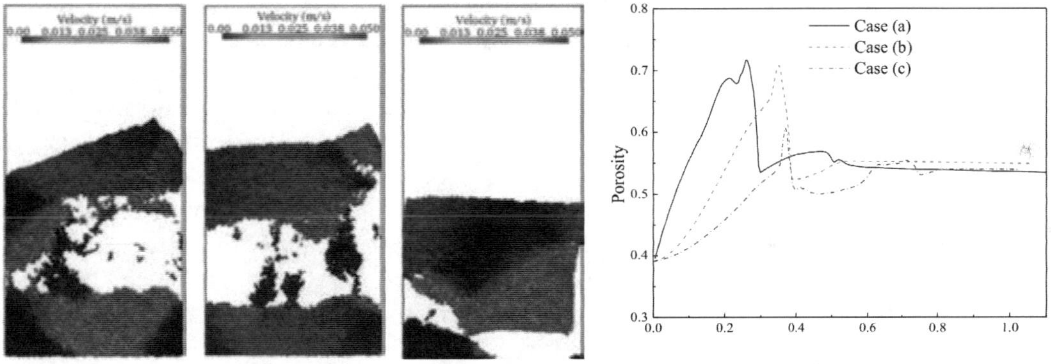

(a) (b) (c) (d)

Fig. 1 Flow patterns in the start-up stage of bed expansion for different gas introduction methods (0.2-0.4s). (a) Instant U_g, (b) Increase to U_g within 0.25s, (c) Increase to U_g within 0.75s, and (d) the variation of porosity with time. (Coloured by particle velocity)

The flow pattern of different Hamaker constant H_a are also compared and shown in Fig. 2. It can be seen that, with the decrease of H_a, the fracture between two bulks becomes obscure and even disappear when there is no van der Waals force. This indicates that the root reason for the formation of the fracture in the expansion of ellipsoids that the cohesive force on ellipsoids are much higher than the spheres, as was shown in Fig. 2.

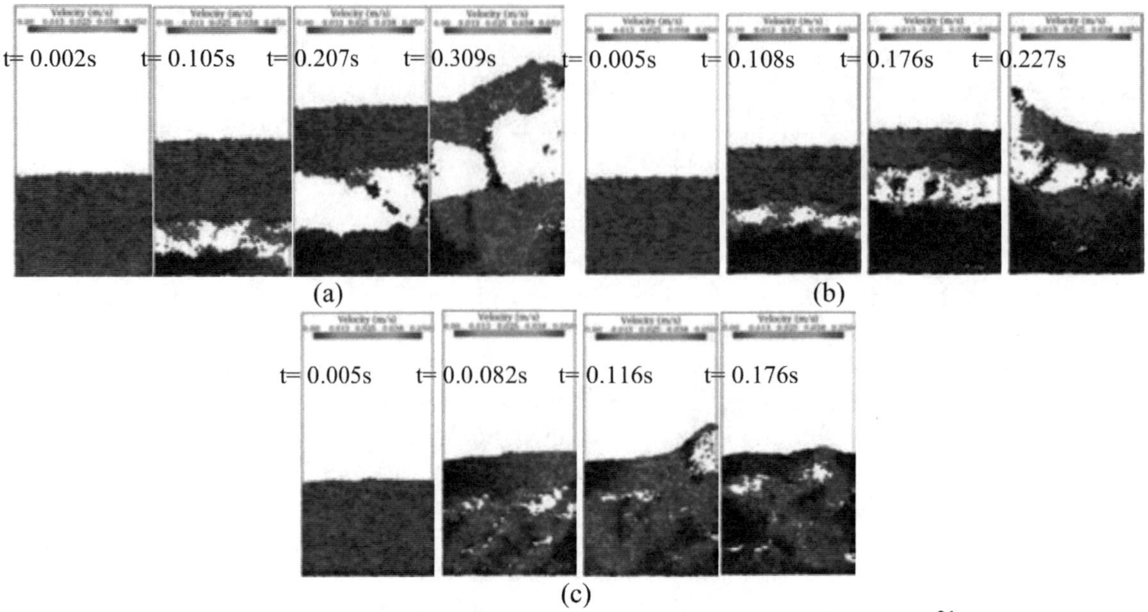

Fig. 2 Start-up stage of bed expansion for different aspect ratios ((a) H_a=2.1×10^{-21}, U_g=0.5U_{mb}=0.004 m/s, (b) H_a=1.05×10^{-21}, U_g=0.5U_{mb}=0.0035 m/s, and (c) H_a=0.0, U_g=0.0035 m/s, α=0.25, U_g=0.004 m/s, coloured by particle velocity)

The variation axial layer-averaged contact force with bed depth at different time in the start-up stage for different aspect ratios is shown in Fig. 3. At the beginning of gas introduction into the beds, the axial contact force increases to the maximum value and then decrease with the bed depth. This is because on one hand, the initial close packed bed will add weight to the underneath particles, and on the other hand, the blowing gas leads to looser bed bottom. When the beds are loose enough, the bulk disconnects, so there shows up a lowest point on the curves. However, there are obvious differences in the variation of lowest contact force with time for different aspect ratios. More specifically, for ellipsoids, with the time increases, the contact forces at the lowest point decreases which implying more weaker contacts, but the position of the lowest point slowly change with time. From Figs. 3(a) and (c), the upper bulk has much higher coordination number (CN) than the bottom bulk. For spheres (Fig. 3(b)), however, with the time increases, the lowest contact force position move from bottom to top, which means that the disconnect position move upwards as the particles in the upper bulk continuously fall to the bottom bulk, giving rise to a smooth expansion. Moreover, the contact forces in the bottom also increases due to the increasing height in the bottom bulk. From the inset of Fig. 3(b), the coordination number in the upper and bottom are quite close.

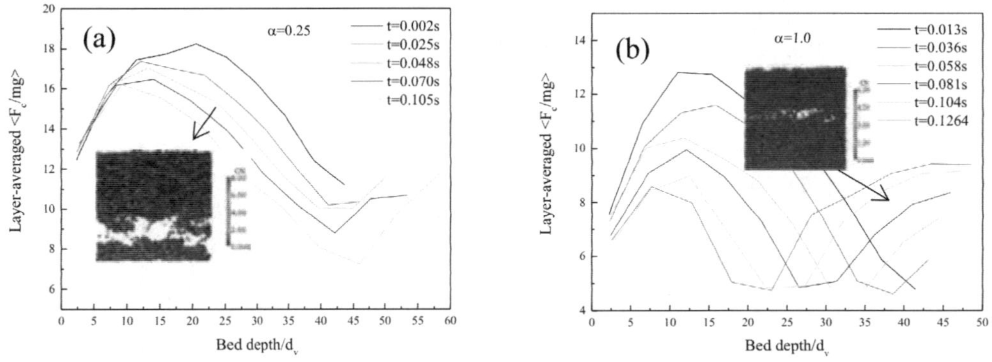

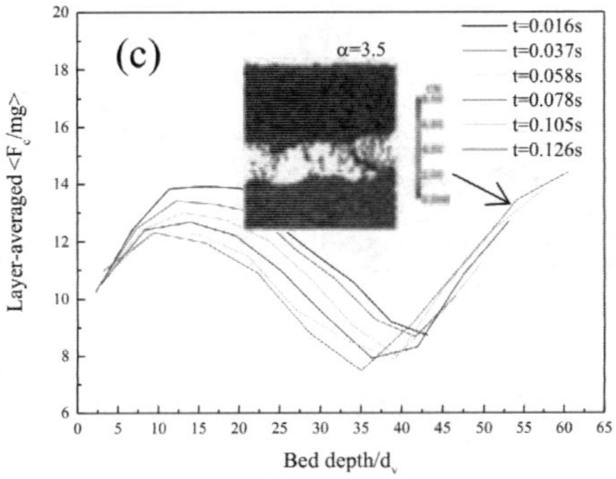

Fig. 3 The axial contact force in loosing and disconnecting stages of bed expansion for different aspect ratios ((a) α=0.25, (b) α=1.0 and (c) α=3.5)

Transition stage

After the bed disconnect into two bulks, the upper bulk continues to ascend under the particle-fluid force. When the upper bulk reaches the highest bed height, the spatial distribution of particle-fluid forces on particles is shown in Fig. 4. In this stage, most particles in the lower bulk have particle-fluid forces close or equal to their gravity, but the upper bulk have particle-fluid forces cannot balance their gravity which makes them falling to the bottom, so most particles have velocities in the downward direction. As shown in Fig. 4(d), when particles in the upper bulk fall to the bottom, there is a pressure drop fluctuation in the bed. For spheres, due to the relative continuous expanding process, the pressure drop fluctuation is quite small, but when aspect ratio deviates from 1.0, the pressure drop fluctuation significantly increases.

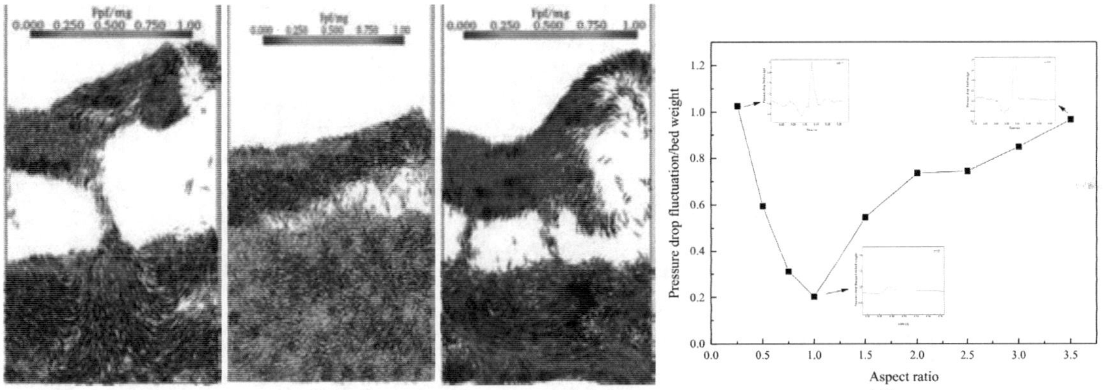

(a) (b) (c) (d)

Fig. 4 Spatial distribution of particle-fluid forces on particles when the upper bulk reaches the highest height in expanded bed (d_p=50 μm, (a) α= 0.25, (b) α=1.0 and (c) α=3.5) and (d) pressure drop fluctuation as a function of aspect ratio

Stable stage

The particle forces spatial distribution for different aspect ratios under the stable state in expanded beds is shown in Fig. 5. It can be seen that all the three forces vary significantly with bed space and particle shape. Generally, in spatial, the large and small contact and van der Waals forces are uniformly distributed in the bed, but the particle-fluid forces near the voids are smaller than inside the bulk. Although some particle-fluid forces are much larger than particles, the bed can still be kept in stable. For different aspect ratio, it shows that when the aspect ratio decreases or increases from 1.0, the contact forces and van der Waals forces increases, and the particle-fluid forces become less uniformly distribution in the beds.

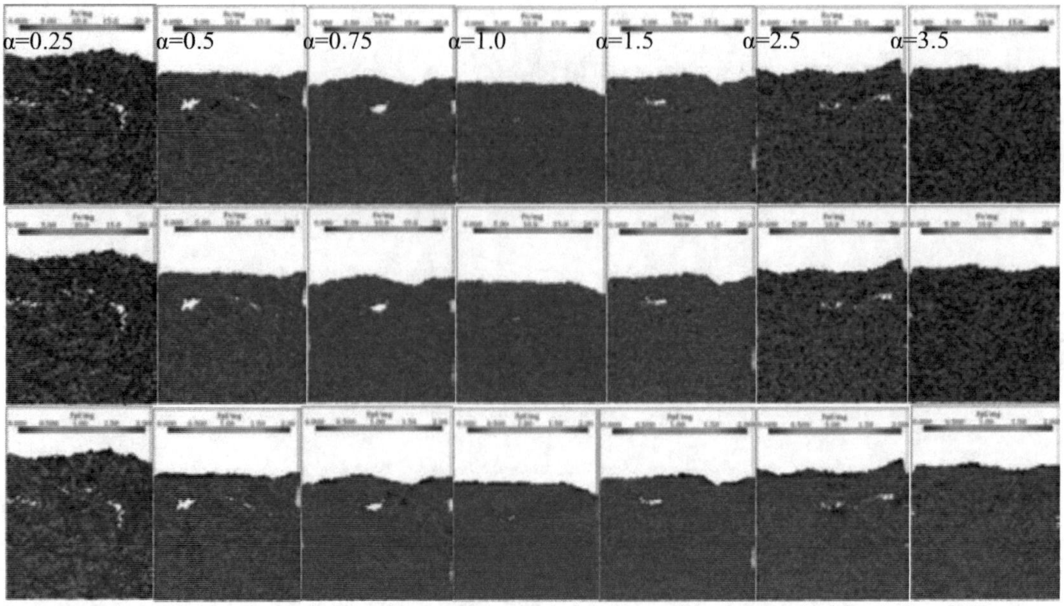

Fig. 5 Spatial distribution of particle forces on particles in stable expanded bed for different aspect ratios (top: $\mathbf{f}_c/m\mathbf{g}$, middle: $\mathbf{f}_{vdw}/m\mathbf{g}$, and bottom: $\mathbf{f}_{pf}/m\mathbf{g}$).

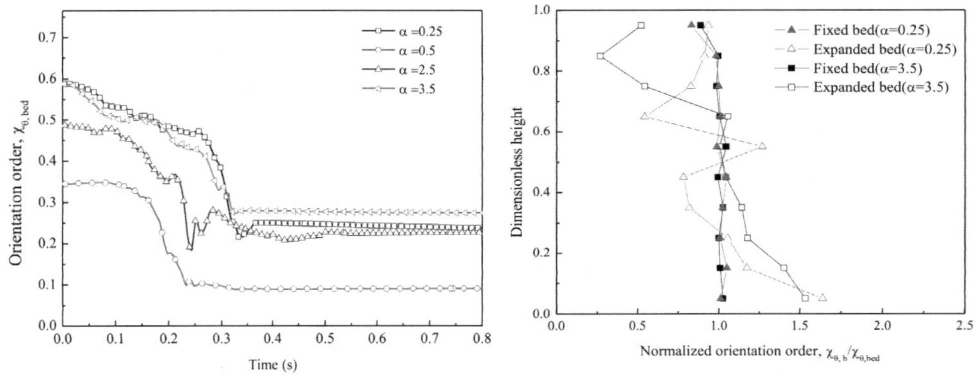

Fig. 6 (left) Variation of bed-averaged orientation order with time for different aspect ratios, and (right) axial distribution of the average orientation order in fixed and expanded beds (d_p=50 μm, fixed bed: $U_g=0.5\ U_{mf}$ and expanded bed: $U_g=0.5\ U_{mb}$)

From the flow patterns mention above, ellipsoidal particles undergo a position rearrangement during the expanding process, and meanwhile there are also changes in particle orientation. Fig. 6(a) shows that variation of orientation order with time for different aspect ratios. Here orientation order χ_θ is defined according to ref. [18, 19], and it becomes zero when the particles are randomly oriented and 1 when all particles lie in the horizontal plane. High orientation order is observed in the packed bed for different aspect ratios. With the introduction of gas, the particles orientation order gradually decreases and finally levels off at the stable stages. For different aspect ratios, the more non-spherical, the higher the orientation order is observed. The axial distribution of the average orientation order in fixed and expanded beds is compared and shown in Fig. 6(b). It can be seen that particles are axially uniform in the fixed beds. However, large orientation order fluctuations show up in the axial direction in the expanded beds, which proves that particles undergo orientation rearrangement during this process.

From the Fig. 5, although particle forces vary spatially, the beds are in force balance to keep a stable expanded bed. In our previous study [15], we have confirmed that the force-balance criterion is valid for the ellipsoidal particles. That is at stable stage, the van der Waals forces balance the particle contact forces, and the particle-fluid forces balance the weight of the bed. In this work, the axial contact forces in the stable expanded beds are studied, and shown in Fig. 7. It can be seen that in stable expanded beds, the forces are quite uniform in the axial direction when aspect ratio is close to 1.0. However, when aspect ratio deviates from 1.0, the layer-averaged contact forces fluctuate in the axial direction, and there is a relative lower contact forces region within the bed. This indicates that for flat or elongated particles, although the two bulks join together at the stable stage, there are still in loose contact near the joint.

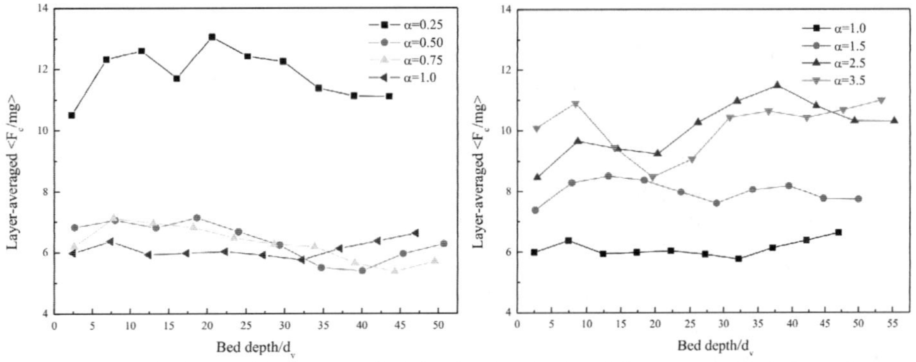

Fig. 7 Axial force ratio in stable expanded beds for (left) oblate and (right) prolate spheroids.

Conclusion

CFD-DEM simulation method is used to analysis the three typical stages (start-up, transition and stable stages) of expanded bed with fine ellipsoidal particles. In the start-up stage, with the gas introduction into the bed, the bulk disconnects into two major bulks with a lowest contact force on the axial direction, especially for flat and elongated particles. The positions of the lowest point of contact force slightly change with time for oblate and prolate particles. For spheres, however, with the time increases, the lowest contact force position move smoothly from bottom to top, giving rise to a smooth expansion. The fractures in the ascending process are not affected by gas introduction method, but gradually disappear when the cohesive forces become smaller. In the transition stage, when aspect ratio deviates from 1.0, the pressure drop fluctuation significantly increases. Compared with the initial packed bed, ellipsoidal particles undergo an orientation rearrangement during the expansion process. In the stable stage, when aspect ratio deviates from 1.0, the layer-averaged contact forces fluctuate in the axial direction, and there is a relative lower contact forces region within the bed.

Acknowledgment

The authors are grateful to the ARC Hub for Computational Particle Technology for the financial support of this work, and the NCI National Facility for the support in computation.

References

[1] Q.F. Hou, Z.Y. Zhou, A.B. Yu. Micromechanical modeling and analysis of different flow regimes in gas fluidization. Chemical Engineering Science. 2012;84(0):449-68.

[2] B.H. Xu, A.B. Yu. Numerical simulation of the gas-solid flow in a fluidized bed by combining discrete particle method with computational fluid dynamics. Chemical Engineering Science. 1997;52(16):2785-809.

[3] K. Suksankraisorn, S.Patumsawad, B. Fungtammasan. Prediction of the minimum fluidization velocity from correlations: An observation. Asian J Energy Environ. 2001;2(2):145-54.

[4] M. ŚciĄżko, J. Bandrowski. Effect of pressure on the minimum bubbling velocity of polydisperse materials. Chemical Engineering Science. 1985;40(10):1861-9.

[5] K.V. Jacob, A.W. Weimer. High-pressure particulate expansion and minimum bubbling of fine carbon powders. AIChE Journal. 1987;33(10):1698-706.

[6] J.M. Valverde, M.A.S. Quintanilla, A. Castellanos, P. Mills. Experimental study on the dynamics of gas-fluidized beds. Physical Review E. 2003;67(1):016303.

[7] M. Ye, M.A. van der Hoef, J.A.M. Kuipers. The effects of particle and gas properties on the fluidization of Geldart A particles. Chemical Engineering Science. 2005;60(16):4567-80.
[8] R. Girimonte, B. Formisani. The minimum bubbling velocity of fluidized beds operating at high temperature. Powder Technology. 2009;189(1):74-81.
[9] J. Wang, M.A. van der Hoef, J.A.M. Kuipers. CFD study of the minimum bubbling velocity of Geldart A particles in gas-fluidized beds. Chemical Engineering Science. 2010;65(12):3772-85.
[10] S. Rapagna, P.U. Foscolo, L.G. Gibilaro. The influence of temperature on the quality of gas fluidization. International Journal of Multiphase Flow. 1994;20(2):305-13.
[11] J.A. Agbim, A.W. Nienow, P.N. Rowe. Inter-particle forces that suppress bubbling in gas fluidised beds. Chemical Engineering Science. 1971 8//;26(8):1293-4.
[12] J.K. Pandit, X.S. Wang, M. Rhodes. Influence of interparticle cohesive force on fluidized bed Behaviour by DEM simulation. Third International Conference on CFD in the Minerals and Process Industries; 2003 10-12 December 2003; CSIRO,Melbourne, Australia; 2003.
[13] Z.Y. Zhou, D. Pinson, R.P. Zou, A.B. Yu. Discrete particle simulation of gas fluidization of ellipsoidal particles. Chemical Engineering Science. 2011;66(23):6128-45.
[14] J.Q. Gan, Z.Y. Zhou, A.B. Yu. CFD–DEM modeling of gas fluidization of fine ellipsoidal particles. AIChE Journal. 2016;62(1):62-77.
[15] J.Q. Gan, Z.Y. Zhou, A.B. Yu. Micromechanical analysis of flow behaviour of fine ellipsoids in gas fluidization. Chemical Engineering Science. 2017 5/18/;163:11-26.
[16] B.H. Xu, A.B. Yu. A new stability criterion for bed expansion of gas fluidisation. *4th world congress on particle technology*. Sydney, Australia. 2002:189.
[17] M.J. Rhodes, X.S. Wang, M. Nguyen, P. Stewart, K. Liffman. Onset of cohesive behaviour in gas fluidized beds: a numerical study using DEM simulation. Chem Eng Sci. 2001 7//;56(14):4433-8.
[18] G.W. Delaney, J.E. Hilton, P.W. Cleary. Defining random loose packing for nonspherical grains. Physical Review E. 2011;83(5):051305.
[19] B.J. Buchalter, R.M. Bradley. Orientational Order in Amorphous Packings of Ellipsoids. EPL (Europhysics Letters). 1994;26(3):159.

MODELING OF HEAT AND MOMENTUM TRANSFER IN THE CONTEXT OF MOVING PARTICLES BY A DOUBLE MRT LATTICE BOLTZMANN SCHEME

Bogdan Kravets, Tony Rosemann, Harald Kruggel-Emden,
Mechanical Process Engineering and Solids Processing (MVTA),
Technical University of Berlin (TU Berlin), Berlin, Germany

Abstract

A lattice Boltzmann method (LBM) framework for fully resolved particle/fluid simulations coupled with the discrete element method (DEM) is derived. The particle motion and particle/particle interactions are covered by the DEM and the fully resolved fluid phase including momentum and heat transfer is prescribed by the LBM. On the LBM side a double distribution function (DDF) approach with multiple-relaxation-time (MRT) collision operators for the density and energy distribution functions is applied. For the boundary conditions interpolated bounce back is used. Particle movement in LBM requires refilling of nodes which switch from solid to fluid as the particle passes by. Different refill algorithms are compared against a case with a static sphere by evaluating the drag force and the Nusselt number. The present study shows, that the investigated refill algorithms provide highly accurate results for a moving sphere. As expected, the numerical error increases with an increase of the particle Reynolds number. Furthermore, sedimentation of two spheres in a finite domain is investigated and the obtained results are compared to results from scientific literature. It can be concluded that the refilling as well as the coupling procedure are sufficiently validated. The derived DEM/LBM framework is capable to simulate dynamic particle/fluid systems and can be regarded as highly accurate.

Introduction

Dynamic particle/fluid systems - like fluidized beds, pneumatic conveying or rotary kilns - passed through by a fluid occur in many areas of technical processes in chemical industry and in energy technology. Modeling of the named systems gained in significance during the past years [1–5]. The most common applicable approach is the discrete element method (DEM) coupled with computational fluid dynamics (CFD), where the solid phase is described by Newton's laws of motion and the fluid phase by the Navier-Stokes equations. However, coupled DEM/CFD usually does not resolve the fluid phase spatially around individual particles.

Particle resolved flow simulations can rely on Navier-Stokes (NS) based approaches like finite volume method (FVM) or finite difference method (FDM) or more alternative schemes like the Lattice-Boltzmann method (LBM). Unlike other direct numerical simulation (DNS) methods, LBM is not solving Navier-Stokes equations but is based on kinetic gas theory and the discrete Boltzmann equation. LBM utilizes a Cartesian mesh and hence does not require a complex mesh derivation or a re-meshing when boundaries are moving. To address heat and momentum transfer in the fluid flow conventional LBM can be extended by thermal LBM (TLBM). The resulting scheme relies on a set of two distribution functions, the so called double distribution function (DDF) approach [6]; one for the fluid density and one for the internal energy.

For the carried out numerical investigations an existing 3D TLBM framework which was utilized for static particle packing simulations in the past [7,8] is extended for moving particles and is coupled with an existing DEM [1–5] framework. The particle movement is taken care of in the context of the DEM. The fluid field and the temperature field are represented by a multiple-relaxation-time (MRT) [9–11] collision operator. For the hydrodynamic boundary conditions, a scheme introduced by Bouzidi et al. [12] is applied with quadratic interpolation. Thermal boundary conditions are prescribed according to Li et al. [13] with linear interpolation. To prescribe unknown distributions occurring during particle movement, different refilling schemes [14–18] are applied.

Numerical Method

Lattice Boltzmann Method

The numerical method in the present work is a three dimensional double distribution functions Lattice Boltzmann model with a double MRT D3Q19 collision operator which is described as follows:

$$f_i(\vec{x} + \vec{e}_i \Delta t, t + \Delta t) = f_i(\vec{x}, t) - \langle M^{-1} \cdot \hat{S} \cdot \left[m_i(\vec{x}, t) - m_i^{eq}(\vec{x}, t)\right] \rangle_i + F_i, \quad (1)$$

$$g_i(\vec{x} + \vec{e}_i \Delta t, t + \Delta t) = g_i(\vec{x}, t) - \langle M^{-1} \cdot \hat{S}_g \cdot \left[m_{i_g}(\vec{x}, t) - m_{i_g}^{eq}(\vec{x}, t)\right] \rangle_i, \quad (2)$$

where f_i and g_i are the density and energy distribution functions respectively; i denotes the $i-th$ discrete direction; $\vec{e}_i$ the discrete velocity; M and $\hat{S}$ are the transformation and collision matrix respectively; F_i is a possible external force; m_i is the moment vector and m_i^{eq} its equilibrium state; the index g denotes the quantities connected to the energy distributions. The diagonal matrixes $\hat{S}$ and $\hat{S}_g$ contain the relaxation rates given by:

$$\hat{S} = diag(0, 0, 0, 0, s_4, s_5, s_6, s_7, s_8, s_9, s_{10}, s_{11}, s_{12}, s_{13}, s_{14}, s_{15}, s_{16}, s_{17}, s_{18}) \quad (3)$$

and

$$\hat{S}_g = diag(s_{g0}, s_{g1}, s_{g2}, s_{g3}, s_{g4}, s_{g5}, s_{g6}, s_{g7}, s_{g8}, s_9, s_{g10}, s_{g11}, s_{g12}, s_{g13}, s_{g14}, s_{g15}, s_{g16}, s_{g17}, s_{g18}). \quad (4)$$

For the correct choice of hydrodynamic quantities like $\hat{S}$ or m_i the reader is referred [9] to and for the thermal quantities like $\hat{S}_g$ or m_{i_g} to [11]. The transformation matrix M in Eqs. (1)-(2) is given in [9].

Hydrodynamic boundary conditions

In the present study interpolated bounce back proposed by Bouzidi et al. [12] is applied. The concept of interpolated boundary conditions in LBM is schematically shown in Fig. 1.

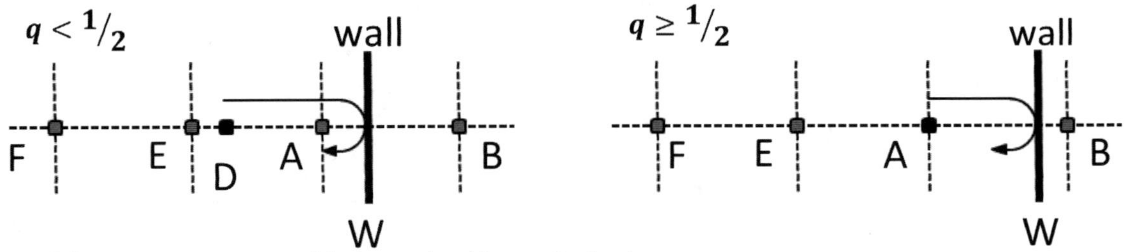

Fig. 1: Schematic representation of the interpolated bounce back scheme.

Depending on the wall position Bouzidi et al. distinguish between two general cases $q < 1/2$

$$f_{\bar{i}}(\vec{r}_i, t + \Delta t) = q(2q + 1) f_i^*(\vec{r}_i, t) + (1 + 2q)(1 - 2q) f_i^*(\vec{r}_i - \vec{e}_i, t) - q(1 - 2q) f_i^*(\vec{r}_i - 2\vec{e}_i, t) + 2\rho_0 w_i (\vec{e}_i \cdot \vec{u}_b) \quad (5)$$

and $q \geq 1/2$

$$f_{\bar{i}}(\vec{r}_i, t + \Delta t) = \frac{1}{q(2q + 1)} f_i^*(\vec{r}_i, t) + \frac{(2q-1)}{q} f_i^*(\vec{r}_i, t) + \frac{(1-2q)}{(1+2q)} f_i^*(\vec{r}_i - \vec{e}_i, t) + \frac{2\rho_0 w_i}{q(2q+1)} (\vec{e}_i \cdot \vec{u}_b), \quad (6)$$

with $q = |AW|/|AB|$ being the fractional wall distance. Here w_i is the weighting factor of the lattice model and $\vec{u}_b$ the boundary velocity. It should be noted that the here applied boundary conditions shown in Eqs. (5)-(6)(4) correspond to quadratic interpolation. For $q = 1/2$ the equations simplify to half way bounce back.

Thermal boundary conditions

Similar to the interpolated boundary conditions by Bouzidi et al. [12], Li et al. [13] proposed interpolated bounce back schemes for the energy distributions. The Dirichlet boundary conditions are given for $q < 1/2$

$$g_{\bar{i}}(\vec{r}_i, t + \Delta t) = (-2q)g_i^*(\vec{r}_i, t) + (2q - 1)g_i^*(\vec{r}_i - \vec{e}_i, t) + 2w_i\Phi_D \tag{7}$$

and $q \geq 1/2$

$$g_{\bar{i}}(\vec{r}_i, t + \Delta t) = -\frac{1}{2q}g_i^*(\vec{r}_i, t) + \frac{(2q - 1)}{2q}g_{\bar{i}}^*(\vec{r}_i, t) + \frac{1}{2q}2w_i\Phi_D. \tag{8}$$

Here Φ_D is the temperature at the solid boundary and all other quantities correspond to those in Eqs. (5)-(6). In contrast to [12] only linear interpolated boundary conditions are proposed; for more details the reader is refereed to [11,13].

Refilling

Due to particle movement several fluid nodes occur which were previously solid nodes (comp. Fig. 2). Therefore, the distribution functions at these nodes are unknown and a refill treatment needs to be applied. Different refill methods were developed in the past [14–18]. The simplest refilling can be performed by setting all distributions f_i to the value of the equilibrium distribution $f_i^{eq}(\rho_0, u_b)$ which in turn is obtained by using ρ_0 as the fluid density and the boundary velocity u_b for fluid velocity [16]. This method is referred as the FEQ refilling.

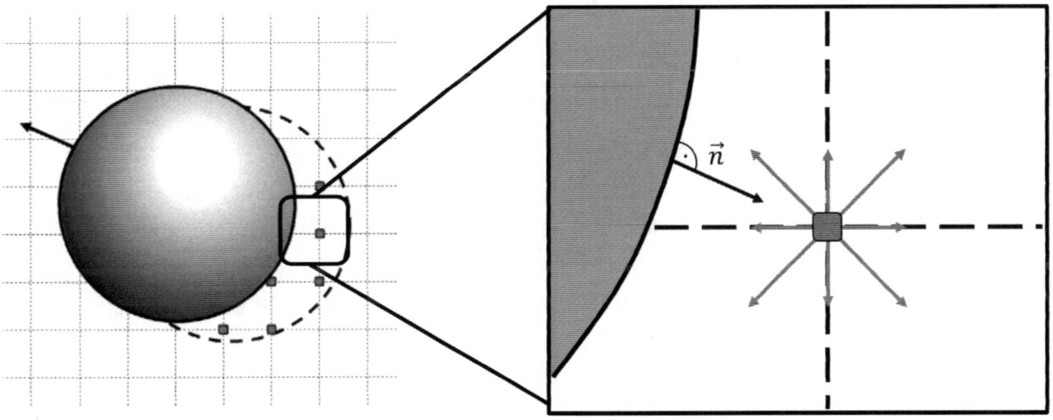

Fig. 2: Schematic representation of particle movement with corresponding refill nodes and unknown distribution functions.

Caiazzo [15] suggested to add a non-equilibrium term f_i^{neq} to the f_i^{eq}. The f_i^{neq} term can be obtained by simply copying it from the closest neighboring node in normal direction. A further scheme is the so called extrapolation (EP) refilling [16], where the unknown distributions are obtained by a second-order normal extrapolation as follows:

$$f_i(\vec{r}_i, t + \Delta t) = 3f_i(\vec{r}_i + \vec{e}_n, t + \Delta t) - 3f_i(\vec{r}_i + 2\vec{e}_n, t + \Delta t) + f_i(\vec{r}_i + 3\vec{e}_n, t + \Delta t), \tag{9}$$

where $\vec{e}_n$ is the direction closest to the normal vector, or in other words where the angle between $\vec{e}_i$ and $\vec{n}$ is minimized (comp. Fig. 2). The considered nodes are the first, second and third fluid node along $\vec{e}_n$. Fang et al. [14] proposed a refill model where an extrapolation for all distributions is conducted, but in contrast to [15] the extrapolation is performed in all directions. Afterwards the computed distributions are averaged over the number of the regarded directions. Different types of extrapolation (e.g. first-order, second-order) can be applied here, but it should be noted that higher orders significantly reduce fluctuations [19]. The above routine is referred to as the average extrapolation refilling (AEP) in the literature. A more sophisticated refill algorithm a so called velocity-constrained (VC) method is proposed by Peng et al. [17]. Where firstly the unknown distributions are extrapolated and afterwards temporary moments are obtained $\vec{m}(\vec{r},t) = M \cdot \vec{f}(\vec{r},t)$. Then the momentum moments (j_x, j_y, j_z), which are connected to the fluid velocity, are modified in such way that the fluid velocity is set to the velocity of the nearest boundary u_b and thus a new moment vector $\vec{m}^*(\vec{r},t)$ is obtained. Finally, new distribution functions are determined $\vec{f}(\vec{r},t) = M^{-1} \cdot \vec{m}^*(\vec{r},t)$. Peng et al. find that their refilling method significantly reduces fluctuations in hydrodynamic forces when compared to the refilling schemes above [17].

All introduced refilling algorithms [14–17] are derived to improve the discrete particle movement and minimize force fluctuations. The introduced refill schemes are proposed for the density distributions. However, refilling is not limited to the hydrodynamic side. In the present study refilling schemes are also applied to energy distribution functions. A reduction of heat transfer fluctuations is expected.

Discrete element method

The translational and rotational particle motion in the DEM is obtained by the integration of Newton's and Eulor equation of each particle given by

$$m_i \frac{d^2\vec{x}_i}{dt^2} = \vec{F}_i^c + \vec{F}_i^{pf} + m_i\vec{g}, \tag{10}$$

$$\hat{I}_i \frac{d\vec{W}_i}{dt} = \sum \vec{M}_i. \tag{11}$$

Here m_i is the particle mass, $d^2\vec{x}_i/dt^2$ particle acceleration, $\vec{F}_i^c$ contact force, $\vec{F}_i^{pf}$ particle/fluid force, $m_i\vec{g}$ gravitational force, $d\vec{W}_i/dt$ angular acceleration, $\vec{W}_i$ angular velocity, $\vec{M}_i$ external moment resulting out of contact or particle/fluid forces, $\hat{I}_i$ the inertia tensor along the principal axis. For detailed description the DEM framework please see [2,3,5]. The DEM hands over particle positions and their translational and rotational velocities to the LBM. The LBM computes particle/fluid forces and rotational moments out of particle/fluid forces and passes these quantities back to the DEM routine. The unit transformation between the two routines is performed using dimensional analysis.

Results

Drag force moving sphere

The implemented LBM framework is sufficiently validated for static particles [7,8]. Therefore, simulations of static particles can be used as a reference. To evaluate the numerical error caused by the particle movement the drag force of a moving sphere is compared to a static case. For this initial test a sphere is placed in the domain center of a rectangular channel with periodic boundaries in x-direction and no-slip walls in all other directions. The sphere is given a constant velocity and reaches a steady state where the acting forces are in balance. After the steady state is reached the drag force is averaged over time. In the static case - the reference case - the sphere is placed in the same manner in the domain but remains motionless and the walls are given a constant boundary velocity. The investigations are performed for different momentum exchange methods (MEM) and different refill algorithms. Two models for particle/fluid force calculation namely the method proposed by Aidun et el. [20] and Wen et al. [19] are studied. For the refilling the velocity-constrained (VC) refilling proposed by Peng et al. [17], FEQ method and the extrapolation (EP) proposed by Lallemand and Luo [16] are applied.

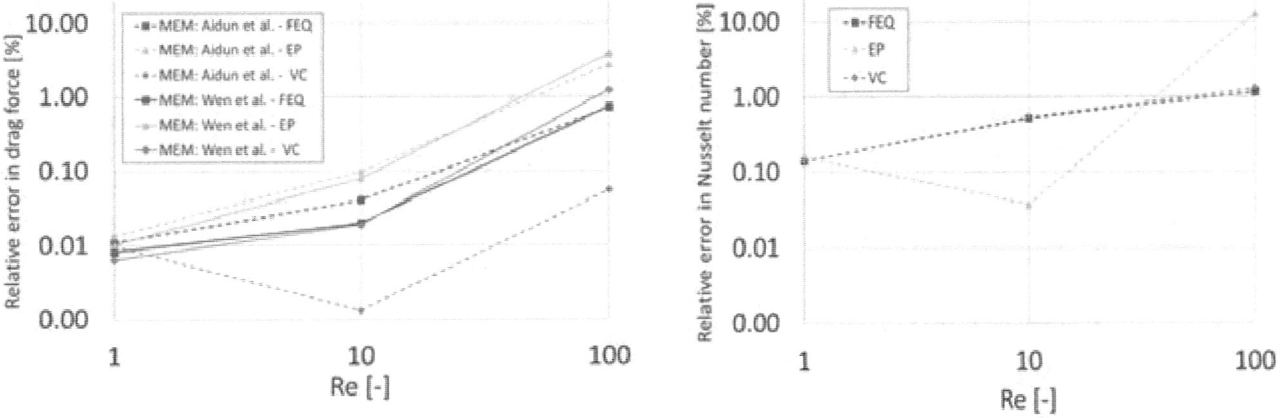

Fig. 3: Relative error in drag fore (left) and Nusselt number (right) computed for moving particles. Comparison of Momentum exchange methods (MEM) by Aidun et al. [20], *Wen et al.* [19] *and refill algorithms proposed by Peng et al. (VC)* [17], *Lallemand & Luo (FEQ,EP)* [16].

Fig. 3 shows the numerical error obtained from the comparison of moving and static particles for Reynolds numbers $Re = 1$, $Re = 10$ and $Re = 100$. It can be seen from the results that the particle movement produces a relatively small numerical error. All drag forces for moving particles are in good agreement with those obtained from static cases. The FEQ and the velocity-constrained (VC) refilling methods provide a slight better result as the normal extrapolation (EP). The two force evaluation methods do not differ significantly from each other. As can be seen, the numerical error increases with growing particle Reynolds number. This was expected, because with increasing velocity the number of refilled fluid nodes rises which leads to an increase of the numerical error.

Two particles sedimentation: Drafting, Kissing, Tumbling

To evaluate particle/particle interactions and the behavior of particles in the wake region a sedimentation of two in line arranged particles is studied. The base area of the numerical domain amounts to $1\ cm \times 1\ cm$ and the height to $4\ cm$. The dynamic fluid viscosity is $\mu = 0.01\ g/cm \cdot s$ and the particle/fluid density ratio is $\rho_s/\rho_f = 1.14$. The particles have a diameter of $d_p = 1/6\ cm$, which are resolved with 8.33 nodes and are arranged in line at $0.5\ cm$ and $0.84\ cm$ from the top. The particles are exposed to gravity force ($g = 9.81\ m/s^2$) and are released at $t = 0$ from their initial positions. The described problem was addressed by Glowinski et al.[21], Apte et al. [22] and Coa et al.[23] in the past.

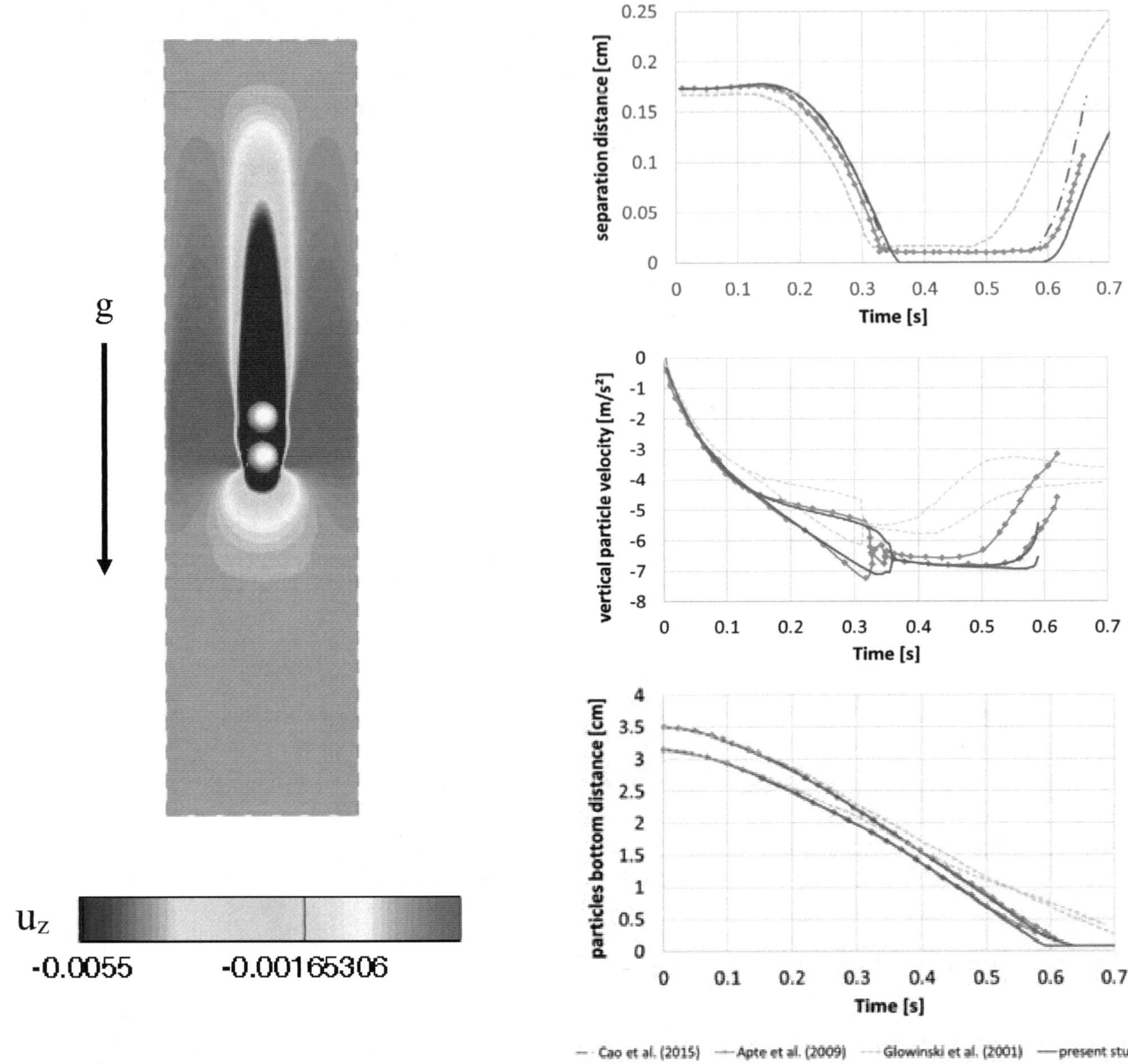

Fig. 4: Sedimentation of two particles: distance of two particles (top), vertical particle velocity (center) and distance from bottom to particle center (bottom) compared to Glowinski et al. [21], *Apte et al.* [22] *and Cao et al.* [23].

The obtained results are shown in Fig. 4 where the distance between the two particles, their vertical velocity and their distance to the domain bottom are plotted over simulation time and compared to [21–23]. For $t < 0.1$ the spheres accelerate equally and the distance between them remains constant. With increasing particle velocities higher viscous forces are acting on the front particle. The second sphere - which is moving through the wake region of the first sphere - experiences less friction and thus accelerates faster. The velocity of the second particle increases and the distance decreases until the two particles collide or "kiss". The process until the first contact at $t \approx 0.35$ is referred as "drafting". After contact the process of "tumbling" begins where the distance between the particles is constant until they reach the bottom and separate again from each other. A very good agreement to Apte et al. [22] and Coa et al. [23] for the whole simulation is observed. However, the particle distance during tumbling in [21–23] differs to our results. In contrast to our study the authors [21–23] applied a short-range repulsive force which leads to a constant interparticle spacing. If the repulsive force is applied "kissing" occurs earlier. With this simple set-up the numerical reliability and accuracy of the derived DEM/LBM framework is demonstrated. It should be furthermore noted the here applied particle resolution ($d_p = 8.33$) is lower compared to those in [22,23].

Conclusions

A coupled DEM/LBM framework for simulating resolved particle/fluid systems including interphase heat transfer is derived and validated. The fluid phase is covered by the LBM and the particle motion by the DEM. At the LBM side a double MRT collision model with interpolating bounce back scheme is applied. Different proposed refilling schemes are tested. A very good agreement for drag force and Nusselt number is achieved in the comparison between a static and moving particle for same Reynolds numbers. Furthermore, sedimentation of two in line arranged particles is studied and compared to results published in scientific literature. It can be stated that the simulation results are in good agreement to other studies. The findings indicate that not only the computation of the resolved flow and the particle/fluid forces (LBM) are done accurately but also validate the DEM routine and the coupling with it. Refilling schemes can be applied to density as well as energy distributions and can help to reduce fluctuations in the drag force or Nusselt number. The derived framework can be applied for resolved simulations of dynamical particle/fluid systems including heat and momentum transfer like fluidized beds, pneumatic conveyance etc..

Acknowledgements

The authors gratefully acknowledge the support by DFG through grant numbers KR3446/6-1 and 3446/13-1.

References

[1] Vollmari K and Oschmann T (2017) Mixing quality in mono- and bidisperse systems under the in fl uence of particle shape : A numerical and experimental study. Powder Technol 308: 101–113.
[2] Vollmari K and Jasevi R and Kruggel-emden H (2016) Experimental and numerical study of fl uidization and pressure drop of spherical and non-spherical particles in a model scale fl uidized bed. 291: 506–521.
[3] Oschmann T and Vollmari K and Kruggel-emden H and Wirtz S (2015) Numerical Investigation of the Mixing of Non-Spherical Particles in Fluidized Beds and during Pneumatic Conveying. Procedia Eng 102: 976–985.
[4] Oschmann T and Schiemann M (2016) Development and veri fi cation of a resolved 3D inner particle heat transfer model for the Discrete Element Method (DEM). Powder Technol 291: 392–407.
[5] Oschmann T and Hold J and Kruggel-Emden H (2014) Numerical investigation of mixing and orientation of non-spherical particles in a model type fluidized bed. Powder Technol 258: 304–323.
[6] He X and Chen S and Doolen GD (1998) A Novel Thermal Model for the Lattice Boltzmann Method in Incompressible Limit. J Comput Phys 146: 282–300.
[7] Kruggel-Emden H and Kravets B and Suryanarayana MK and Jasevicius R (2016) Direct

numerical simulation of coupled fluid flow and heat transfer for single particles and particle packings by a LBM-approach. Powder Technol 294: 236–251.
[8] Kravets B and Kruggel-Emden H (2017) Investigation of local heat transfer in random particle packings by a fully resolved LBM-approach. Powder Technol 318: 293–305.
[9] D'Humières D and Ginzburg I and Krafczyk M and Lallemand P and Luo L-S (2002) Multiple-relaxation-time lattice Boltzmann models in three dimensions. Philos Trans A Math Phys Eng Sci 360: 437–451.
[10] Yoshida H and Nagaoka M (2010) Multiple-relaxation-time lattice Boltzmann model for the convection and anisotropic diffusion equation. J Comput Phys 229: 7774–7795.
[11] Li L and Mei R and Klausner JF (2017) Lattice Boltzmann models for the convection-diffusion equation: D2Q5 vs D2Q9. Int J Heat Mass Transf 108: 41–62.
[12] Bouzidi M and Firdaouss M and Lallemand P (2001) Momentum transfer of a Boltzmann-lattice fluid with boundaries. Phys Fluids 13: 3452–3459.
[13] Li L and Mei R and Klausner JF (2013) Boundary conditions for thermal lattice Boltzmann equation method. J Comput Phys 237: 366–395.
[14] Fang H and Wang Z and Lin Z and Liu M (2002) Lattice Boltzmann method for simulating the viscous flow in large distensible blood vessels. Phys Rev E 65: 1–11.
[15] Caiazzo A and Itwm F (2008) Analysis of lattice Boltzmann nodes initialization in moving boundary problems. Prog Comput Fluid Dy 8:3-10.
[16] Lallemand P and Luo LS (2003) Lattice Boltzmann method for moving boundaries. J Comput Phys 184: 406–421.
[17] Peng C and Teng Y and Hwang B and Guo Z and Wang L (2016) Implementation issues and benchmarking of lattice Boltzmann method for moving rigid particle simulations in a viscous flow. Comput Math with Appl 72: 349–374.
[18] Krithivasan S and Wahal S and Ansumali S (2014) Diffused bounce-back condition and refill algorithm for the lattice Boltzmann method. Phys Rev E 33313: 1–15.
[19] Wen B and Zhang C and Tu Y and Wang C and Fang H (2014) Galilean invariant fluid – solid interfacial dynamics in lattice Boltzmann simulations. J Comput Phys 266: 161–170.
[20] Aidun CK and Lu Y and Ding E-J (1998) Direct analysis of particulate suspensions with inertia using the discrete Boltzmann equation. J Fluid Mech 373: 287–311.
[21] Glowinski R and Pan TW and Hesla TI and Joseph DD (2001) A Fictitious Domain Approach to the Direct Numerical Simulation of Incompressible Viscous Flow past Moving Rigid Bodies : Application to Particulate Flow. 426: 363–426.
[22] Apte S V and Martin M and Patankar NA (2008) A numerical method for fully resolved simulation (FRS) of rigid particle – flow interactions in complex flows. J Comput Phys 228: 2712-2738
[23] Cao C and Chen S and Li J and Liu Z and Zha L and et al. (2015) Simulating the interactions of two freely settling spherical particles in Newtonian fluid using lattice-Boltzmann method. Appl Math Comput 250: 533–551.

508346 Combining High and Low-Order Computational Models to Simulate Biomass Fast Pyrolysis Reactors

Abstract:

Fast pyrolysis is a leading candidate technology for the thermochemical conversion of solid biomass into liquid bio-oils which can be used as feedstocks for producing biofuels and high-value chemicals. Bio-oils are commonly generated in fluidized, circulating, or entrained flow reactors in which biomass particles rapidly devolatilize in the absence of oxygen into mixtures of light gases, condensable bio-oil vapors, and char. To maximize bio-oil yields, the reactors typically operate at temperatures near 500°C and must maintain particle residence times in the range of 2-5 seconds and gas residence times less than 1 second. Deviations from these conditions can result in significant production and quality penalties, therefore optimal reactor design and control become crucial to achieving commercially viable bio-oil production.

In this presentation, we describe preliminary results produced with a hybrid modeling approach for simulating laboratory-scale biomass pyrolysis reactors. By exploiting the fact that the biomass solids loadings and heats of reaction are both relatively low, we can separate the complex gas-particle hydrodynamics from particle-scale and homogeneous gas-phase chemical reaction kinetics and thereby greatly accelerate the simulation process. Our approach is based on initially modeling the detailed particle-gas hydrodynamics in MFiX using the discrete element method (DEM) and the two-fluid model (TFM) to determine the residence time distributions (RTDs) of the different size biomass particles in the reaction zone. We then use the resulting particle RTDs in low-order reactor models written in Python to account for particle-scale and gas-phase reaction kinetics to estimate the trends in net bio-oil pyrolysis yield for a range of reactor conditions.

EXPERIMENTAL AND NUMERICAL INVESTIGATION OF A THREE-DIMENSIONAL PRISMATIC SPOUTED BED WITH LIQUID INJECTION

Swantje Pietsch[1], Paul Kieckhefen[1], Michael Müller[2], Michael Schönherr[2], Frank Kleine Jäger[2], Stefan Heinrich[1]*
[1]Hamburg University of Technology, Institute of Solids Process Engineering and Particle Technology, Hamburg, Germany
[2]BASF SE, Ludwigshafen, Germany

Abstract

In this contribution, a three-dimensional prismatic spouted bed for coating purposes is investigated experimentally and numerically. Spouting behavior was experimentally investigated by means of Fourier transform of the pressure drop signal. The spouting stability could be increased by two parallel draft plates and influenced by using binary or ternary mixtures of different particle sizes. Simulations were performed with a coupled CFD-DEM approach, whereby droplets were injected in a post-processing step, which allowed the estimation of the liquid surface coverage of the used γ-Al_2O_3 particles. It was found that the insertion of draft plates results in a broader liquid distribution in the first ten seconds of the process. By applying the recurrence CFD (rCFD) method, the broader liquid distribution could be confirmed for a process duration of 1 hour.

Introduction

Spouted bed technology

In contrast to conventional fluidized beds, the fluidization gas enters the process chamber of spouted beds via a tube or a slit instead of a distribution plate over the whole cross-sectional area. The high gas velocities in the slit or tube result in an entrainment of particles, whereby the typical spout is formed. Due to the over the height increasing cross-sectional area and the decreasing gas velocity, the particle velocity is reduced in the so-called fountain region and the particles fall back into the annulus zone. In this area with high particle packing, the particles move up into the gas inlet zone and are again entrained by the fluidization gas. In general, spouted bed technology has been applied in several industrial applications since its invention in 1954 [1], whereby different geometries have been developed. One distinguishes between axisymmetric, asymmetric and slot-rectangular spouted beds, which again differ in terms of gas inlet geometry. In this contribution, a slot-rectangular spouted bed in its prismatic variation with a shallow static bed height, which is lower than the prismatic region of the apparatus, is investigated. Former investigations with this type of apparatus were mostly focused on pseudo two-dimensional beds in order to reduce the number of particles in numerical investigations. Salikov et al. [2] investigated a pseudo two-dimensional prismatic spouted bed and determined different spouting regimes by high frequency pressure measurements and Fast Fourier Transform (FFT) of pressure drop signal. The resulting FFT power plots showed the main frequency of characteristic pressure fluctuations or, in case of instable spouting, a distribution of peaks. By means of that, the different spouting regimes in a spouted bed were detected depending on the gas velocity: After the fixed bed state, first bubbles are formed, which are transported to the bed surface. After the minimum spouting velocity, which determines the start of spouting behavior, a stable regime with a homogeneous bed expansion is obtained. By further increasing the gas velocity, the instable regime is reached, which is characterized by spout deflections and irregular bed expansions. At very high gas velocities another stable regime with lower particle concentrations (dilute spouting) was described by Salikov et al. [2]. In general, spouted beds are well applicable for processes with liquid injection, e.g. for spray granulation or coating, due to the more structured flow pattern and the resulting improved heat, mass and momentum transfer compared to classical bubbling fluidized beds.

Numerical investigations

Due to a lack of empirical correlations and the complex granular dynamics, numerical simulations of gas-solid processes are required. Generally, two main approaches for modeling of gas-solid processes can be distinguished: Euler-Euler and Euler-Lagrange simulations. In Euler-Euler simulations, as e.g. the Two Fluid Model with the Kinetic Theory of Granular Flow [3], both the fluid and the particle phase are treated as continua. In contrast, in Euler-Lagrange simulations, the motion of individual particles is tracked. The tracking of particle velocities and forces allows the analysis of bed dynamics on the scale of single particles, which gives information about e.g. collision frequencies and rotational velocities that cannot be obtained by experimental investigations. One of the most often applied approaches is the CFD-DEM-simulation. Computational fluid dynamics (CFD) describes the fluid phase by solving the Navier Stokes equations and via discrete element method (DEM) the particle forces, trajectories and velocities are calculated with Newton's equations of motion. With today's computational power and the use of coarse-graining approaches, millions of particles can be modeled, which allows the investigation of processes on laboratory or pilot plant scale. Nevertheless, the efficiency is not sufficient for modeling of real-time industrial processes. For this purpose, approaches with statistical functions, as e.g. recurrence CFD (rCFD) [4,5], are in development. Most of the work regarding simulation of fluidized or spouted beds was focused on conical or pseudo two-dimensional spouted beds mainly without any liquid injection. Recently, Sutkar et al. [6] investigated a pseudo two-dimensional spouted bed with injection of droplets. They modelled the collision of droplets and particles and the resulting transfer of mass, volume and momentum from droplets to particles. Simulations were verified by visual observations and measurements with infrared thermography. Additionally, in case of fluidized beds, for example Börner et al. [7] investigated particle residence times and crossing length in a cylindrical top-spray fluidized bed by applying a similarity model. Hilton et al. [8] investigated spray coating in a fluidized bed, whereby spray droplets were modelled as discrete Stokesian particles within the fluid flow field. Toschkoff and Khinast [9] summarized different approaches for simulation of coating processes in fluidized beds and drum coaters. They found analytical methods to be the fastest and easiest to apply approaches though drawbacks in detailed investigations were reported. The methods differ in the acting forces that are taken into account. While Shi and McCarthy [10] tracked the moisture content for each particle by the simplification of a uniform layer around the particles but the inclusion of viscous and cohesive forces, liquid transfer and viscous forces were ignored in the proposed model by Muguruma et al. [11], who focused on cohesion forces due to liquid bridges. The aim of this work was the finding of a connection between the spouting stability in a three-dimensional spouted bed and the homogeneity of a liquid layer induced by spray coating.

Material and Methods

The apparatus investigated was lab-scale ProCell 5 (Glatt, Germany). In Figure 1 the whole apparatus and a sketch of the process chamber with the two-fluid nozzle (970-S4, Schlick, Germany) in bottom-spray configuration is shown. As the observation window in the process chamber is small and only in two dimensions, a transparent replica of the apparatus made of plexiglass was installed (Figure 1c) in order to validate the spouting stability.

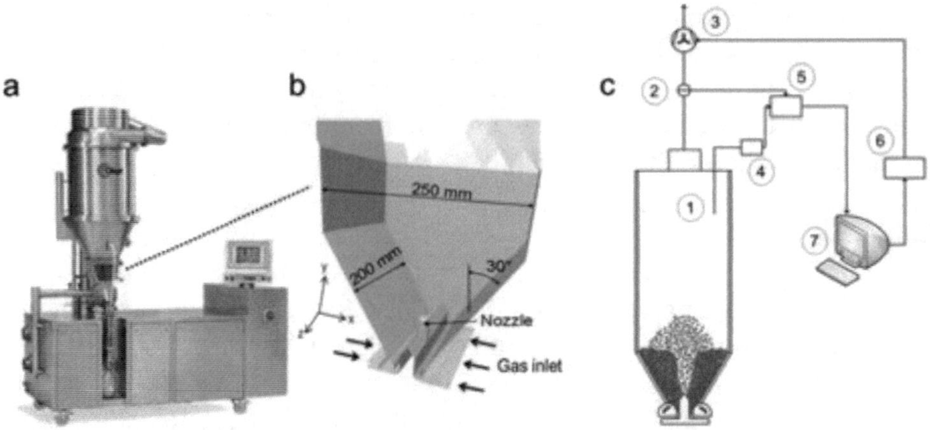

Figure 1. ProCell 5 (Glatt, Germany): (a) Laboratory plant, (b) Geometry of the process chamber with nozzle in bottom-spray configuration.

The fluidization gas is sucked through the apparatus, meaning that it is operated at under-pressure. In the plexiglass version, a high frequency pressure detector was installed. As influencing parameter the fluidization gas volume flow rate was varied. Nearly monosized γ-Al_2O_3 particles (Sasol, Germany) with a mean diameter (d_p) of 656 µm were used as model core particles. Additionally, γ-Al_2O_3 particles with a diameter of 1.8 and 3.0 mm were available for the investigation of spouting stability in experiments. The particles were assumed to be ideally monosized and spherical in the simulations. Former investigations [12,13] showed a stabilizing effect of the flow pattern by inserting draft plates into the process chamber. Therefore, two parallel plates (60 mm x 10 mm x 20 mm) with a horizontal distance of 45 mm and a distance in vertical direction from middle profile of 10 mm were inserted in part of the simulations. A bed mass of 1 kg particles was used. For first investigations of liquid injection, pure water was injected.

Simulations

CFD-DEM coupling

Simulations were performed with open source software CFDEMcoupling, which couples the CFD tool OpenFOAM with DEM software LIGGGHTS [14]. The gas phase dynamics in presence of a granular phase are solved by applying the PISO (Pressure-Implicit with Splitting of Operators) algorithm in OpenFOAM to the incompressible Navier Stokes equations. The calculated fluid velocity field is transferred to DEM software LIGGGHTS, which calculates the particles' interactions, velocities and forces by solving Newton's equations. Although computational performance is steadily increasing, coarse-graining approaches are necessary for simulation of industrial or even laboratory scale. In this contribution, the approach by Bierwisch et al. [15] is applied with a scaling factor of four, which reduces the number of particles from more than 6 million to around 80,000 parcels. The geometry of the process chamber, as shown in Figure 1 b, was drawn in CAD software SolidWorks and the simulation mesh was created by OpenFOAM meshing tool SnappyHexMesh. For the fine inlet geometry a mesh cell size of 3.125 mm was used, which corresponds to the particle diameter size times 1.7. The process chamber area above the inlet region had a mesh cell size of 6.25 mm ($3.5 \cdot d_p$) and in the upper part a cell size of 12.5 mm ($7 \cdot d_p$) was found to be sufficient as the particle concentration in this area is low. The mesh around the nozzle was calibrated with spray pattern at different heights that were determined in experimental investigations. In total, the mesh contained about 700,000 cells of mostly hexahedral type.

Simulation setup

The particle properties were measured in several experimental set-ups and are given in Table 1 with additional simulation parameters. For example, Young's modulus was determined in single particle compression test with Texture Analyzer (Stable Micro Systems, UK) and restitution coefficient was measured with free-fall device as described in [16].

Table 1. Fluid and particle properties and simulation parameters.

Parameter/Setting	Value	Unit	Parameter/setting	Value	Unit
Slit height h	3.5	mm	Wall:		
Time step:			Young's modulus Y_w	2.5	GPa
CFD Δt_{CFD}	$5 \cdot 10^{-5}$	s	Poisson ratio ν_w	0.3	-
DEM Δt_{DEM}	$2.5 \cdot 10^{-7}$	s	Restitution coefficient:		
Liquid injection Δt_{post}	10^{-4}	s	Particle-particle $e_{p\text{-}p}$	0.72	-
Particle diameter d_p	656/1800	µm	Particle-wall $e_{p\text{-}w}$	0.73	-
Scaling factor δ	4	-	Static friction:		
Particle density ρ_p	1328	kg/m³	Particle-particle $\mu_{s,p\text{-}p}$	0.06	-
Fluid density ρ_f	1.225	kg/m³	Particle-wall $\mu_{s,p\text{-}w}$	0.05	-
Fluid dynamic viscosity μ_f	$1.7894 \cdot 10^{-5}$	kg/(m·s)	Field sampling frequency f	10^2	Hz
Particles:			Liquid injection:		
Young's modulus Y_p	3.06	GPa	Parcel injection rate	10^5	Hz
Poisson ratio ν_p	0.3	-	Mass flow rate	5	g/min

Liquid injection

Liquid injection was performed after the resource-intensive CFD-DEM simulations. The post-processing solver, which is based on a combination of the sprayFoam-solver and the cfdemPostproc-post-processing tool, loads the velocity fields and particle positions. Droplets are injected as a Lagrangian phase in the loaded velocity fields with a frequency of $2.5 \cdot 10^6$ Hz at the nozzle inlet patch with a mass flow rate of 5 g/min. A particle-droplet capturing efficiency of 100 % was assumed, whereby the area coated by a single parcel was set equal to the droplets crossectional area.

Recurrence CFD (rCFD)

As the flow pattern in spouted beds is characterized by a high periodicity, they show a high applicability for the recurrence CFD method. For a detailed description of the mechanism, the reader is referred to the work of Lichtenegger and Pirker [4]. For performing the recurrence simulation a data base is needed which is generated by detailed CFD-DEM data. The length of the sampling period is important as only the mixing patterns present during this period can be reproduced in rCFD. In this study, the 10 s simulation data was used for the recurrence plot.

Results and Discussion

Spouting stability

In order to investigate the spouting stability, experiments with different bed fillings and varying gas volume flow rates (20 – 170 m³/h) for 1 kg γ-Al_2O_3 particles (d_p= 656 µm) were performed. Up to a gas flow rate of 30 m³/h the bed remained completely static and measured pressure drop fluctuations were irregular with low amplitudes. With increasing gas velocity small cavities were observed, which turned into bubbles that ascended to the bed surface. The minimum spouting velocity, which corresponds to the point of initial spouting was determined at a gas flow rate of 40 m³/h. Stable spouting occurred at gas flow rates between 40 and 70 m³/h. The corresponding pressure drop signals show regular fluctuations, which results in single peaks in the power plots. From the power plot of a volume flow rate of 50 m³/h a main frequency of about 5 Hz can be determined. The power plot and the corresponding snapshots obtained by high speed camera measurements are shown in Figure 2.

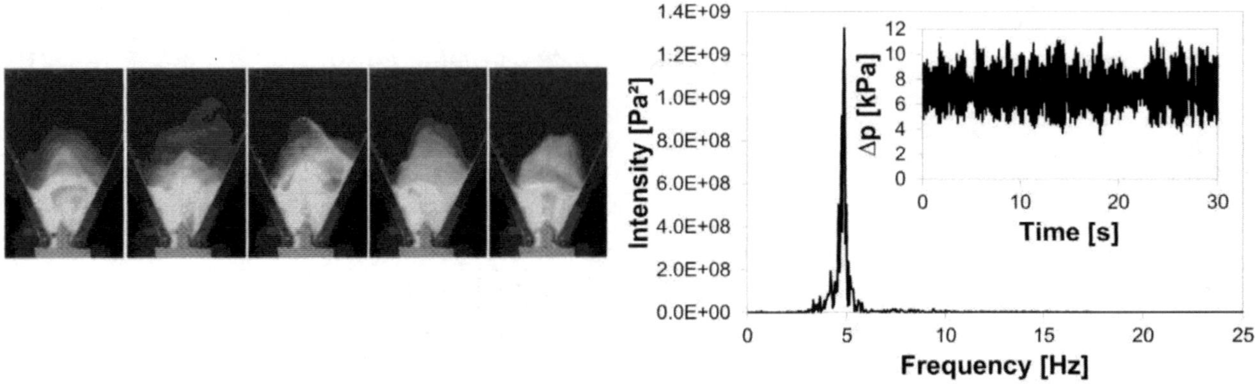

Figure 2. Snapshots obtained from experiment at a gas flow rate of 50 m³/h (stable regime) in time interval of 0.1 s and corresponding FFT power plot with measured pressure fluctuations.

Further increase of the gas velocity results in irregular bed expansion heights and spout deflections. This inhomogeneity is also visible in the irregular pressure fluctuations and the broad peak distribution in the power plot indicating the instable regime. In summary, a dependency of the spouting stability on the gas volume flow rate for the given bed material and spouted bed apparatus can be obtained by plotting the amplitude expressed as standard deviation of the pressure drop fluctuations against the gas volume flow rate. The transition from bubbling to stable regime and from stable to instable regime can be detected by a sudden increase in the standard deviation. During stable spouting, the amplitude is almost constant as shown in Figure 3 [17].

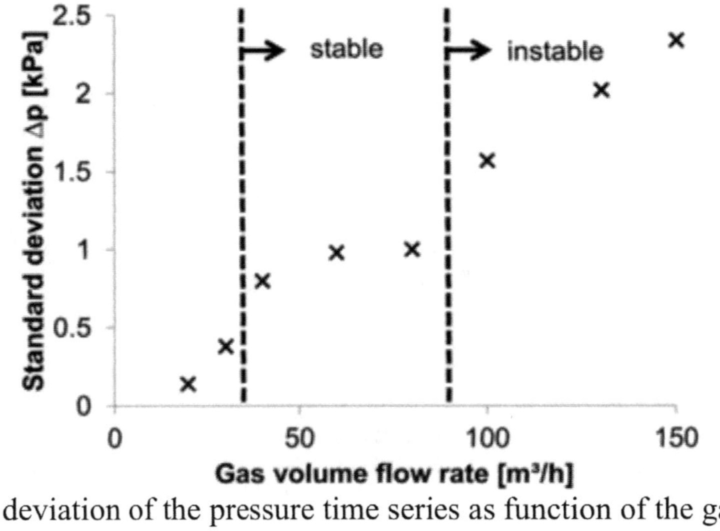

Figure 3. The standard deviation of the pressure time series as function of the gas volume flow rate for a bed mass of 1 kg γ-Al_2O_3 particles (d_p= 656 μm) indicating the transition between different spouting regimes.

Previous investigations were performed with an almost monodisperse particle system. When performing processes with liquid injection, the particle size is changing with time due to the layer building and the size is distributed over the entire particle batch. In order to judge the influence of different particle sizes in the bed on the spouting stability, experiments with additional γ-Al_2O_3 particles of 3 mm and of 1.8 mm were performed. In case of a mixture of 50 wt.-% 0.656 mm and 50 wt.-% of 1.8 mm sized particles, the range of stable spouting remained almost the same (40-80 m³/h) as for the small particles. But when using three different sized particles in the same ratio, the spouting behavior is noticeable changed and no single slim peak over the whole gas volume flow rate was observed. Exemplarily, the FFT signals for 50, 80 and 100 m³/h are shown in Figure 4. The absence of the stable regime comes along with segregation effects observed in experiments with colored particles (yellow: 0.656 mm, red: 1.8 mm, white: 3.0 mm).

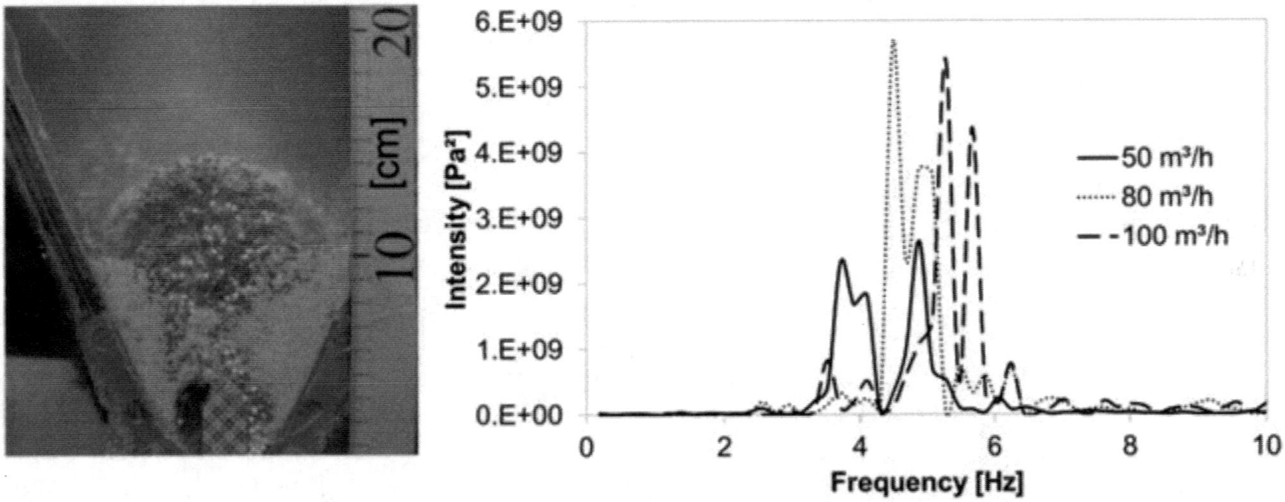

Figure 4. Snapshot obtained from experiment at a gas flow rate of 60 m³/h and FFT power plots for three different spouting regimes indicating the absence of a stable regime.

Liquid injection

As the liquid distribution on the scale of single particles is not accessible in experiments, it was investigated numerically. For simplification only monodisperse systems were simulated. In order to quantify the influence of the different gas volume flow rates with their related liquid deposition rates on the particle surface coverage, the respective distributions for a simulation time of 10 s are given in Figure 5a based on the approach of Kariuki [18].

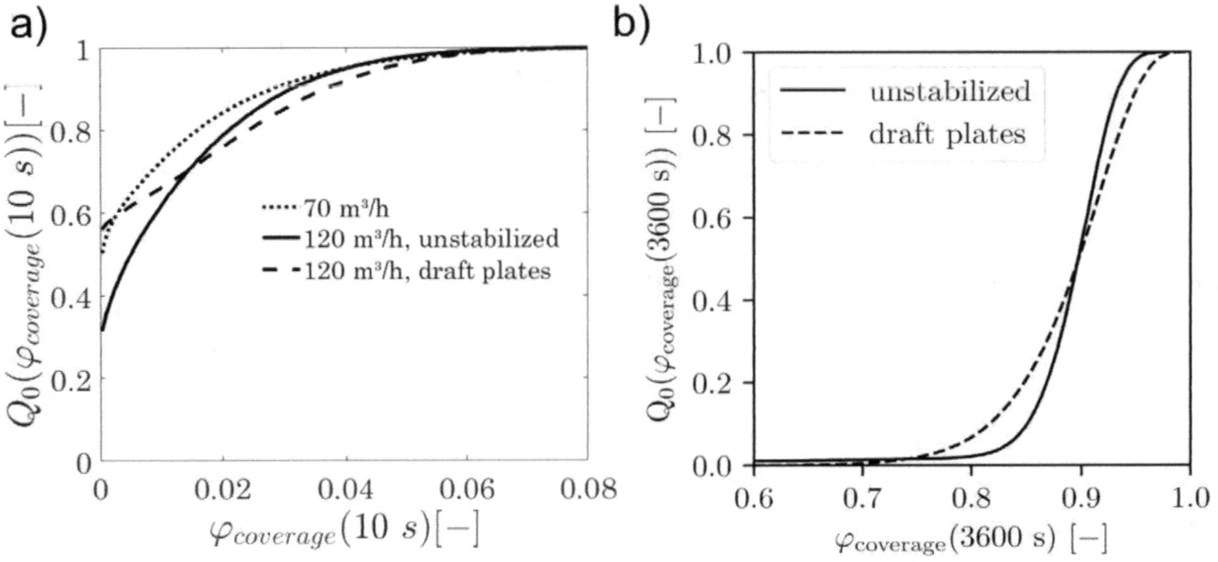

Figure 5. Cumulative number distribution of particle surface coverage, 1 kg γ-Al_2O_3 particles, a) 10 s CFD-DEM simulation; 70, 120 m³/h and 120 m³/h with draft plates, b) 3600 s rCFD simulation; 120 m³/h and 120 m³/h with draft plates.

A comparison for the original chamber geometry shows that in case of stable spouting (70 m³/h) around 50 % of the particles are uncoated after 10 s, whereas only about 32 % are completely uncoated in instable regime (120 m³/h). Nevertheless, the number of particles with a surface coverage of almost 6 % is comparable for both regimes. The curve of the optimized chamber geometry with draft plates is similar to that of the stable spouting, which shows their stabilizing effect. Similar to the stable regime, a lot of uncoated particles (around 56 %) exist after 10 s, which indicates a high distribution of coverage during stable spouting. In order to judge whether the draft plates results in a more inhomogeneous droplet distribution, a rCFD simulation based on the CFD-DEM data was performed for 1 hour [19]. The resulting distribution is given in Figure 5 b. As can be seen, the distribution with draft plates is broader compared to the original chamber geometry. Nevertheless, it is assumed that especially for sticky particles and those tending to agglomeration, the draft plates are still advantageous as the spray zone is physically separated from the annulus zone.

Summary

In this contribution, the stability range of a three-dimensional prismatic spouted bed was determined experimentally. The stability could be increased by the insertion of two parallel draft plates. When considering ternary systems of γ-Al_2O_3, the stability is decreased, which will influence granulation processes. Liquid injection for coating purposes was analyzed during post-processing of coupled CFD-DEM simulations of a three-dimensional prismatic spouted bed with monodisperse particles. By means of additional rCFD simulations it was shown that particle coating is more homogeneous in case of the unstabilized process chamber, which can be explained by the more degrees of freedom that the particles have resulting in a better distribution of the droplets on the particle bed. Nevertheless, the draft plates seem promising for particles, which are difficult to handle in the process, as e.g. very electrostatically charged particles.

References

[1] K.B. Mathur, P.E. Gishler, A technique for contacting gases with coarse solid particles, AIChE J. 1 (1955) 157–164.

[2] V. Salikov, S. Antonyuk, S. Heinrich, V.S. Sutkar, N.G. Deen, J.A.M. Kuipers, Characterization and CFD-DEM modelling of a prismatic spouted bed, Powder Technology 270 (2015) 622–636.

[3] D. Gidaspow, R. Bezburuah, J. Ding (Eds.), Hydrodynamics of circulating fluidized beds: Kinetic theory approach, 1991.

[4] T. Lichtenegger, E. Peters, J.A.M. Kuipers, S. Pirker, A recurrence CFD study of heat transfer in a fluidized bed, 5th International Granulation Workshop 172 (2017) 310–322.

[5] T. Lichtenegger, S. Pirker, Recurrence CFD – A novel approach to simulate multiphase flows with strongly separated time scales, Chemical Engineering Science 153 (2016) 394–410.

[6] V.S. Sutkar, N.G. Deen, A.V. Patil, V. Salikov, S. Antonyuk, S. Heinrich, J.A.M. Kuipers, CFD–DEM model for coupled heat and mass transfer in a spout fluidized bed with liquid injection, Chemical Engineering Journal 288 (2016) 185–197.

[7] M. Börner, A. Bück, E. Tsotsas, DEM-CFD investigation of particle residence time distribution in top-spray fluidised bed granulation, Chemical Engineering Science 161 (2017) 187–197.

[8] J.E. Hilton, D.Y. Ying, P.W. Cleary, Modelling spray coating using a combined CFD–DEM and spherical harmonic formulation, Chemical Engineering Science 99 (2013) 141–160.

[9] G. Toschkoff, J.G. Khinast, Mathematical modeling of the coating process, Progress in Film Coating 457 (2013) 407–422.

[10] D. Shi, J.J. McCarthy, Numerical simulation of liquid transfer between particles, Powder Technology 184 (2008) 64–75.

[11] Y. Muguruma, T. Tanaka, Y. Tsuji, Numerical simulation of particulate flow with liquid bridge between particles (simulation of centrifugal tumbling granulator), Powder Technology 109 (2000) 49–57.

[12] V. Salikov, S. Heinrich, S. Antonyuk, V.S. Sutkar, N.G. Deen, J.A.M. Kuipers, Investigations on the spouting stability in a prismatic spouted bed and apparatus optimization, Advanced Powder Technology 26 (2015) 718–733.

[13] S. Pietsch, S. Heinrich, K. Karpinski, M. Müller, M. Schönherr, F. Kleine Jäger, CFD-DEM modeling of a three-dimensional prismatic spouted bed, Powder Technology 316 (2017) 245–255.

[14] C. Goniva, C. Kloss, N.G. Deen, J.A.M. Kuipers, S. Pirker, Influence of rolling friction on single spout fluidized bed simulation, Particuology 10 (2012) 582–591.

[15] C. Bierwisch, T. Kraft, H. Riedel, M. Moseler, Three-dimensional discrete element models for the granular statics and dynamics of powders in cavity filling, Journal of the Mechanics and Physics of Solids 57 (2009) 10–31.

[16] B. Crüger, V. Salikov, S. Heinrich, S. Antonyuk, V.S. Sutkar, N.G. Deen, J.A.M. Kuipers, Coefficient of restitution for particles impacting on wet surfaces: An improved experimental approach, Particuology 25 (2016) 1–9.

[17] S. Pietsch, P. Kieckhefen, S. Heinrich, M. Müller, M. Schönherr, F.K. Jäger, CFD-DEM modelling of circulation frequencies and residence times in a prismatic spouted bed, Chemical Engineering Research and Design (2018).

[18] W.I.J. Kariuki, B. Freireich, R.M. Smith, M. Rhodes, K.P. Hapgood, Distribution nucleation: Quantifying liquid distribution on the particle surface using the dimensionless particle coating number, Chemical Engineering Science 92 (2013) 134–145.

[19] P. Kieckhefen, T. Lichtenegger, S. Pietsch, S. Pirker, S. Heinrich, Simulation of Spray Coating in a Spouted Bed using Recurrence CFD, Particuology (2018), accepted.

507132 CFD-DEM Simulation for Drying of Food Grain in a Fluidised Bed

Abstract:

Drying is a common practice for food grain processing, mainly to prevent postharvest deterioration, to restrict microbial growth and to extend storage time. Drying of food grains by gas consumes huge thermal energy and it is of importance to improve its energy efficiency by optimization and integration of drying processes. Numerical approach is a cost-effective alternative way of experimental approach for optimisation and modification of the existing drying processes. A combined computational fluid dynamics (CFD) - discrete element method (DEM) numerical model has been developed to provide an accurate and detailed predictions for both fluid and solid phases of fluidised bed drying. The model is incorporated with a simple drying model to consider water evaporation as a resemblance to chemical reaction. The developed model can provide both macroscopic and microscopic drying characteristics such as drying kinetics and moisture content of individual particle. First, the model is validated by comparing the predicted drying rate curve and temperature profile with the experimental data, showing good agreement. Then, a parametric study has been conducted based on the validated model to examine the effects of different operational conditions and grain properties on drying rate, thermal efficiency and moisture distribution of particles. The model follows a general understanding of drying that drying rate increases with the increase of air velocity and air temperature, and the decrease of initial moisture content and particle size. Thermal efficiency of dryer increases with increasing temperature and decreasing air velocity, initial moisture content and particle size. However, the standard deviation of moisture distribution of particles (obtained at a given mean moisture content) decreases with the increase of air velocity or particle size and the decrease of temperature or initial moisture content. Based on particle scale information, new finding is revealed that is, higher air velocity, higher particle size, lower temperature or lower initial moisture content gives more uniform moisture distribution of particles. Such a particle scale model would be useful for designing and controlling drying processes.

507969 Importance of Cohesion Strength Measurements As Flowability Indicator of Powders in a Fluidized Bed

Abstract:

It is well known that the difference in bulk material properties such as particle size, shape and density as well as surface moisture influences the flowability of fine powders [1]. Fluidized beds are commonly treated as i) at "incipient fluidization" mode whilst the packed-bed of particles are about to "break" solid-solid surface contact and ii) in "suspension mode" whilst the particles are fully-suspended in the carrier fluid which supports their weight and is subject to fluid-solid surface interactions [2]. This study links cohesion strength with the flowability of commonly used industrial fine powders (i.e. Cement, Barite, Bentonite, Alumina and Silicon carbide). The solid like structures observed in cohesive powders can be broken by pre-consolidating and initiating shearing using an Anton Paar MCR Rheometer with Powder Cell and a fluidization set with controlled gas flow [3].

Cohesion strength is determined via shearing the sample following a protocol where the sample is compressed at low consolidations with an air-permeable piston and then a spoke-shaped geometry penetrates the sample and shears it. Warren Spring cohesion is calculated as the maximum value.

Furthermore comparisons on shear rate dependent viscosity and Brookfield Powder Flow Tester (using annular shear cell) contribute to the calculations which determine the flowability of the powders and allow the characterization of harsh cohesive powders.

[1] Valverde J.M., Castellanos A. Effect of particle size and interparticle force on the fluidization behaviour of gas fluidized beds. Phys. Rev. E 67, 051305 (2003)

[2] Rietema, K The dynamics of fine powders, Elsevier, London (1991)

[3] Salehi H, Schutz D, Romirer R, Barletta D, Poletto M. On the use of a powder rheometer to characterize the powder flowability at low consolidation with torque resistances. AIChE Journal http://dx.doi.org/10.1002/aic.15934

507965 Investigation into the Attrition Behaviour of FCC Catalysts

Abstract:

Fluidized bed operations have a key importance in fluid catalytic cracking units. Here the intensive mechanical stress causes attrition and consequently important material loss. The distributor region, the jets and the cyclone are the most common contributors to this phenomenon since attrition of friable particles is most likely to occur at impact points where the impact forces are high. Therefore, attrition can be minimized by minimizing the impact angle Î,, and minimising stream velocity.

In this work, the attrition of fluid cracking catalyst is considered in detail with emphasis on the link between single particle properties and bulk attrition behaviour, by studying two different sample types and considering particle-wall impacts. The experiments involve the bench scale centrifugal attrition tester (developed by the Wolfson Centre for Bulk Solids Handling Technology, UK) which simulates a particle-wall impact scenario. This apparatus has more recently been used for the evaluation of particle attrition by impact [1] and comprises a rotating disc surrounded by an array of targets. The particulate samples are fed into the centre of the rotating disc and radially accelerated via the tubes and impacted against the target array. Typically, as the impact velocity and impact angle increase, the level of attrition is also observed to increase. Good correlation was found at high impact velocities/gas velocities considering real conditions found in cyclones.

This paper presents a degradation model for particles in fluidized bed conditions. The model utilises a breakage matrix approach. Where the breakage matrix is empirically-obtained from the centrifugal attrition tester, and transforms the input particle size distribution in fluidizing conditions, to the output particle size distribution after the test. Particle size analysis and particle shape analysis were carried out using Sympatec and Malvern Morphologi G3 respectively.

Regarding the operational capacity of the apparatus, the centrifugal tester allows greater control of particle impact velocity; particle-gas interactions are lower than those of the gas-blast tester and steady state conditions are more readily defined as recently summarised in a review paper [3].

[1] B. Kotzur, M.S. Bradley, R.J. Berry, R.J. Farnish, Influence of Solids Loading Ratio on Particle Attrition within a New Centrifugal Accelerator Impact Tester, in: PARTEC Int. Congr. Part. Technol., 2016.

[2] B.A. Kotzur, M.S.A. Bradley, R.J. Berry, R.J. Farnish, Breakage Characteristics of Granulated Food Products for Prediction of Attrition during Lean-Phase Pneumatic Conveying, Int. J. Food Eng. 12 (2016) 835â€“850

[3] B.A. Kotzur, R.J. Berry, S. Zigan, P. Garcia-Trinanes, M.S.A. Bradley, Particle attrition mechanisms, their characterisation and application to lean phase pneumatic conveying systems: A Review. (submitted to Powder Technology 05/05/2017)

508693 Particle Tracking Velocimetry Measurements for CFD-DEM Validation

Abstract:

The coupled Eulerian-Lagrangian method known as CFD-DEM is a powerful modeling tool that provides particle-scale, detailed insight into gas-solid multiphase flow devices. However, since CFD-DEM tracks the motion of every particle in a given system, many of the validation-grade experiments conducted previously for Eulerian-Eulerian (TFM) methods are too large (particle count too high) for CFD-DEM. By validation-grade, we mean high-quality experimental data where all particle properties and flow boundary conditions relevant to the computational model have either been well-characterized or the uncertainties quantified. Some more recent works, see e.g., Gopalan et al. (2016), have collected validation-grade experimental data specifically targeted at CFD-DEM. Previous works focused on simple systems and simple particle physics, most commonly in rectangular fluidized beds with a single gas inlet, leaving a gap between the existing CFD-DEM validation database and the TFM validation database and more realistic, engineering data. We aim to help bridge this gap by providing validation-grade data of increasingly more complex problems and physics which are still amenable to CFD-DEM--both current capabilities and projected short-term future capabilities.

In this presentation, we summarize recent experiments conducted at Particulate Solid Research, Inc. and the University of Colorado which utilize particle tracking velocimetry (PTV). At one end of the Geldart spectrum, large Group D particles are studied in a semicircular fluidized bubbling bed that is operated slightly above and slightly below minimum fluidization. Additional fluidization is provided near the flat, front face of the unit via four horizontal air jets. The two pairs of jets are submerged, subsonic but high-velocity and set at the right and left hand sides of the bed directed inwards towards each other at elevations of approximately 5 and 15 cm above the distributor. High speed video (HSV) imaging is taken of the face of the semicircular bed to capture the dynamics of the opposing jets. PTV is used to collect data from the HSV images which is then averaged into bins (grid) to create 2-D time averaged solids velocity profiles and ultimately used to calculate momentum-based jet penetration depth measurements. At the other end of the Geldart spectrum, experiments have also been performed for mildly cohesive Group A and A/C particles in the dilute riser section of a circulating fluidized bed. The entrained fines tend to form agglomerates: groupings of a few particles in size held together by interparticle forces. Local regions of the riser flow can be imaged with HSV and a new PTV method has been developed that can distinguish between primary particles and agglomerates. With the new PTV method, the hydrodynamic behavior of particles and agglomerates are extracted, such as the size, concentration, mean velocity, solids volume fraction and kinetic energy associated with relative particle and agglomerate motion (e.g. granular temperature).

CFD-DEM simulations using the open source MFiX code have also been conducted for some of the experimental conditions and we will briefly provide preliminary results, both successes and challenges. In general, the numerical simulations of the air jets in the semicircular bed compare quite favorably to the experimental data. The most noticeable bias that exists is that the simulated jets tend to penetrate further into the bed than observed experimentally, perhaps a consequence of the (momentum) point source modeling of the jets. For simulations of the cohesive particles (Geldart Group A), the CFD-DEM predictions are complicated by the fact that interparticle forces need to be modeled. We briefly report on simulations which utilize a recently developed adhesion model (LaMarche et al. 2017). Corresponding with the increase in complexity of the physics of the problem, the numerical results are degraded compared to the more idealistic (spherical, monodisperese and non-cohesive) Group D particle results, although some good agreement is still observed. Notably, when the uncertainty in local superficial gas velocity is considered, the agreement in the entrainment flux improves substantially.

Gopalan, B., M. Shahnam, R. Panday, J. Tucker, F. Shaffer, L. Shadle, J. Mei, W. Rogers, C. Guenther, M. Syamlal, Measurements of pressure drop and particle velocity in a pseudo 2-D rectangular bed with Geldart Group D particles, Powder Technology, 291, 299-310 (2016).
LaMarche, C.Q., S. Leadley, P. Liu, K. M. Kellogg, C. M. Hrenya, "Method of quantifying surface roughness for accurate adhesive force predictions," Chemical Engineering Science, 158, 140-153 (2017).

508034 Experimental Verification of Solid-like and Fluid-like States in Homogeneous Fluidization Regime of Geldart a Particles

Abstract:

Experimental Verification of Solid-like and Fluid-like States in Homogeneous Fluidization Regime of Geldart A Particles

Although homogeneous fluidization for Geldart A particles has received considerable attention for a long time, the underlying mechanism has not yet been fully understood.[1] Some ascribed the stability to inter-particle forces,[2, 3] while others insisted a purely hydrodynamic explanation.[4] Valverde et al.[5] fluidized fine particles with the average size of 8.53 ¦Ìm (i.e., Geldart C particles) by the addition of fumed silica nanoparticles and found that even during homogeneous fluidization both solid-like and fluid-like behaviors can be distinguished, indicating that the stability of homogeneous fluidization may have two distinct origins: in solid-like state, the particle contact forces dominate; while in fluid-like state, the hydrodynamic factors dominate. However, it is evident that the above conclusions were drawn based on the modified Geldart C particles. To our best knowledge, there is no experimental report on whether both solid-like and fluid-like states exist for true Geldart A particles.

It should be stressed that, however, fluidization behaviors of the modified Geldart C particles and true Geldart A particles should be different from each other: besides some apparent fluidization characteristics such as the expansion ratio, maximum free volume and the ratio of the minimum bubbling velocity to the minimum fluidization velocity, the modified Geldart C particles undergo a dynamical aggregation process,[6] while for Geldart A particles, it is still questioned whether or not aggregation presents in homogeneous expansion region.[7] Homogeneous fluidization of Geldart A particles is of great relevance: for fundamental research, it is closely associated with the instability and origin of meso-scale structures in fluidized beds;[8] for industrial application, the optimal operation for particles circulation in fluidized bed reactor-regenerator systems, such as in fluid catalytic cracking (FCC) and methanol to olefins (MTO) process, is achieved when particles flow downwards a standpipe in a bubble-free condition,[9] which is actually homogeneous fluidization. Therefore, it is significant to investigate if both solid-like and fluid-like states exist in homogeneous expansion regime for true Geldart A particles as well.

In this work, incipient fluidization characteristics of FCC catalyst, which is a typical Geldart A powder, were carefully investigated by the use of electrical capacitance tomography (ECT), pressure fluctuation and camera recording. It was found that the dependency of the standard deviation of the overall solid concentration measured by ECT on superficial gas velocity shows three platforms, which were further verified by pressure fluctuation and camera recording to correspond to fixed bed, solid-like homogeneous and fluid-like homogeneous regimes, respectively, and the gas velocity in the terminal point of each platform is the minimum fluidization velocity, the minimum bubbling velocity and the critical velocity between the solid-like and fluid-like homogeneous fluidization states, respectively. Once the existence of both states is verified, it is reasonable to speculate that the mechanism of homogeneous fluidization for Geldart A particles have two distinct factors: one dominated by inter-particle forces in solid-like state and the other dominated by fluid dynamics in fluid-like state.

Figure 1. Average and standard deviation of the overall solid concentration against superficial gas velocity.

Literature Cited

1. Oke O, Lettieri P, Mazzei L. An investigation on the mechanics of homogeneous expansion in gas-fluidized beds. *Chem Eng Sci*. 2015;127:95-105.

2. Rietema K. The effect of interparticle forces on the expansion of a homogeneous gas-fluidised bed. *Chem Eng Sci*. 1973;28:1493-1497.

3. Menon N, Durian DJ. Particle motions in a gas-fluidized bed of sand. *Phys Rev Lett*. 1997;79:3407-3410.

4. Cody GD, Goldfarb DJ, Storch GV, Norris AN. Particle granular temperature in gas fluidized beds. *Powder Technol*. 1996;87:211-232.

5. Valverde JM, Castellanos A, Quintanilla MAS. Self-diffusion in a gas-fluidized bed of fine powder. *Phys Rev Lett*. 2001;86:3020-3023.

6. Valverde JM, Castellanos A. Effect of vibration on agglomerate particulate fluidization. *AlChE J*. 2006;52:1705-1714.

7. Sande PC, Ray S. Fine mesh Computational Fluid Dynamics study on gas-fluidization of Geldart A particles: Homogeneous to bubbling bed. *Ind Eng Chem Res*. 2016;55:2623-2633.

8. Sundaresan S. Instabilities in Fluidized Beds. *Annu Rev Fluid Mech*. 2003;35:63-88.

9. Xie H-Y, Geldart D. Fluidization of FCC powders in the bubble-free regime: effect of types of gases and temperature. *Powder Technol*. 1995;82:269-277.

508190 Magnetic Resonance Imaging and Modeling of Liquid-Solid Fluidization

Abstract:

Magnetic Resonance Imaging and Modeling of Liquid-Solid Fluidization

Christopher M. Boyce[1,2], Alexander Penn[2,3], Klaas P. Pruessmann[3], Christoph R. Müller[2]

[1]Department of Chemical Engineering, Columbia University

[2]Department of Mechanical and Process Engineering, ETH Zürich

[3]Institute for Biomedical Engineering, University of Zürich and ETH Zürich

Although less studied than gas-solid fluidized beds, liquid-solid fluidized beds have a variety of applications ranging from particle separation to thermal energy storage to bioreactors[1]. On a fundamental level, liquid-solid fluidized beds typically exhibit homogeneous fluidization, making them an ideal candidate for studying particle motion in a granular gas[2]. Studies of liquid-solid fluidized beds have been somewhat limited by difficulties in studying particle and liquid motion in the interior of 3D opaque systems.

Here, we use rapid magnetic resonance imaging (MRI) to study both liquid and particle dynamics inside of a 3D liquid-solid fluidized bed. MRI techniques are used to create instantaneous maps of particle and liquid velocity over time[3], as well as the distribution of particle and liquid velocities[4] in the system at the molecular level. The findings are compared with various theories for fluid and particle motions in a granular gas. Additionally, the results are compared with computational fluid dynamics – discrete element method (CFD-DEM)[5] simulations to understand the abilities and deficiencies of such a model in capturing liquid and solid dynamics in a liquid-solid fluidized bed.

References:

(1) Epstein, N. Applications of Liquid-Solid Fluidization. *Int. J. Chem. React. Eng.* **2003**, *1*, 1–16.

(2) Rouyer, F.; Menon, N. Velocity Fluctuations in a Homogeneous 2D Granular Gas in Steady State. *Phys. Rev. Lett.* **2000**, *85* (17), 3676–3679.

(3) Penn, A.; Tsuji, T.; Brunner, D. O.; Boyce, C. M.; Pruessmann, K. P.; Müller, C. R. Real-Time Probing of Granular Dynamics with Magnetic Resonance. *Sci. Adv.* **2017**, *Accepted.*

(4) Boyce, C. M.; Rice, N. P.; Ozel, A.; Davidson, J. F.; Sederman, A. J.; Gladden, L. F.; Sundaresan, S.; Dennis, J. S.; Holland, D. J. Magnetic Resonance Characterization of Coupled Gas and Particle Dynamics in a Bubbling Fluidized Bed. *Phys. Rev. Fluids* **2016**, *1* (7), 074201.

(5) Tsuji, Y.; Kawaguchi, T.; Tanaka, T. Discrete Particle Simulation of Two-Dimensional Fluidized Bed. *Powder Technol.* **1993**, *77* (1), 79–87.

507036 Improving Circulating Dehydrogenation Technology through Optimization of Fluidization

Abstract:

The dehydrogenation of light alkanes to corresponding light olefins is industrially an important process. One of the commercial alkane dehydrogenation circulating fluidized bed reactors was underperforming for quite some time. In addition, the catalyst losses were significantly higher than the design value. Several aspects related to fluidization, catalysis and process engineering were investigated to understand and resolve these issues in the Plant. The co-authors of this study focused on the fluidization aspects of the process, and benchmarked equipment design as well as operating conditions against industrial best practices. The team extensively studied the design of feed distributors, cyclones, internals, strippers, etc. Furthermore, attrition properties were also assessed for different types of catalyst used in the unit. Several recommendations and improvement modifications were proposed by the team in order to increase olefin production and to reduce catalyst losses significantly. This study provides insights into the improvement of fluidization behavior and systematic approach followed by the team to determine underlying root causes in the circulating fluidized bed unit.

508509 Hydrodynamic Characteristics of High Pressure High Temperature Counter-Current Moving Bed Chemical Looping System Circulated with Geldart Group D Oxygen Carrier Particles

Abstract:

The chemical looping combustion processes developed at the Ohio State University (OSU), which uses Fe-based oxygen carrier and a counter-current moving bed reducer design, have the capability of achieving nearly 100% carbon capture and the flexibility of co-generating high purity hydrogen and electricity. To accommodate the moving bed operation, the oxygen carriers are classified as Geldart Group D particles with an average particle diameter of 1.5 mm and a density of 2500 kg/m^3. The system consists of a reducer operated as counter-current moving bed reactor, and a combustor operated as a fluidized bed reactor, to perform reductionâ€"oxidation reaction cycles for converting fuels carbon dioxide (CO_2). An oxidizer can also be integrated into the system, between the reducer and combustor reactors, to produce a nearly pure stream of hydrogen (H_2). The oxidizer is operated in the counter-current moving bed mode similar to the reducer reactor. An L-valve is used in the chemical looping systems to transport the solids between the moving bed reactors and the fluidized bed combustor where the aeration gas flow rate is directly correlated to the global solid circulation rate. An iso-kinetic device is installed on the standpipe of the moving bed side to measure the solids circulation rate of the system. Thousands of hours of operations with different solids circulation rates and gas processing capacities have been performed on sub-pilot and pilot scales of the systems under high operating pressures (>10 atm) and temperatures (>1,000 °C). The hydrodynamic characteristics of oxygen carrier particles operated in the systems, such as pressure drops and fluidization regimes, under different operational conditions are obtained. The operational ranges of gas flow rates for the reducer and the combustor under different pressures and temperatures are obtained. The characteristic relationship between solids circulation rate and aeration gas of L-valve under different temperatures is investigated. The gas demands for fluidizing the combustor and transporting solids through the riser under varing operating temperatures and pressures are also investigated. Fluidization regimes of the combustors are identified. The slugging behavior of the combustor with high height-to-diameter ratio under elevated temperature conditions is also investigated with the aid of ECVT measurement.

509052 Â Defluidization Behaviour of Industrial Reactive Powders at High Temperature

Abstract:

Fluidized bed reactors are used in a wide range of industrial chemical processes, and they are particularly useful in high temperature systems thanks to their ability to provide high heat transfer rates and rapid solids mixing which lead to isothermal and more controllable conditions. Their applications spread the environmental, chemical, energy and process industries. For example, they are a key technology in the petroleum industry, catalytic chemicals synthesis processes, combustion and gasification of solid fuels (coal, wastes and biomass), metals productions and many more.

Operation below incipient fluidising conditions in combustion â€" type operations can lead to particle agglomeration and sintering which change the characteristics of the bed, leading to possible bed defluidisation unscheduled process downtime and additional costs. However, the details of the physical processes under such conditions are still not well understood. Despite the countless researches that have been done so far, such processes still disclose unrevealed features and challenge the researchers worldwide.

The objective of this work is to assess the effect of high temperature on the flow properties and fluidization behaviour of such systems using carbon coke and two different titanium ores. To this end, a simple approach for the early detection of agglomeration and defluidisation in the regime below incipient fluidization - when the carbon combustion initiates the defluidization- will be followed. Such an approach is based on the simultaneous measurements of local temperatures and the pressure drop across the particle bed.

508544 Fluidization of Cohesive Materials Using Modified Solids Mechanics Models

Abstract:

Fluidization of cohesive materials is difficult to achieve. These materials tend to form channels during fluidization. They also form structures within the fluid bed that lift surround bulk material and can develop pressures and pressure gradients that are larger than the hydrostatic gradient normally experienced in fluid beds. We want to approach this phenomenon from a solid mechanics point of view. One can model the channel formed in a fluid bed as a pipe or a rathole in material that has strength and treat it as a slope stability problem â€“ except that the bulk material has pressure gradients that modify the stability of the channel surface. The basis of these equations is part of a previous work. The size of stable ratholes with gas pressure gradients can be computed using this past work. However, the motion of material in the channel during fluidization can induce a shear that is not compatible with the traditional assumptions for calculation of rathole stability from a solids mechanics point of view. Traditional theory assumes an unconfined state at the surface of the rathole and implies unique conditions on principle stress at this location. This assumption is not valid when a shear stress exists on this surface. We have reformulated the rathole stability equations with the assumption of a local shear stress at this boundary and also included gas pressure gradients. The resulting equation was designed to predict the stability of a channel formed in a bulk solid subject to movement in the channel and local gas pressure gradients. This is closer to the conditions that one would expect in the fluidization of a cohesive material.

We constructed a fluid bed and measured the standard fluidization curves as well as the size of channels formed in the bed. We also measured the unconfined yield strength of the material at low solid contact pressures and then compared this data to the values predicted from the modified rathole theory, including movement of material. An FEM analysis of gas pressures and solids stresses surrounding the channel was coupled with the results of the rathole equations to predict the channel size. We also measured the settlement time of cohesive materials as a function of the unconfined yield strength, and then created a settlement model that included gas loss through channels that formed. Finally, we correlated this behavior with the expected aerated rathole stability predictions. The results are presented in this paper.

507254 Evaluating the Influence of Dry Granulation Processing Variables on the Rheological Properties of Granules

Abstract:

Granulation is often employed in the production of pharmaceutical formulations for oral solid dose delivery. It is most often required when constituent materials in a formulation have poor flow properties, enabling these components to be combined to improve process performance and ensure content uniformity.

Dry granulation, based around the use of a roller compactor and integral mill/screen, can have significant advantages over more common wet granulation techniques. Dry granulation can reduce processing costs and can also be used with moisture/solvent/heat sensitive materials. Unlike a wet granulate, the dry granulate does not require additional drying and milling before encapsulation or tabletting.

Dry granulation is not suitable for all types of material, however, there is little research into which properties make a formulation suitable for this process. In addition, there is little to indicate which process parameters produce optimal granulate quality in order to achieve interruption free processing and high quality final products.

In this study an FT4 Powder Rheometer® (Freeman Technology, Gloucestershire, UK) was used to demonstrate how processing parameters influence the properties of a placebo formulation granulate. The granulate was produced using a roller compactor (MiniPactor, Gerteis Maschinen+Processengineering AG, Switzerland) on which the roll gap and compaction force were varied. The dynamic flow, bulk and shear properties of the resulting granulates were then evaluated using the FT4 Powder Rheometer.

The data clearly demonstrates that several rheological properties, known to influence die filling efficiency and Critical Quality Attributes (CQAs) of the resulting tablets, could be controlled by changes in roller compactor settings.

This study emphasises the importance of assessing powder properties using techniques that closely simulate process conditions and how it is possible to identify Critical Process Parameters to optimise a dry granulation process and ensure high final product quality.

507489 Assessment of the Homogeneity of an Alumina Powders Mixture Using Rheological Parameters

Abstract:

This study has been carried out in order to improve the comprehension of powder mixing in the nuclear fuel manufacturing process. Thus, a binary mixture of two Alumina (Al_2O_3) powders, GE15 and CR6, was used to simulate the oxides powders actually involved in the process. Different mixing technics and parameters were tried and compared in terms of homogeneity of the mixtures produced. The concept of intensity of segregation [1] was used for assessing mixture homogeneity through a sampling protocol and classical statistical analysis tools [2]. One of the difficulties of working with these compounds is that, due to the similarities between CR6 and GE15 powders (same chemical composition, same colour, etc.), the direct measure of the composition of a given sample is not possible. Since homogeneity of a mixture is assessed from the variation of the composition between samples, it is important to find an indirect way to measure the composition of an Alumina powders mixture. Therefore, different correlations between measurable parameters (specific surface area, Sauter diameter, cohesion and flow function coefficient) and the composition of the mixtures were found empirically allowing indirect measurement of the composition of a given sample.

The four studied parameters: specific surface area (, measured by nitrogen adsorption using BET method), Sauter diameter (, measured by laser granulometry, dry transportation with air at 0.1 bar), cohesion and flow function coefficient (and , measured by FT4 shear test at 9 kPa on a 10 mL cell), were compared in terms of precision, relevance and feasibility.

It appears that the homogeneity assessment methods involving rheological parameters (cohesion and flow function coefficient) show many advantages compared to the other methods. Not only are they the most accurate correlations found but both parameters are measured simultaneously, meaning that it is possible to apply both methods on the same samples while all the tests (N_2 adsorption, laser granulometry and FT4) are destructive. Those two methods also allow the estimation of the homogeneity at a more interesting scale (around 7 g of powder), indeed, specific surface area and Sauter diameter methods give a homogeneity at a scale of less than one gram which is not relevant for the end-use properties of the final product.

[1] P. Danckwerts, The definition, measurement and some characteristics of mixtures, Applied scientific research. Section A, 3 (1952) 279-296

[2] S. Massol-Chaudeur, H. Berthiaux, S. Muerza, et J. Dodds, A numerical model to identify the structure of a high-dilution powder mixture, *Powder Technol.*, 128 (2002) 131-138

508516 Characterising the Grain Shape: In Search of Size Independent Shape Descriptors

Abstract:

Granular materials contain a large number of grains that differ in size and shape. While the characterisation of the size distribution is typically done using sieve analysis, the characterisation of the shape of the grains remains a difficult task and is one of the major bottlenecks in the characterisation of the granular material. In this study, we attempt to characterise the shape of three different types of grains commonly used in iron and steel making industry, namely pellet, sinter and iron ore lump. While the pellet particles can be considered nearly spherical in shape, imposing the spherical shape on the iron ore lumps and highly irregular sinter particles seems a very poor choice. We consider two different size ranges of pellet particles and four different size ranges for sinter and iron ore lumps. Imaging of the grains is done using an iPhone 7 Plus camera and sufficient number of grains of each type in all the size ranges are considered. We perform the image analysis and obtain various boundary parameters (such as projected area, minor and major axis, perimeter, etc.) of the grains using MATLAB image processing toolbox. In addition, we define two additional boundary parameters to characterise the deviation of the projected area from that of a circle and calculate their values for all the different type of grains considered in this study. Using these length scales, we define a total of nine different shape descriptors and obtain the distribution of these shape descriptors for each type of grain for all the size ranges. Interestingly, for a given type of grains, the frequency distribution curves for these nine shape descriptors turn out to be nearly independent of the size range of the grains. This suggests that the chosen shape descriptors indeed represent properties related to the shape of the grains alone, irrespective of the size of the grains considered.

We investigate the correlation among these shape descriptors by evaluating the values of the correlation coefficients between each pair of the shape descriptors using the Pearson method, Kendall method and Spearman method. Our results indicate that out of the nine shape descriptors considered in this study, five shape descriptors are strongly correlated with each other and only the remaining four shape descriptors need to be considered in characterising the shape of the grains. Out of these four uncorrelated shape descriptors, two shape descriptors, namely the aspect ratio of the grains and the root of form factor, have been considered important by researchers in previous studies. The two other shape descriptors proposed in this study, which are used to characterise the deviation of the projected grain shape from that of a circle, are not measured previously to the best of our knowledge. We compare these four shape descriptors for the pellets, sinter and iron ore lumps having the same sieve size range and show that the shape descriptors for these three different types of grains are different.

We next calculate the values of these four shape descriptors for relatively larger coke particles using 2d image analysis. We also perform 3d image analysis of some of these coke particles and calculate the values of the analogous shape descriptors in 3 dimensions. We investigate if the identified 2d shape descriptors (obtained from 2d imaging) can be correlated to the analogous 3d shape descriptors (obtained from 3d imaging) and explore if it is possible to use the 2d imaging technique to estimate the 3d shape descriptors, thus eliminating the need of 3d imaging.

508543 Investigating Mechanical Properties of Surfactant Films at the Solid-Liquid Interface Using AFM

Abstract:

Corrosion is a well-known problem in the oil and gas sector and also in many other industries. Surfactant based inhibitors have been used for corrosion control in oil pipelines but their inhibition mechanism is still not fully known. Our group is studying surfactant corrosion inhibition mechanism by various characterization techniques (AFM, Electrochemical and Spectroscopic techniques). A better understanding of corrosion inhibition mechanism will lead to the design and development of not only cost-effective but also greener surfactants system for corrosion inhibition purposes.

Atomic force microscopy (AFM) is an effective tool to study the barrier properties of surfactant films at the solid-liquid interface. Determination of mechanical and topographical properties of single and mixed surfactant coatings by atomic force microscopy (AFM) can delineate the role of physical vs. electrochemical barrier by such films in the corrosion inhibition process. . So far, the AFM data, obtained using a closed cell, indicate the effect of surfactant concentration and that of hydrocarbon chains interactions on the mechanical properties of cationic surfactant films. Ultimately, integration of the AFM and electrochemical results will reveal the correlation between the surfactant structures at the solid-liquid interface and their corrosion inhibition efficacy.

508545 Deposition Rate Consequences of the Formation of Multi-Spherule Cluster Aggregates in Gases "Role of Momentum Shielding™"

Abstract:

For engineering flow systems characterized by particulate suspensions in gases, will the formation of *cluster aggregates* (CAs) in the mainstream accelerate deposition rates on immersed targets or confining walls ---or is aggregate formation more likely to suppress capture rates? Recent research on the drag properties of large cluster aggregates (e.g., stimulated in part by the Sorensen (2011) review) has opened the door to answering these important mechanism- (and pressure-) dependent questions. Sub-100 nm "primary" spherules, which may find themselves in even tenuous 'fractal-like' N-spherule clusters, can take advantage of what we call "momentum shielding"---analogous to the drag advantages experienced by birds who elect to fly in formation. We show here that the extent of this "momentum shielding" can be conveniently quantified (in a dimensionless function we call: $S_{mom}(N;Kn_1;\text{aggregate structure}))$, and used to predict the deposition-rate consequences of aggregation in aerosol flow systems when the cluster deposition mechanism is dominated by either: i) isothermal convective diffusion (C-D), ii) thermo-phoresis (TP-) or iii) inertial impaction (II). As expected, we show that for any type of aggregate structure (including 'fractal-like'), momentum shielding increases with cluster size (e.g. N) and gas pressure (via the reciprocal Knudsen number based on spherule radius R_1). We also show how S_{mom} can be calculated for the limiting case of a N>>1 random close-packed-D_f=3 cluster --- which is nearly gas impermeable. For fractal-like aggregates, including DLCAs (D_f=1.8) and RLCAs (D_f=2.1) we have recently shown that the abovementioned function: $S_{mom}(N, Kn_1; D_f)$ is actually sufficient to make useful engineering estimates of CA Brownian diffusion coefficients (Rosner and Tandon (2017b), thermophoretic diffusivities (Rosner and Tandon (2017a) and aggregate "stopping times" (Rosner and Tandon (2017c)----each in terms of their previously known values (D_1, $\alpha_T D_1$, and $t_{p,1}$) for isolated individual spherules in the same environment.

This information now enables instructive engineering calculations of the deposition-rate consequences of spherule aggregation to form large cluster aggregates-----i.e. at the same mainstream spherule volume fraction we can now determine how much aggregation will increase or decrease the total spherule acquisition rate on an immersed target. We present here representative results for each of the abovementioned particle transport mechanisms----i.e.: i) isothermal convective-diffusion (C-D), ii) thermo-phoresis (TP-) or iii) inertial impaction (II)----- including a brief account of the consequences of dealing with Kn1-dependent coagulation-'aged' aggregate size distributions (ASDs) in the mainsteam. We summarize our current re-examination of the canonical problem of inertial impaction on a circular cylinder in Re>>1 crossflow (Rosner and Tandon (2017c), dealing now with CA-"projectiles" but in the absence of spherule loss from the much larger target. We conclude by briefly outlining our planned work on the efficient "inversion" of sampled cluster-ASD-data to reconstruct, using the abovementioned $S_{mom}(N,Kn_1;D_f$ information, what ASD must have existed in the mainstream---at least in the absence of appreciable supercritical aggregate break-up.

References:

1. Rosner, D. E. and P. Tandon (2017a). Knudsen Transition Effects on the Thermophoretic Properties of Fractal-like Aggregates; Implications for Thermophoretic Sampling of High-Pressure Flames, Aerosol Science and Technology (in press).

2. Rosner, D. E. and P. Tandon (2017b). Aggregation- and Knudsen Transition Rarefaction- Effects on Particle Mass Deposition Rates by Convective-Diffusion, Thermophoresis or Inertial Impaction; Consequences of Multi-spherule 'Momentum Shielding', Aerosol Science and Technology (in press).

3. Rosner, D. E. and P. Tandon (2017b). Mainstream Aggregation and Polydispersity Effects for Inertially-Dominated Deposition Rates on Cylinders (or Leading 'Edges') in Crossflow"---- Consequences of 'Momentum Shielding' for the Impaction of Large, "Compact" ($2<D_f<3$) Multi-spherule Aggregates, Aerosol Science and Technology (prepared for submission, August 2017).

4. Sorensen, C. M. (2011). The Mobility of Fractal Aggregates: A Review, Aerosol Science and Technology. 45: 765-779.

508609 Experimental and Numerical Studies on the Thermal Diffusivity of PackedÂ Powder Beds

Abstract:

Several industrial processes, such as calcination and drying, involve thermal treatment of particles. In order to achieve efficient processes and good quality products, the thermal properties of the particles and that of the bulk powder must be well defined. Previously, some models, such as the Krischer and Maxwell-Eucken models, have been proposed to relate particle thermal properties to that of the bulk powder. Though these models have been shown to work well, it is not very clear how to account for changes in the packing density, moisture content, and arrangement of particles. In addition, the particle thermal properties of most materials are not readily available, which limits the application of these models. To get a better understanding of the relationship between the particle thermal properties and that of the bulk powder and also to quantify the effects of packing density and moisture content on the bulk powder thermal properties, we run several experiments and discrete element method (DEM) based simulations. In both the experiments and simulations, a pipe is filled with particles where the temperature of the pipe is kept constant. We measured the temperature change in the powder bed and compute the actual thermal diffusivity of the powder bed. We tested several types of powders and glass beads in the experiments. The packing fraction, the moisture content, and size of particles were varied in the experiments. In the simulations, in addition to the packing fraction and particle size, the thermal properties of the particles were also varied. In this talk, we will present the details of the experimental and simulation results.

509556 Determination of Size, Size Distribution and Refractive Index of Artificial and Biological Microparticles

Abstract:

For the approval of clinical studies, it is mandatory to validate the production process of artificial oxygen carriers by reliably specifying the properties of the microparticles used like the newly developed blood substitute HbMP-700 (1-3). In particular, besides shape, size and size distribution, the amount of hemoglobin (Hb) and methemoglobin (metHb) content is essential.

We present a method based on the simultaneous measurement of collimated and diffuse transmission spectra in the visible and ultraviolet range and numerical analysis to infer the wavelength-dependent refractive index, size and size distribution of microparticles.

For the validation of the method, we used aqueous suspensions of monodisperse polystyrene microspheres of known concentration, determined by flow cytometric measurements. The related data analysis is carried out for synthetic and experimental data. For homogeneous spheres, Mie theory is applied to compute the solution for the scattering of a plane electromagnetic wave by a particle and to calculate the extinction cross section C_{ext}, i.e. the shadow cast by the particle as a function of the vacuum wavelength λ, the particle size, and the complex refractive index $n(\lambda)$. For dilute suspensions of particles, the measured quantity is an ensemble-averaged extinction cross section C_{ext}. This average is obtained by integrating over the particles' size distribution, which may not be known a priori. From the resulting spectral extinction cross sections, the complex refractive index and the mean and standard deviation of the size distribution can be inferred by means of nonlinear least-squares optimization. In order to reduce the number of free parameters of the problem to a feasible amount, $n(\lambda)$ is represented in a set of harmonic-oscillator-like resonance functions. The refractive index of polystyrene microspheres obtained is in good agreement with literature values for the bulk material, the mean and standard deviation of the distribution of particle *size D* are determined with an uncertainty in the range of a few nanometers. For example, for microspheres of mean diameter mean(D) = 2µm and distribution width std(D) = 9.4nm the uncertainty of the mean is $u\{\text{mean}(D)\}$ = 1.2nm and the uncertainty of the distribution width amounts to $u\{\text{std}(D)\}$ = 0.4nm.

In a second step, we applied the method to characterize HbMP-700 microparticles, approximately featuring a prolate spheroidal shape with 700nm length and 400nm width. Besides shape, size and size distribution, the amount of hemoglobin (Hb) and methemoglobin (metHb) content is essential. The relative concentrations of oxyHb, deoxyHb, metHb are accessible from spectral extinction and diffuse transmission measurements, since the different Hb variants exhibit different wavelength-dependent refractive indices. The method to infer size and refractive index described above can be applied to these non-spherical particles, since the deviation of their light scattering properties from those of spheres turns out to be small. The results based on the optical method will be compared to alternative procedures for HbMP particle characterization with respect to their size and size distribution, i.e. electron microscopy, centrifugation and dynamic light scattering. In addition, x-ray spectrometry will be used to detect the ratio between Fe^{2+} (oxy/dexyHb) and Fe^{3+} (metHb) as well as the iron concentration.

References

1) Novel hemoglobin particles--promising new-generation hemoglobin-based oxygen carriers. Bäumler H, Xiong Y, Liu ZZ, Patzak A, Georgieva R. Artif Organs. 2014 Aug;38(8):708-14. doi: 10.1111/aor.12331.

2) Nonvasoconstrictive hemoglobin particles as oxygen carriers. Xiong Y, Liu ZZ, Georgieva R, Smuda K, Steffen A, Sendeski M, Voigt A, Patzak A, Bäumler H. ACS Nano. 2013 Sep 24;7(9):7454-61. doi: 10.1021/nn402073n.

3) Hemoglobin-based oxygen carrier microparticles: synthesis, properties, and in vitro and in vivo investigations. Xiong Y, Steffen A, Andreas K, Müller S, Sternberg N, Georgieva R, Bäumler H. Biomacromolecules. 2012 Oct 8;13(10):3292-300. doi: 10.1021/bm301085x.

517131 FlowCam NanoÃ‚Â® Provides Counts, Sizes and Images of Nano-and Microparticles: Application to a Therapeutic Protein Pumping Study

Abstract:

Sub-visible particle characterization is becoming a more important method in assessing drug substance and drug product stability. However, the current USP protocol does not require any characterization of particles below 10 Î¼m. This greatly hampers our ability to properly assess protein stability as a function of particulate formation and obscures most sub-visible particles and various morphological properties.

In this study, we assess the capabilities of the new FlowCam Nano from Fluid Imaging Technologies, in conjunction with a pumping study that focuses on particulate formation generated during fill operations. Initial experiments show that the new FlowCam Nano can generate highly resolved images of nanoparticulates, allowing the simultaneous assessment of particulate morphology for nano- and microparticulates. This is extremely useful, as currently unpublished data (for the pumping study) show that as a function of post-pumping agitation, a significant increase in microparticle concentration coincides with a significant depletion of the nanoparticle concentration. This suggests that agglomeration (and probably not nucleation) is the more likely mechanism for larger microparticulate formation.

The ability to characterize the nano- and microparticulate morphology simultaneously, will allow us to better assess the cause of particulate formation (such as observing the transition of nanoparticles to microparticles and seeing if specific morphological properties of larger nanoparticulates are conserved in microparticulates). In addition, the FlowCam Nano is also sensitive enough to assess differences in the colloidal stability of a drug product in different formulations. During filling operations, and accelerated degradation studies, FlowCam Nano showed the differences in size distributions of IVIG in PBS and 0.25 M glycine buffer. Since IVIG is less colloidally stable in PBS, the size distribution varies significantly more per size bin and the overall mean effective spherical diameter is also ~ 600 percent greater after 24 hours of agitation in comparison to the size distribution data of IVIG in glycine. Thus, Flow Cam Nano has capabilities that are very useful in assessing drugs in the substance (e.g. optimizing drug formulations) and product (e.g. assessing particulates post fill & finish operations) phases.

Numerical and experimental estimation on the normal and tangential capillary bridge force adhered to two spheres

Kazuo Murase*, Keisuke Arai, Takato Ootsuka, Daiki Sakamoto and Futa Egawa,
Department of Applied Chemistry, Chuo University
1-13-27, Kasuga, Bunkyo-ku, Tokyo,1128551, Japan
**hmurase@kc.chuo-u.ac.jp*

Introduction

Liquid bridges extensively occur as a binder liquid adhered to solid surfaces in a variety of engineering applications such as crystal growth, ink-jet printing, material processing and powder technology. Especially the strength of adhesion and cohesion of liquid bridges adhered to particles contribute to the consolidation of an agglomerate in wet granular materials. Estimation of the attractive or repulsive force between particles is important in controlling kinetics of agglomeration and handling of granular materials. Characteristics of a liquid bridge e.g. the equilibrium bridge geometry [1], stability [2], static and dynamic binding force [3-6], rupture energy [7,8], and wetting hysteresis [9] etc., have been experimentally and numerically investigated.

The liquid distributions within a particles bed are classified into four types of categories, i.e. pendular, funicular, capillary and slurry states, according to the liquid contents [10]. The static normal capillary force at pendular liquid state have elaborately been investigated as Fisher [11], toroidal model [8] and the exact solution of Yung-Laplace equation [12]. The dynamic normal capillary forces [13] are experimentally estimated with lubrication theory. The tangential force of a liquid bridge [14] also contributes to the primary factor for the granular consolidation with the shear stress of wet granular materials but have not yet substantially been investigated.

This work aimed to estimate experimentally and numerically the normal and tangential capillary force of a static and dynamic liquid bridge adhered to two spheres. The asymmetry static liquid bridge with the shear stress demand for the exact solution of Young-Laplace equation for the equilibrium liquid drop including the effect of gravity.

1. Theory

1.1 Formulation of a asymmetry static liquid bridge

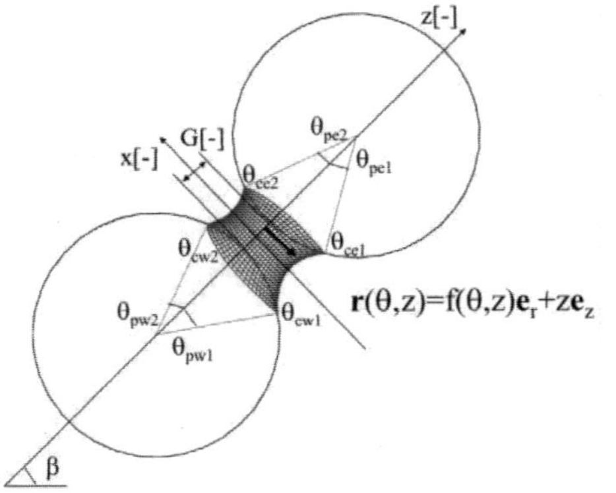

Figure 1 Geometry of a liquid bridge adhered to two spheres with arbitrary inclination in cylindrical coordinate system.

The asymmetry static bridge shape demands for the exact solution of the Young-Laplace equation due to the limit of the toroidal approximation. A prevalent geometry of a liquid bridge adhered to two spheres is drawn in Figure 1. The dimensionless position vector of a liquid surface in cylindrical coordinates is described as $\mathbf{r}(\theta,z) = f(\theta,z)\mathbf{e_r} + z\mathbf{e_z}$ divided by spherical radius, R', which $\mathbf{e_r}$, $\mathbf{e_z}$ are unit basis vectors in r-,z-directions. The symbol (') means the dimensional value. Fig.1 indicates the two dimensional front view with four half-filling (θ_{Pw1}, θ_{Pw2}, θ_{Pe1}, θ_{Pe2}) and contact angles (θ_{Cw1}, θ_{Cw2}, θ_{Ce1}, θ_{Ce2}).

The unit outward normal vector of a bridge surface, $\mathbf{n}_L$, is presented as follows [15],

$$\mathbf{n}_L \equiv \frac{r_\theta \times r_z}{|r_\theta \times r_z|} = \frac{f\mathbf{e}_r - f_\theta \mathbf{e}_\theta - ff_z\mathbf{e}_z}{[f^2(f^2+f_z{}^2)+f_\theta{}^2]^{1/2}} \quad \ldots(1)$$

,where low suffixes (r,θ,z) describe partial derivatives.

The dimensionless effective potential energy, $E=E'/\sigma R'^2$ with Lagrangian multiplier, $K=K'/\sigma R'^2$, on the constrained condition of a constant volume, $V=3V'/4\pi R'^3$, is described as follows [16],

$$E = \int_{-\pi/2}^{\pi/2}\int_{z_{min}}^{z_{max}} 2\left[f^2\left(f^2+f_z{}^2\right)+f_\theta{}^2\right]^{1/2} d\theta dz + \frac{Bo}{2}\int_{-\pi/2}^{\pi/2}\int_{z_{min}}^{z_{max}} (z\sin\beta + x\cos\beta) f^2 d\theta dz + K\left[V - \frac{1}{2}\int_{-\pi/2}^{\pi/2}\int_{z_{cp1}}^{z_{cp4}} f^2 d\theta dz\right] \quad \ldots(2)$$

Bond number (Bo) is a dimensionless number defined by $(\rho_L-\rho_G)gR'^2/\sigma$, where is the ratio of gravitational force to surface tension force equally as Etovoes number (Eoe).

The integral domain of a bridge surface is determined by the trapezoidal section of the upper and lower surface lines, [$z_{w2}<z<z_{e2}$ at $\theta=\pi/2$, $z_{w1}<z<z_{e1}$ at $-\pi/2$] and contact lines, [$z_{w1}<z<z_{w2}$, $z_{e1}<z<z_{e2}$ at $-\pi/2<\theta<\pi/2$] with four contact points settled by four half-filling angles ($\theta_{Pe1},\theta_{Pe2},\theta_{Pw1},\theta_{Pw2}$) in Fig. 1. Two contact lines on the upper and lower spheres were approximated by ellipse lines with parametric aspect ratios, which pass through respectively pairs of two contact points settled with half filling angles, ($\theta_{Pw1},\theta_{Pw2}$) and ($\theta_{Pe1},\theta_{Pe2}$). Filling angles, ($\theta_{PW}(\theta)$, $\theta_{PE}(\theta)$) on the lower and upper spheres have variable properties along the circumferential direction. The formulation of a static bridge shape at equilibrium state results in the extremum problem, which corresponds to the condition that the first order variation equals to zero, $\delta E=0$.

1.2 Estimation of the capillary force of a liquid bridge

The dimensionless mean curvature (H) and contact angles (θ_c) of a bridge surface are respectively calculated by eqs. (3),(4) once the outline of a liquid bridge is numerically evaluated.

$$-2H = \nabla_s \cdot \mathbf{n}_L = \left[(\alpha^2 + f_\theta^2) - f(f^2 + f_\theta^2)f_{zz} - f(1+f_z^2)f_{\theta\theta} + 2ff_z f_\theta f_{\theta z}\right]/\alpha^3 \quad \text{...(3)}$$
$$,where\ is\ \alpha^2 = f^2(1+f_z^2) + f_\theta^2$$

$$\theta_c(\theta) = \cos^{-1}(\mathbf{n}_s \cdot \mathbf{n}_L) \quad \text{...(4)}$$

,where contact angles also depend on the circumferential angle.

According to the Young-Laplace equation, the dimensionless pressure difference, $\Delta P = \Delta P'R'/\sigma$, across a liquid surface is related with the mean curvature of the surface.

$$\Delta P = (-2H) \quad \text{...(5)}$$

The capillary force on a liquid bridge adhered to two spheres must be evaluated only on the contact line as known by the boundary method due to unspecified neck radius of a bridge and asymmetric condition. Dimensionless averaged capillary forces, $F_{Ci} = F'_{Ci}/(2\pi R'\sigma)$ on the upper and lower sphere (i=1,2 respectively correspond to lower, upper sphere) are presented as follows by three terms, which correspond to the surface tension, hydrostatic pressure and buoyancy terms along the longitudinal angle (θ). The sign of the capillary force set to be positive for attractive direction.

$$F_{ci} = \frac{1}{\pi}\int_{-\pi/2}^{\pi/2} \sin(\theta_{Pi})\sin(\theta_{Pi}+\theta_{Ci})d\theta - \frac{1}{2\pi}\int_{-\pi/2}^{\pi/2} \sin^2(\theta_{Pi})\Delta P d\theta \mp \frac{2}{3}\mathrm{Bo}\cdot V_{bi} \quad \text{...(6)}$$

, which V_b is sphere volume submerged in a bridge

Finite Element Method is applied for the solution of the $\delta E=0$ equation. Isoparametric elements of a biquadratic polynomial shape function were selected for the integral domain and unknown radius. The set of non-linear algebraic equations for unknown radius at element nodes and Lagrangian multiplier was resolved by the Newton-Raphson method [9].

2. Experiment

Tensile strength of a static and a dynamic pendular liquid bridge

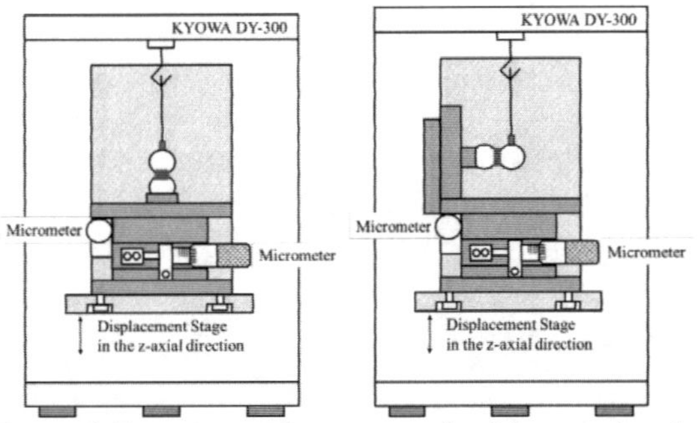

Figure 2 Experimental apparatus for (a) normal and (b) tangential tensile strength of a liquid bridge adhered to two spheres

Figure 2 shows the surface tensiometer (Kyowa interface Science Co., Ltd. DY-300) for measuring (a) the normal and (b) the tangential tensile strength of a water liquid bridge adhered to two stainless steel spheres (radius 3.0mm). The rod attached with one sphere is hanged up to the tensiometer and the other sphere fixed on the displacement stage is moved upward and downward by the tensiometer. The discrete and continuous displacement of the stage are performed for the normal and tangential deformation of a liquid bridge for the static and dynamic states. Two micrometers adjust the appropriate position of the lower spheres for the upper sphere. Water droplets are poured into the spheres aperture and substantially wetted by the displacement of the stage. The tensile strength at static and dynamic state and snapshots of liquid droplets by the digital camera and movie are recorded. The outlines of droplets are evaluated by a digitizer and estimate the half filling angles (θ_{Pe1},θ_{Pe2},θ_{Pw1},θ_{Pw2}) and contact angles (θ_{Cw1},θ_{Cw2},θ_{Ce1},θ_{Ce2}) at four contact points. The sphere configuration (radius, gap), the droplet volume, the half filling angles and the contact lines on spheres initiate parameters of numerical simulation.

3. Results and discussions

3.1 The normal tensile strength of a liquid bridge adhered to two vertical spheres.

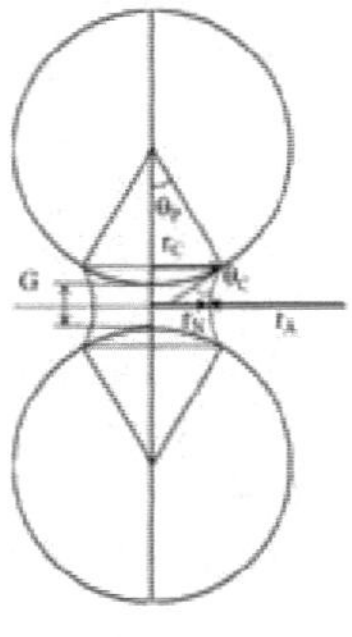

Figure 3 Geometry of toroidal model

The symmetric geometry of a liquid bridge adhered to two vertical spheres without gravity has been extensively investigated by toroidal approximation in Figure 3. The boundary method overestimates the capillary for of a liquid bridge rather than the gorge method but is feasibly and distinctly defined for the asymmetric bridge outline. Symmetric condition simplifies eq.(3) for the mean curvature and eq.(6) for the capillary force into eq.(7),(8) [7,16,17].

$$-2H = \nabla_s \cdot \mathbf{n}_L = \frac{1}{\alpha}\left(\frac{1}{f} - \frac{f_{zz}}{\alpha^2}\right), \textit{where} \quad \alpha^2 = 1 + f_z^2 \qquad \ldots(7)$$

$$F_{Boundary} = r_C \sin(\theta_{Pi} + \theta_{Ci}) - \frac{r_C^{\ 2}}{2}(\Delta P) \mp \frac{2Bo}{3} V_{bi}, \quad r_C = \sin\theta_P \qquad \ldots(8a)$$

$$F_{Borge} = r_N - \frac{r_N^{\ 2}}{2} \sin\theta_{cpi}(\Delta P) \mp \frac{2Bo}{3} V_{bi} \qquad \ldots(8b)$$

,where $r_N - r_A = \frac{A}{\tan(\pi/2 - \theta_P - \theta_C)} + \sin\theta_P, r_A = \left[A^2 + \{\sin\theta_P - (r_N - r_A)^2\right]^{1/2} A = G/2 + 1 - \cos\theta_P$

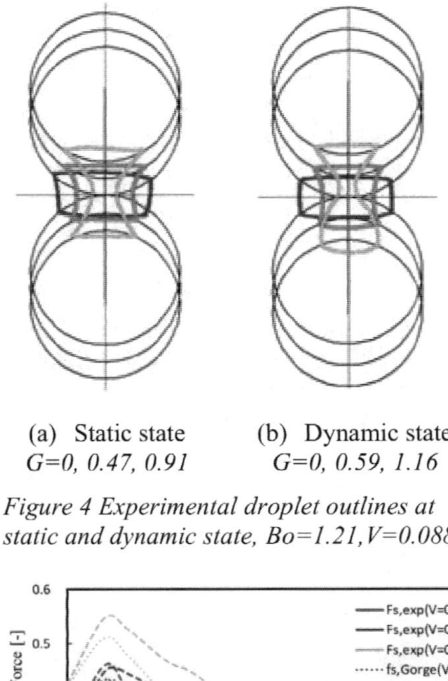

(a) Static state $G=0,\ 0.47,\ 0.91$

(b) Dynamic state $G=0,\ 0.59,\ 1.16$

Figure 4 Experimental droplet outlines at static and dynamic state, Bo=1.21, V=0.088

Figures 4 show symmetric outlines of liquid bridges around the vertical axis respectively at the static and dynamic state. Static state results in the shorter rupture gap and the larger dimensionless normal tensile strength from Figures 5a. Sphere gaps at the peak static capillary force depend on liquid volume. The contact angle on the upper sphere corresponds to the advancing contact angle at the higher Bo number due to gravity effect.

Fig. 4b indicates the downward transition of a liquid immediately before the rupture due to the wettability a liquid. The toroidal approximations over-estimate experimental data at the lower sphere gap but the sphere gaps at the peak dynamic capillary force depend on the velocity of stage displacement. The velocity reduces the peak value of dynamic capillary due to the downward transition of a liquid in Fig. 5b. The stepwise profiles of capillary force are due to the movement sensitivity of tensiometer for the stage displacement.

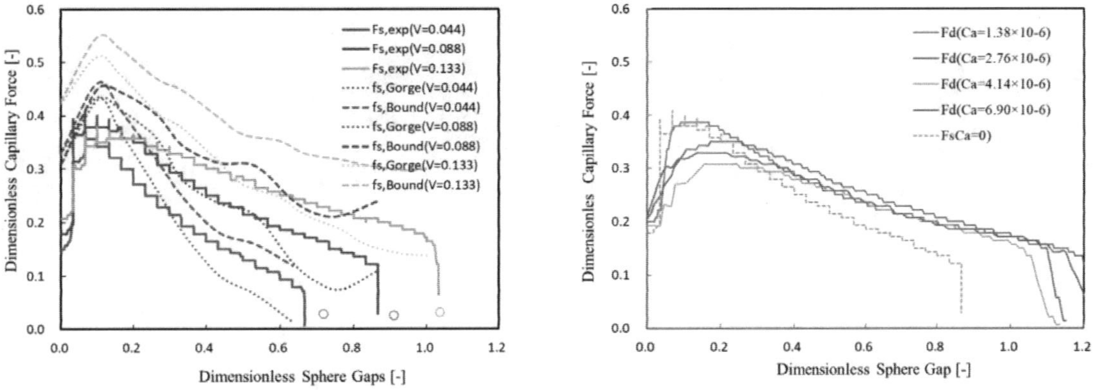

(a) at the static state (b) at the dynamic state

Figure 5 Effect of sphere gap on dimensionless capillary force

3.2 Estimation of boundary condition at contact line by LBM

The Young's equation describes the force balance among three interface tensions (solid-liquid, solid-gas and liquid gas) at three phase contact point and also defines the relationship between interface tension and contact angle. The solution of the Young-Laplace equation at the static state demands for one unknown boundary condition of either a contact angle or a contact line. The simulation for the wettability on three dimensional contact line demands for the elaborate process. Our experiment determines the contact lines on spheres under the constant liquid volume and is utilized for the boundary condition. Besides the perfect wettability the smooth transition of three phase contact lines is unexpected due to the advancing and receding contact angle. Lattice Boltzmann Method (LBM) is expected to estimate the boundary condition at static and dynamic state. The geometry of a liquid bridge is numerically estimated by FEM with half-filling angles predicted by 2 dimensional Shan & Chen model of LBM, which is used by D2Q9 model for transition of distribution function. The estimation for the liquid geometry by FEM has been verified. Figures 6 show the predicted liquid outlines by LBM and FEM. at a given liquid volume.

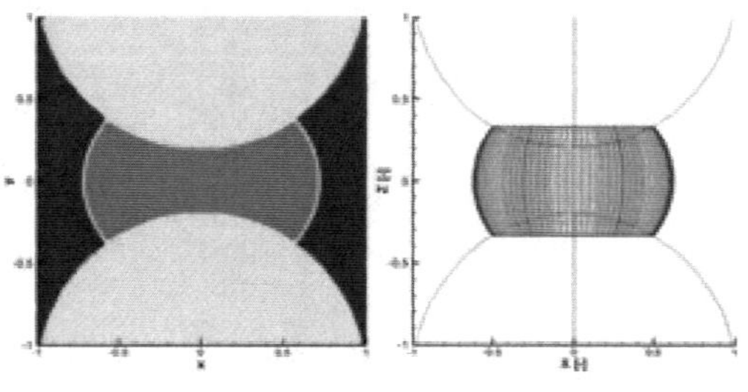

(a) Repulsive interparticle force

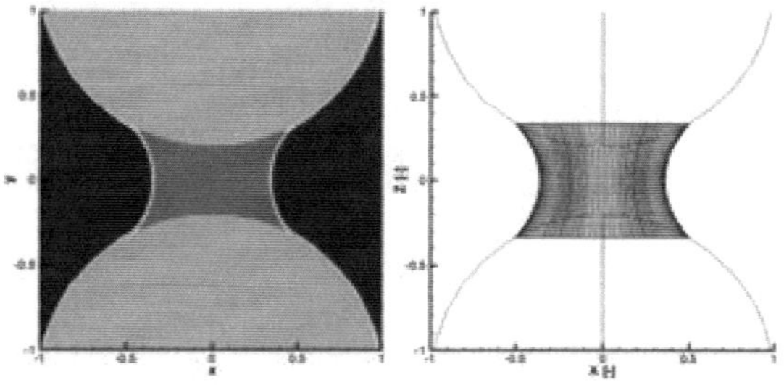

(b) Attractive interparticle force

Figure 6 Snapshots of liquid geometries at static state by LBM and FEM

Figures 7 show snapshots of dynamic liquid outlines at a given liquid volume up to the rapture state by LBM. The transition of boundary conditions, i.e. filling and contact angles, is smoothly described. Gravity and imperfect wettability, i.e. advancing and receding contact angles, are excluded in Figs. 7 but the effect of gravity on the liquid outline is avairable.

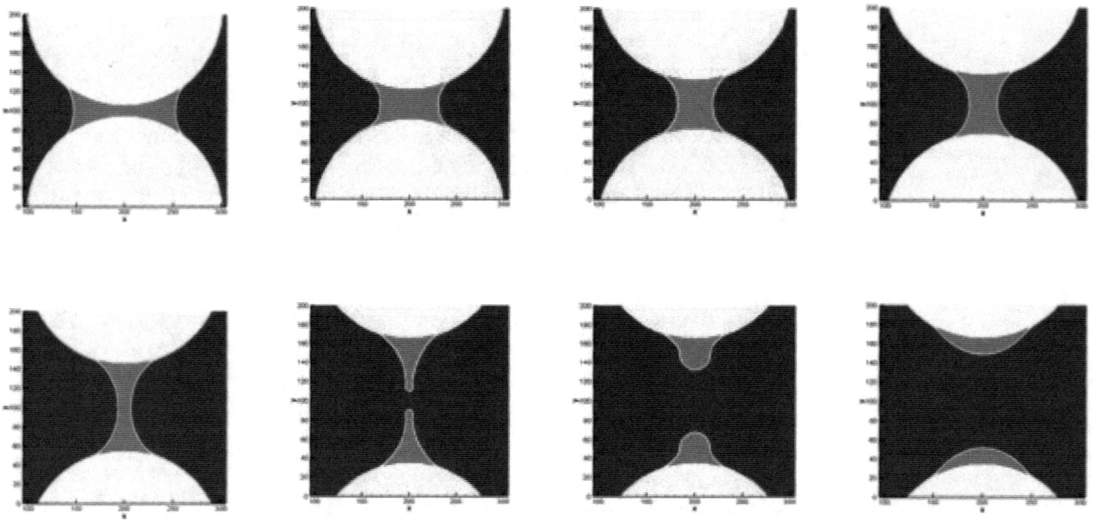

Figure 7 Snapshots of snapshots of dynamic liquid outlines at a given liquid volume

3.3 The tensile strength of a liquid bridge with shear deformation.

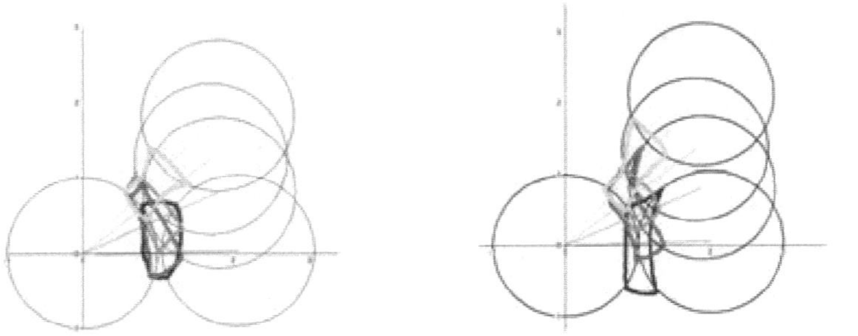

(a) Static state (G=0, 0.59, 0.98) (b) Dynamic state (G=0, 0.59, 0.98)

Figure 8 Liquid outlines at static and dynamic state (Bo=1.21, V=0.088)

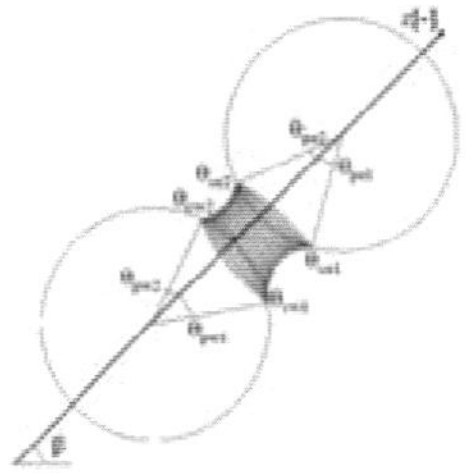

Definitions of half-filling and contact angles.

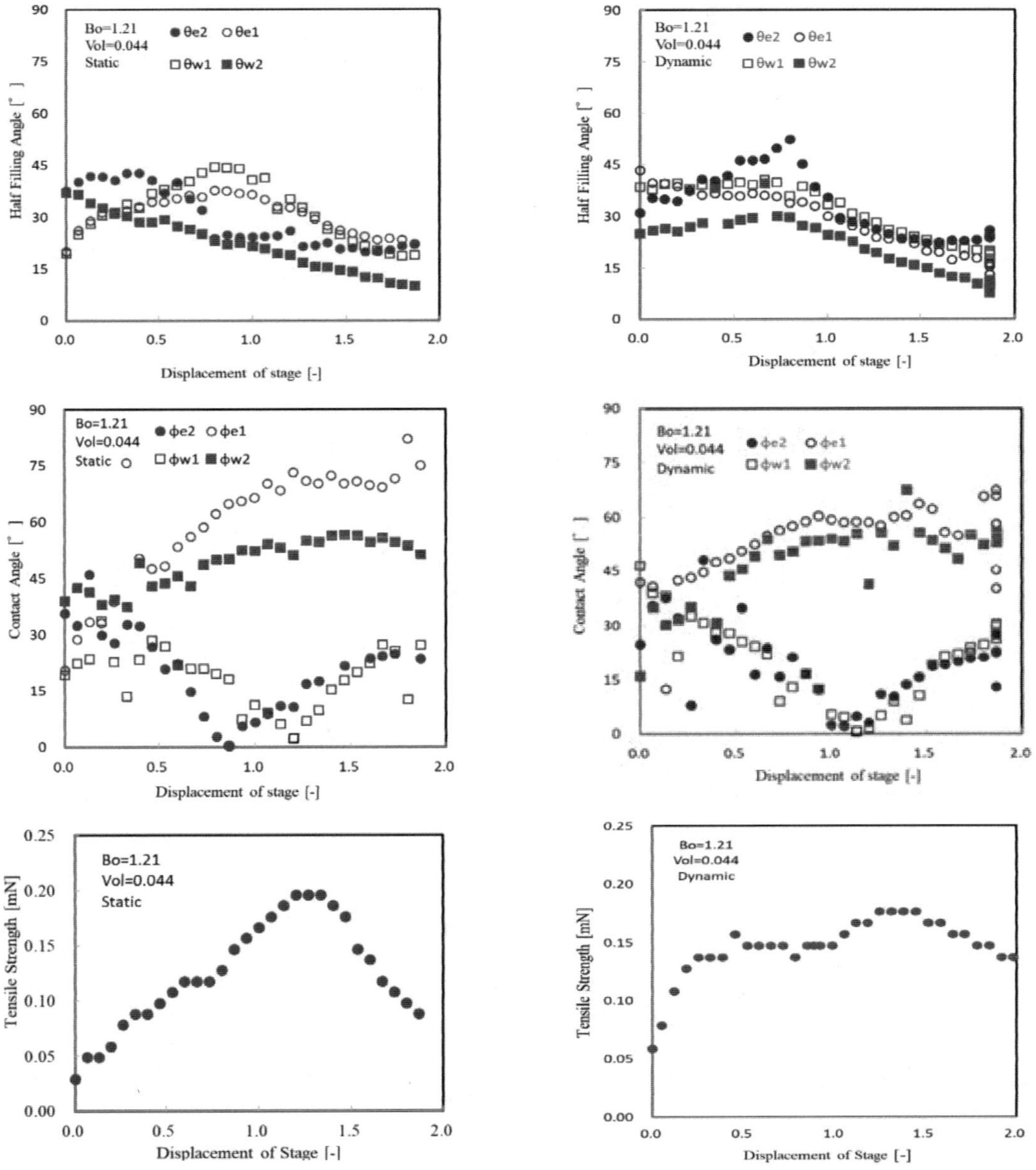

(a) Static state (b) Dynamic state

Figure 9 Profiles of half-filling angles, contact angles and tensile strength , Bo=1.21,V=0.088

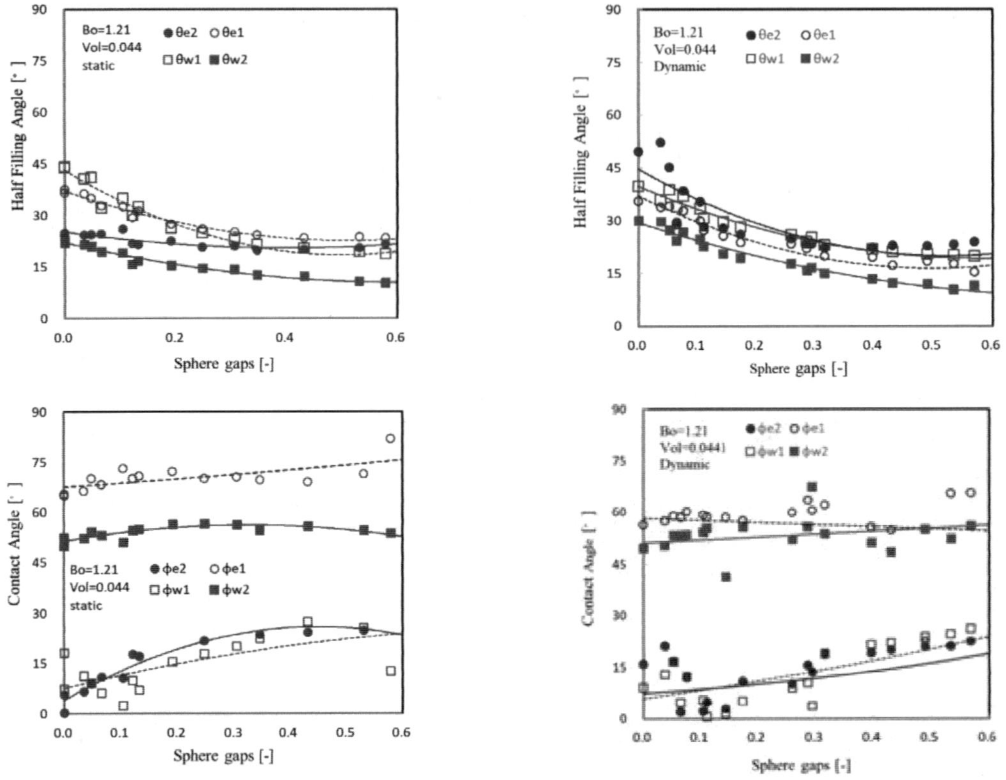

(a) Static state (b) Dynamic state

Figure 10 Profiles of half-filling angles, contact angles with sphere gaps , Bo=1.21,V=0.088

Figures 8 show four types of liquid outlines at static and dynamic state. Magenta and green plots are drawn respectively at the extreme values of half filling angles, contact angles and tensile strengths in Figures 9. During the lower stage displacement two spheres keep at the contact state and the half-filling angles reached to the peak values. Half-filling angles sifted with displacements of a stage have the peak values except the static half-filling angle at the top point of the lower sphere (θ_{Pw2}), which monotonously decrease with stage displacement. The increase and decrease of contact angle respectively corresponds to the advancing and receding contact angle. The contact angles at the bottom point of the lower sphere (θ_{Cw1}) and at the top point of the upper sphere (θ_{Ce2}) transit from the receding to the advancing contact angle due to the infinitesimal values. Figures 10 redraw the half-filling and contact angles with respect to sphere gap. Contact angles contribute to the peak tensile strength, which is measured as resultant force (F_T) of normal (F_n) and tangential (F_t) tensile strength in Figure 11. The normal tensile strength conventionally corresponds to the capillary force. The stage displacement at the peak tensile strength minimize the sum of half-filling and contact angle due to the positive hydrostatic pressure across liquid surface. The larger dynamic tangential strength occurs during the sphere contacting due to the asymmetric deformation. The adhesion force of liquid to sphere at the static state contribute to the larger peak tensile strength.

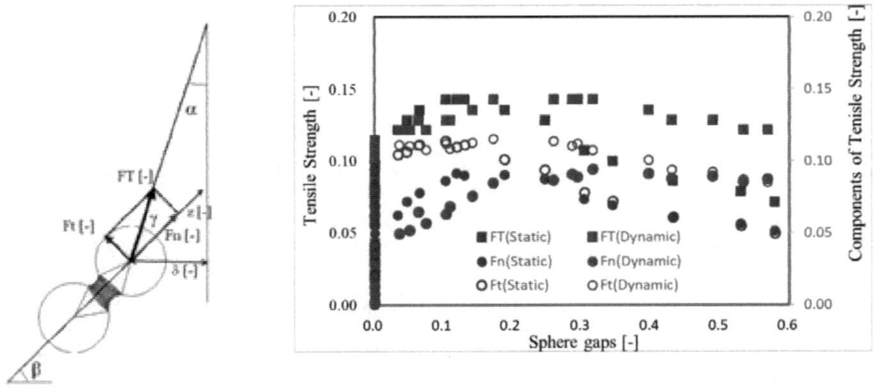

Fig.11 Normal and tangential components of tensile strength at static state

3.4 Numerical simulation at the static state

The solution of Young-Laplace equation demands for one unknown necessary condition on the contact point, i.e. half-filling or contact angle. Liquid surfaces simulated by FEM are drawn with solid lines in Figures 12. The simulation reveals the asymmetric liquid geometry with half-filling and contact angles along longitudinal angle. Disagreements of liquid interfaces results from the approximation of contact lines on spheres. The tangential strength influences depend on the mean curvature. Especially the second figure at the peak half-filling angle has the negative hydrostaic pressure in the direction of relative sphere displacement and positive pressure in the oppsite direction.

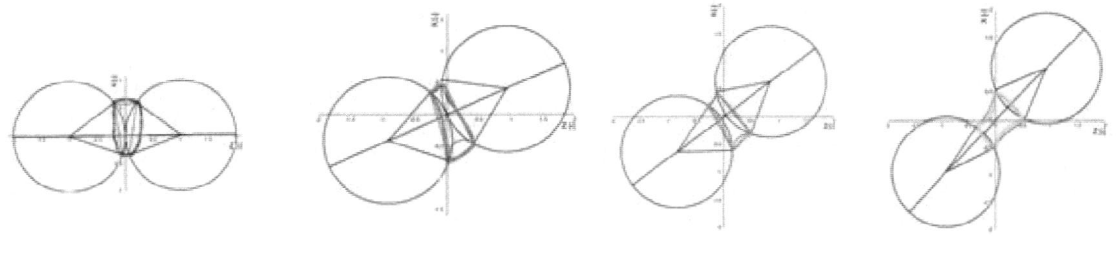

At the initial geometry At the peak halffilling angle At the peak tensile strength At the maximum shpere gap

Figure 12 Numerical estimation at the static state

Figure 13 displays transitions of measured normal tensile strengths ($F_{exp,n}$) and simulated capillary force (F_{cal}) to sphere gap. The initial droplet is attached to the above aperture in the direction of sphere movement due to gravity. The stage displacement drifts parts of liquid volume into the lower aperture of spheres. The sphere on displacement stage keep contacting with the other sphere hang up to the tensiometer up to the peak tensile strength. Fcal and FcalB are numerically estimated by outlines of liquid surfaces respectively with Young-Laplace equation and boundary toroidal model. The sign of hydrostatic pressure with deformation of liquid droplets results in the repulsive capillary force. The normal tensile strength corresponds to the capillary force during the range of symmetric liquid bridge. The angle of beta is defined by the inclination of center distance of a sphere pair. The lower normal tensile strength than the capillary force results from the orthogonal component of capillary force at the low range of beta.

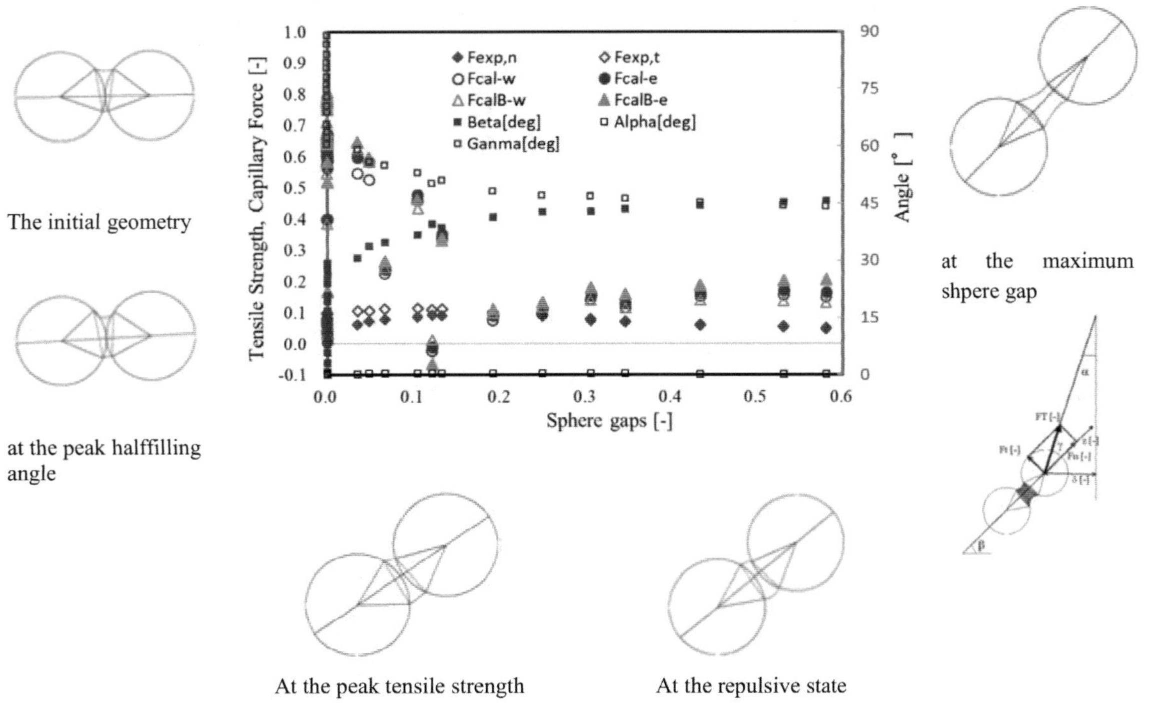

Figure 13 Normal tensile strength and numerical capillary force

Conclusions

Liquid bridges adhered to particles are deformed with the normal and tangential stress in wet granulator. The interparticle force with asymmetric droplet by shear stress must be estimated by the three dimensional solution of Young-Laplace equation. A surface tensiometer and numerical simulation contribute to the normal and tangential tensile strength of liquid bridge adhered to two spheres.

1. The normal strength of symmetric liquid bridge with a surface tensiometer is verified by the conventional toroidal model and the solution of Young-Laplace equation.
2. Sphere gaps at the peak capillary force in the vertical sphere configuration depend on liquid volume and displacement velocity of sphere.
3. The lattice Boltzmann Method is expected to numerically predict the boundary conditions on the contact lines, which are half-filling angle and contact angle at the static and dynamic state.
4. The tensile strength depends on transition of contact angles, which correspond to advancing or receding angles. The larger sum of contact and half-filling angles than the orthogonal angle corresponds to the positive hydrostatic pressure across liquid surface.
5. The normal tensile strength is related with the solution of Young-Laplace equation by FEM. The tangential tensile strength depends on the liquid transition within droplets and the variable mean curvature.

References

[1] K.Murase, H.Shibata, Y.Yasuda and S.Toyama, *Jpn J. Multiphase Flow,* **5**, 322(1991)
[2] A.E.Saez and R.G.Carbonell, *J. Colloid Interface Sci.,* **140**, 408(1990)
[3] D.N.Mazzone, G.I.Tardos and R.Peffer, *Powder Tech.*, **51**, 71(1987)
[4] C.D.Willet, M.J.Adama, S.A.Johnson and J.P.K.Seville, *Langmuir*, **16**, 9405(2000)
[5] M.J.Adama, S.A.Johnson, J.P.K.Seville and C.D.Willet, *Langmuir*, **18**, 6180(2002)
[6] R.K.Niven, *J Petroleum Sci. Eng.*, **52**, 1(2006)
[7] O.Pitois, P.Moucheront and X.Chateau, *Eur. Phys. J.B*, **23**, 79(2001)
[8] S.J.R.Simons, J.P.K. Seville and M.J. Adams, *Chem. Eng. Sci.*, **49**, 2331(1994)
[9] C.D.Willet, M.J.Adama,S.A.Johnson and J.P.K.Seville, , *Powder Tech.*, **130**, 63(2003)
[10] S.M.Iveson, J.A.Beathe and N.W.Page, *Powder Tech.*, **127,** 149(2002)
[11] R.A.Fisher, *J. Agri. Sci.*, 16,492(1926)
[12] F.M.Orr, L.E.Scriven and A.P.Rivas, *J. Fluid Mech.*, **67**, 723(1975)
[13] O.Pitois, P.Moucheront and X.Chateau, *J. Colloid Interface Sci.,* **231**, 26(2000)
[14] U.El Shamy and T.Groeger, *Int. J. Numer. Anal. Meth. Geomeh.*, **32**, 1763(2008)
[15] F.M.Orr, R.A.Brown and L.E.Scriven, *J. Colloid Interface Sci.,* **60**, 137(1977)
[16] K.Murase, T.Mochida, Y.Sagawa and H.Sugama, *Adv. PowderTech.,* **19**, 349(2008)
[17] C.D.Willet, M.J.Adama,S.A.Johnson and J.P.K.Seville, *Handbook of powder technology*, **11**, chap.28, ed. A.D.Salman, M.J.Hounslow and J.P.K.Seville,Elsevier (2007)

NUMERICAL SIMULATION OF WIRE-PLATE ELECTROSTATIC PRECIPITATOR - EFFECT OF PARTICLE CONCENTRATION

Ding Yang[1], *Baoyu Guo*[2*], *Jun Guo*[1], *Aibing Yu*[2]

[1] *Fujian Longking Co. Ltd, Longyan, Fujian, 364000, China*
[2] *Department of Chemical Engineering, Monash University, Clayton VIC 3800, Australia*

Introduction

Electrostatic Precipitators (ESP) have been widely used to separate fine particles from large scale industrial exhaust gases, such as coal fired power plants. In-depth knowledge and careful design optimization are critical to meeting the increasingly strict emission standard in terms of $PM_{2.5}$ emission limit, e.g., 5-10 mg/Nm^3. Numerical modelling has been employed as a powerful tool in this area. Complicated interactions among electric field, gas flow and particle transport are summarized by Adamiak [1]. To model such a system, various levels of simplifications have been made in published papers.

Among them, Lagrangian particle tracking technique is a widely accepted approach. The mass concentration of ash particles in coal-fire flue gas is typically 10-50 g/Nm^3, corresponding to volume fraction in the order of 10^{-5}. Such a small value is often regarded as sufficiently dilute particle concentration in numerical modeling of multiphase flows. For the case of ESP, complicated interactions among the e-field, gas flow and charged particles are involved, even if the deposited dust layer is ignored. Full coupling between the continuous gas phase and discrete particle phase is computationally expensive to achieve a converged steady solution. To avoid such a high cost, the model for ESP was simplified as so called one-way coupling in many previous works, so that the electric field is solved without the particle phase (Guo et al. [2,3]; Farnoosh et al. [4]). This simplification may be acceptable in the end stage of ESP, but will introduce significant error in the early collection stage. Limited studies reported in the literature have considered the particle charges as a source term in the Poisson's equation of electrostatics. For example, Choi and Fletcher [5] predicted a reduced current for the particle-loaded case (d_p = 1-5μm) compared with the particle-free case, but the corona ion charge density somehow shows an increasing trend along the stream. When fine dust particles are treated, the large specific surface area of the particle phase, combined with reduced collection efficiency, implies high space charge density in the collection region. Its effect was understood as a partial suppression of the corona current [6]. As was experimentally proved by Podlinski et al. [7], space charge of charged particles can contribute to the EHD flow pattern and particle transport. In present work, particle space charge is considered through two-way coupling and interesting results are generated in terms of ESP performance as particle loading changes.

Integrated ESP Modelling

Mathematical Equation

Coupled electric field, gas-particle flow and particle charging processes are simulated using

Eulerian-Lagrangian method in Computational Fluid Dynamics (CFD). Major governing equations of vector format are listed below for completeness. For electric field, Poisson equation of potential

$$\nabla^2 V = -(\rho_{ion} + \rho_c)/\varepsilon_0 \quad \mathbf{E} = -\nabla V \tag{1}$$

Current continuity equation

$$\nabla \bullet ((b\mathbf{E} + \mathbf{U}_f)\,\rho_{ion} - D\nabla\rho_{ion}) = S_c \tag{2}$$

where ρ_{ion} is corona charge density, b the ion mobility, D is ion diffusivity, ρ_c is particle space charge density, evaluated by averaging the charges of particles that pass through a control volume (cell),

$$\rho_c = \frac{1}{V_{cell}} \sum_{n=1}^{N} \dot{n}_p q_p \Delta t \tag{3}$$

Where $\dot{n}_p$ is the number flow rate of a parcel of particles tracked, Δt is the residence time in the cell. The source term, S_c, in Equation (2) is transfer rate of ion charges to particles, which is not considered at the moment, i.e., $S_c = 0$. For the gas flow, the momentum equation is,

$$\begin{aligned}\nabla \bullet (\rho_g \mathbf{U}_g \otimes \mathbf{U}_g) &= \nabla(p + \tfrac{2}{3}\rho_g k) + \nabla \bullet [(\mu + \mu_t)(\nabla \mathbf{U}_g + (\nabla \mathbf{U}_g)^{\mathrm{T}})] + \rho_{ion}\mathbf{E} \\ &- \frac{3\pi n_p d_p \mu}{\mathrm{C_C}}(1 + \mathrm{Re}_p^{0.687})(\mathbf{U}_g - \mathbf{U}_p)\end{aligned} \tag{4}$$

For the particle flow, the force balance of a single particle is

$$m_p \frac{d\mathbf{U}_\mathrm{p}}{dt} = \frac{3\pi n_p d_p \mu}{\mathrm{C_C}}(1 + \mathrm{Re}_p^{0.687})(\mathbf{U}_g - \mathbf{U}_p) + \tfrac{1}{6}\pi d_p^3 \mathbf{g}(\rho_p - \rho_g) + n_e q_e \mathbf{E} + \mathbf{F}_{dis} \tag{5}$$

k-ω SST turbulence model is chosen for gas flow, and the dimensionless Lawless charging rate equation is used for particle charging [8].

Model Conditions

An integrated ESP model developed comprises two parallel modules or submodels, namely, electric field model and coupled gas-particle flow model. One-way coupling between the two modules is used earlier is upgraded to two-way in the present work. The electrostatic field caused by charged particles in suspension is considered in the solution of electric field. The two submodels are coupled through a small under-relaxation factor, in views of strong interaction between the electric field and particles flow, and the stochastic nature of the discrete particle phase. In this way all the volume averaged quantities related to the particle phase (such as the particle space charge density) are smoothed out over time in the simulation.

A single channel, 4-wire and flat plate configuration is considered. The dimensions are 3.5mm in wire diameter, 0.24m for wire-wire spacing, 0.2m for wire-plate spacing. The plate length is 1.2m. The simulation domain is defined as 1.74m long by 0.4m wide box with the 4 holes inside. The grid is smoothly refined at places where higher gradient in electric quantities is expected, such as the proximity of the wires and middle of the adjacent wires. Other base case conditions are as follows.

- Applied voltage on the wire electrode: Va = -60kV. The charge density on the wire surface is determined using an injection law combined with Peek equation (Guo et al.[9]).
- Gas temperature: 120°C. As the inlet boundary condition for the gas phase, the velocity is set as 1 Nm/s; relative turbulence intensity 3.7%, turbulent length scale $0.3D_h$ = 0.24m.
- Dust particle size is mono-dispersed at d_p=2.5 μm. Particle density 2000 kg/m^3; ash dielectric constant of κ_p=6.6, Particle loading at the inlet, $(\rho_p)_{in}$, is varied with default value of 10 g/Nm^3.

Typical Results

Electric Field without Particles

When the inlet/outlet boundaries are moved sufficient long distance away from the electrodes in the simulation domain (referred to as "natural boundary condition"), the electric field around the boundary wires show an asymmetric distribution pattern that differ from that around the inner wires. Electric field-lines/current can go out of the limit of the collection region (Figure 1). Electric field intensity on the collection plate is of interest as it is considered a factor to determine the upper limit of operation before spark-over [10]. Normally the field intensity on the plate is highest at the point that faces exactly the wire center with the least distance. This maximum field intensity on the plate is 3.80×10^5 V/m. However, when the natural boundary condition is used, the highest value (1.06×10^6 V/m) predicted in field intensity appears at the endpoint of the plate due to its sharpness. This is a likely point that would trigger electric spark-over prematurely. Practically such a sharp edge should be avoided. Generally, the field strength at the middle wires is smaller than that at a boundary wire, due to the counteraction of adjacent wires. Then the corona charge density at the boundary wires should be greater than that at the inner wires (Figure 2), which is similar to the simulation result of Kasdi [11].

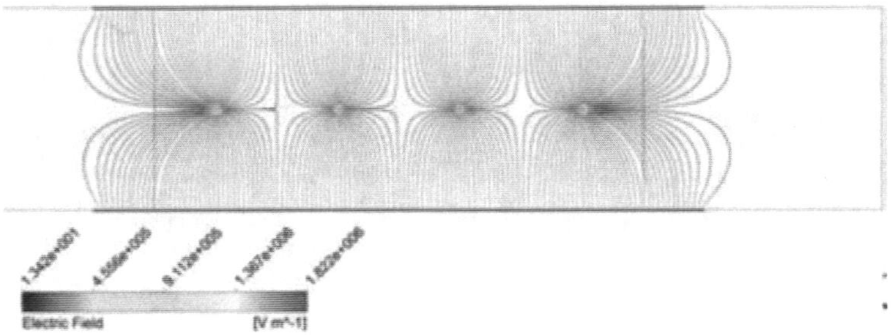

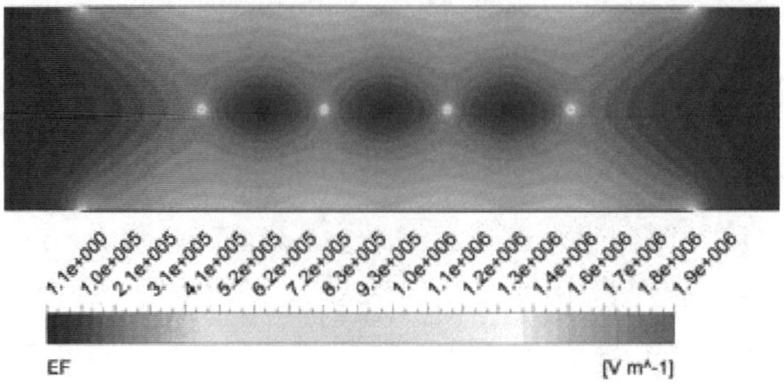

Figure 1. Field streamlines and intensity distribution (no particles).

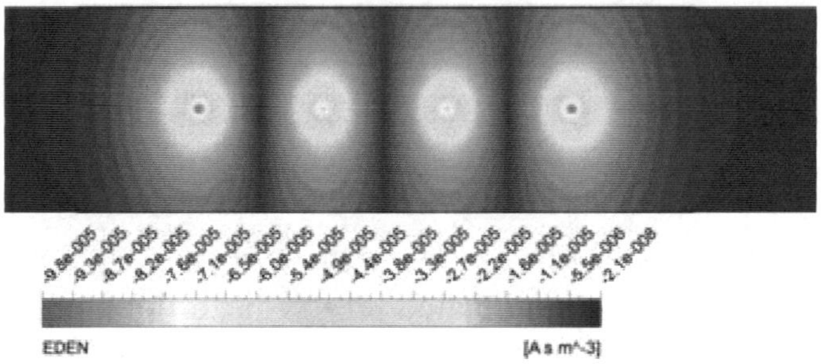

Figure 2. Corona charge density (no particles).

Effect of Particle Space Charge

Particles are randomly injected from the inlet plane with uniform probability in locations and no-slip for velocity (same as the gas velocity). In using random number generator, it is ensured that the injection points are not repeated for every coupling cycle. Figure 3 shows sampled particle trajectories with color scaled to particle charge. Figure 4 shows a pattern of particle deposition rate on the plate. It is seen that the end points of the plate, particularly the downstream one, are special spots to accumulate more dust.

Figure 3. Particle trajectories with colour scaled to particle charge.

Figure 4. Indicative pattern of particle deposition rate on the plate.

Figure 5 shows the averaged space density of charges carried by particles. Highest space charge density appears near the first wire electrode, where electric current density is highest. It diffuses gradually from the centre line with particles drifting toward the plate while moving forward. Due to the radiating nature of the field-lines around each wire, the lateral component of Coulomb's force in the region close to the centre line may not be as large as that off the centre line. Not collected charged particles of negative polarity are conveyed towards the outlet, lowering the potential downstream outside the collection region (Figure 6), causing distortion of the overall electric field distribution.

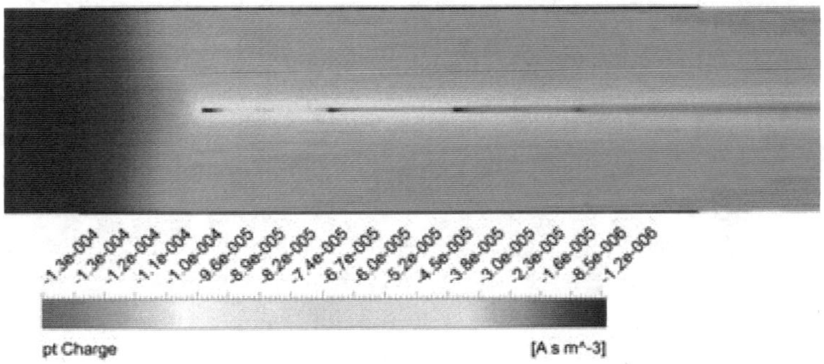

Figure 5. Distribution of space charge density of particles.

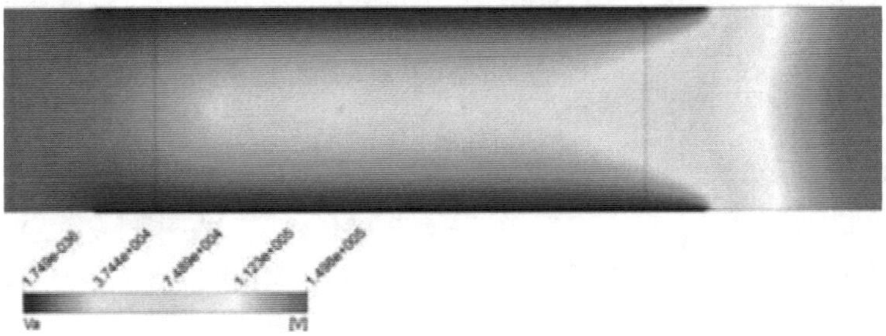

Figure 6. Distribution of electric potential (in absolute value).

Effect of Particle Loading

On Electric Current

Figure 7(a) shows the mean corona current, based on unit length of wire. As can be seen, the corona current drops quickly as particle loading increases, although a fixed voltage is applied. This is caused by the secondary electrostatic field created by the space charges of particles that cancels out the primary electric field at the wire surface and subsequently suppressed the corona generation.

On Particle Collection Efficiency

Predicted particle collection efficiency, as particle loading changes, is given in Figure 7(b). The collection efficiency, increases initially with particle loading (<3 g/m^3), peaks at particle loading of 4-5 g/m^3 and then decreases as particle loading increases further. When ignoring particle space charge, the electric field would be independent of particle loading. This is the case for very low particle loading, where the fractional collection efficiency is constant, or in other words, the quantity of dust collected as well as that penetrated increases proportionally with particle loading.

The observed enhancement of collection by a slight amount of particles is due to the redistribution of electric field so that the near-plate and volume-averaged field intensity happens to increase, which favor the charging of most particles. The asymptotic lower limit of the curve should correspond to the situation when the corona current is so small that turbulent dispersion dominates over the drift force.

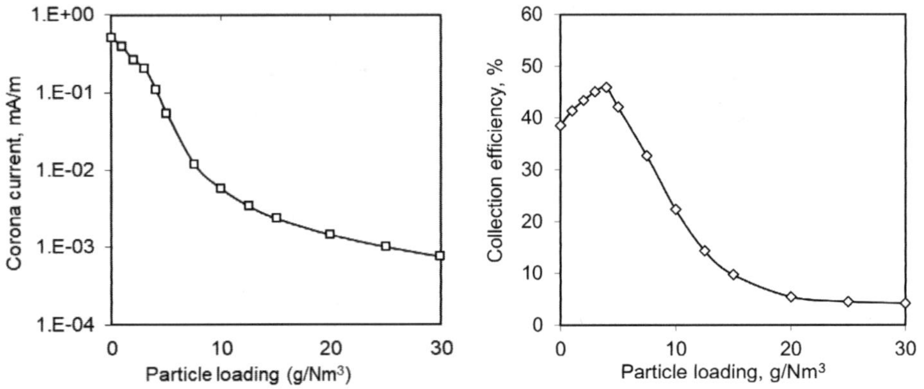

(a) (b)

Figure 7. Mean corona current (a) and collection efficiency (b) as function of particle loading.

On Electric Field Distribution

Figure 8 visualizes the predicted corona charge density and field-lines, which shows how the electric field distribution evolves with particle loading. As the electric field is distorted starting first from the end side, corona density at 3 of the 4 downstream wires is suppressed faster than the 1st wire. When the particle loading is above 7.5 g/m^3, all wires except the first one stop emitting. Meanwhile, reversed field-lines appear around the last wire, and the affected region expands as the loading increases further. Otherwise the field-lines become smoother with increasing loading. Since the field-lines are extended to the outlet, some escaped particles can be pulled back into the collection region, a factor that contributes to significantly greater deposition rate around the end-point of the plate.

Overall, the particle charges affect the particle collection adversely, as they tend to suppress the corona emission or reduce the corona current. Particularly the penetrated particles, which carried significant charges, have prolonged effect as they float along. The corona at the downstream electrodes even extinguish unless the applied voltage is raised individually. When highly charged, more particles precipitate to the plate, reducing the number of particles floating ahead (or less space charge density). The charges carried by these remaining particles exert a negative feedback to the electric field and particle charging, which compromises the collection efficiency.

Electric Quenching

Corona quenching is a common phenomenon in ESP when fine particles are treated. Riebel et al. [12] experimentally studied the corona discharge characteristics of ESP with glycerol aerosol, which found corona onset voltage increases and the corona current significantly decreases after the glycerol steam was injected into the system. Under fixed voltage in the present simulation, corona currents from all the wire electrodes drop with increasing solid loading. The downstream coronas (from the second one) will eventually extinguish at some point. An interesting phenomenon arises when the particle loading increases above 10 g/Nm^3, i.e., the electric streamlines in a small zone around the last wire reverse direction locally. This implies that a charged particle, once falls accidently within the zone, could be trapped and attracted to the same wire, as seen from the particle trajectories in Figure 9. Thus this wire is subject to heavier dust deposition.

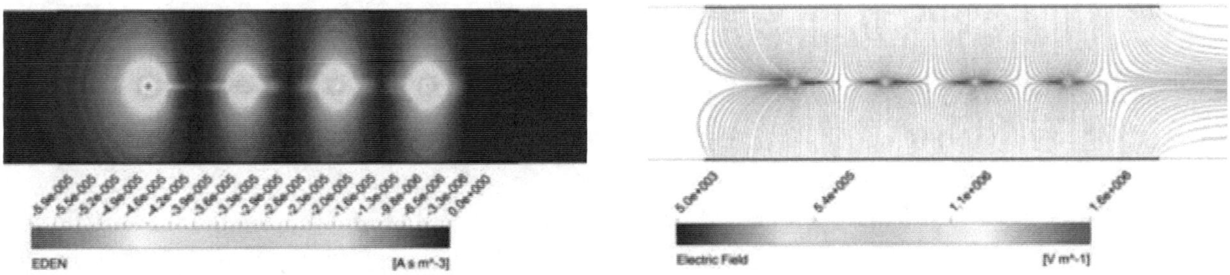

$(\rho_p)_{in}$ = 3 g/Nm^3

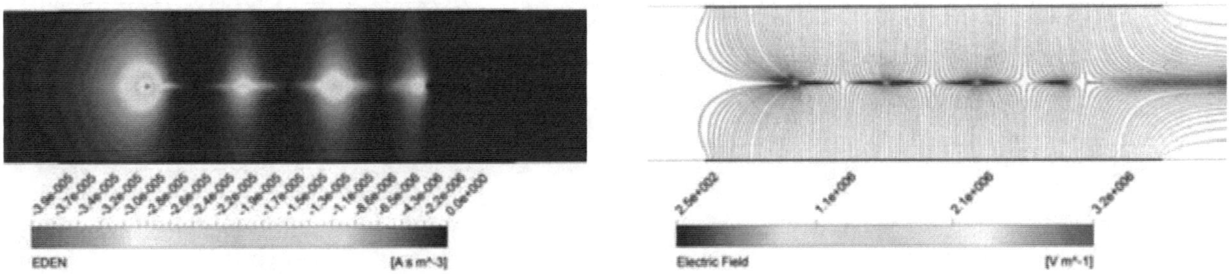

$(\rho_p)_{in}$ = 5 g/Nm^3

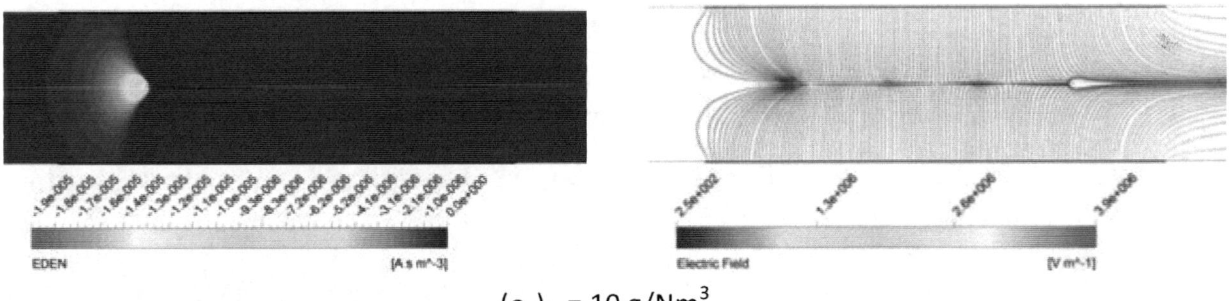

$(\rho_p)_{in}$ = 10 g/Nm^3

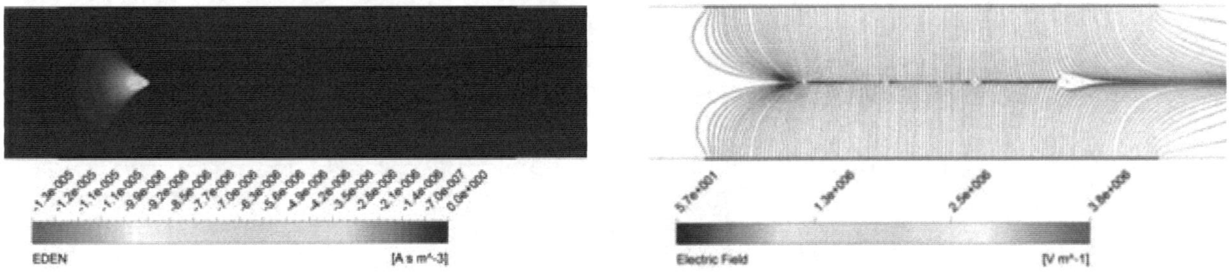

$(\rho_p)_{in}$ = 20 g/Nm^3

Figure 8. Visualizations of corona charge density and field streamlines for different particle loadings.

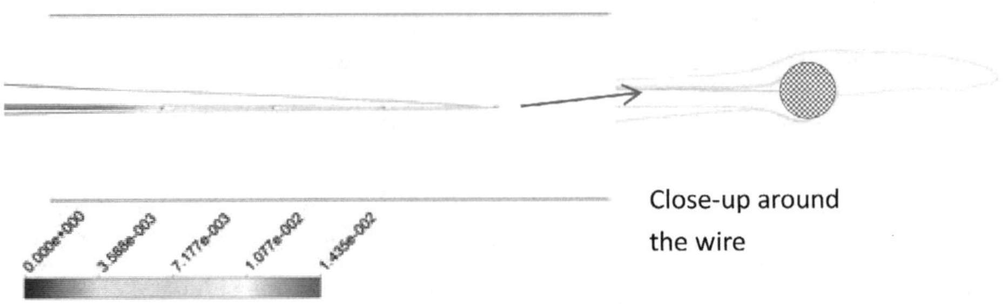

Figure 9. Trajectories of particles falling on the corona wire ($(\rho_p)_{in}$ = 20 g/Nm^3).

Concluding Remarks

CFD based ESP model has been advanced, with an enhanced coupling between electric field and particle flow. So the model can be readily applied to the early collection stage of the ESP, where dust particle loading is too high to neglect. Space charge of particles is considered, and its significance is clarified in a broad range of particle mass concentration in the gas. Dust loading significantly affects the dust collection efficiency for fine particles considered. Maximum collection efficiency exists at a critical particle loading that depends on particle size. However, as particle loading increases, collection efficiency decreases and eventually approaches asymptotically to a lower limit. The phenomenon of electrical quenching is observed, where downstream corona is extinguished or one of the emitters is polluted when the particle loading is high.

References

1. Adamiak K. (2013), "Numerical models in simulating wire-plate electrostatic precipitators: A review", *J. Electrostatics*, 71, pp.673-680.
2. Guo B.Y., Yu A.B. and Guo J. (2014), "Numerical modelling of electrostatic precipitation: effect of gas temperature", *J. Aerosol Science*, 77, pp.102-115.
3. Guo B.Y., Yu A.B. and Guo J. (2015), "Numerical modelling of ESP for design optimization", *Procedia Engineering*, 102, pp.1366-72.
4. N. Farnoosh, K. Adamiak, G.S.P. Castle (2010), "3-D numerical analysis of EHD turbulent flow and mono-disperse charged particle transport and collection in a wire-plate ESP", *J. Electrostatics*, 68, pp.513-522.
5. Choi B.S. and Fletcher C.A.J. (1998), "Turbulent particle dispersion in an electrostatic precipitator," *Applied Mathematical Modelling*, 22, pp.1009-1021.
6. Parker K.R., Applied Electrostatic Precipitation, Blackie Academic & Professional, 1997.
7. Podlinski J., Niewulis A., Mizeraczyk J. and Atten P. (2008), "ESP performance for various dust densities", *J. Electrostatics*, 66, pp.246-253.
8. Lawless P.A. (1996), "Particle charging bounds, symmetry relations, and an analytic charging rate model for the continuum regime", *J. Aerosol Sci.*, 27, pp.191-215.
9. Guo B.Y., Guo J. and Yu A.B. (2014), "Simulation of the electric field in wire-plate type electrostatic precipitators", J. Electrostatics, 72, pp.301-310
10. White H.J. (1974), "Resistivity problems in electrostatic precipitation", *J. Air Pollut. Control Assoc.* 24, pp.313–338.
11. Kasdi A. (2016), "Computation and measurement of corona current density and V–I characteristics in wires-to-plates electrostatic precipitator", *J. Electrostatics*, 81, 1-8.
12. Riebel U., Radtke R. and Loos R. (2002), "An experimental investigation on corona quenching", *J. Electrostatics*, 54, pp.159-165.

Triboelectric Charge of Spherical Glass Particles against Metal Pipeline

HoSu CHOI [1], Teruo SUZUKI [2] and Kwangseok CHOI [3,]*
[1] Christian Academy in Japan, [2] Kasuga Denki Inc.,
[3] National Institute of Occupational Safety and Health, Japan
**choiks@jniosh.go.jp*

Introduction

Insulating powders are used in many ways such as in the food, pharmacy, and chemical industries. Usually, these powders are highly charged when transported, loaded, dried, and mixed, due to repeated contact and separation between powder particles and the equipment. This results in the powders acquiring electrostatic charges. If not dissipated quickly, electrostatic charge buildup can cause inter-particle cohesion and adhesion to operating equipment, causing a production efficiency decrease. Incendiary electrostatic discharges such as spark discharges, cone discharges, and propagation brush discharges can also take place during industrial operations and may lead to fires and even explosions. Riedewald reported on an industrial fire in a pharmaceutical powder transfer operation [1]. A large number of studies involving electrostatic charge generation on powders, as well as electrostatic discharges and hazards during operation, have been reported. For example, Kaialy reviewed the main factors affecting the electrostatic charging of pharmaceuticals such as particle resistivity, size, shape, and surface roughness [2]. Gibson also reviewed industrial hazards related to electrostatic charges and discharges [3]. Our previous studies [4–6] investigated the electrostatic charging of polymer powder in mixing processes and the ignitability of aluminous, magnesium and polymer powders due to electrostatic discharges involving the effect of the ratio of N2 in the air on the ignitability.

In order to easily measure and confirm the electrostatic charge level of powders, a novel tribocharging system that utilizes the Faraday cup is presented, and the charging behavior of the sample powders is investigated in this paper.

Experimental Apparatus and method

Figure 1 shows the spiral air tribocharge apparatus presented in this study. The apparatus was divided into four main parts: the air supply part, the powder supply part, the tribocharge part, and powder collection part. The air supply part consisted of a compressor (Hitachi, PO3.7MA) that included an auto dryer (CKD, F3-217199), air controller, electronic valve, and switch. In this study, dried air was used because moisture in the air may have an effect on the electrostatic characteristics of the powder sample. The powder supply and tribocharge parts consisted of a powder port, a spiral air flow device, and a pipeline. The spiral air flow device (photo 1) was attached in front of the sample powder port in order to create spiral air inside the pipeline. The spiral air flow device had 8 holes that were 6 mm in diameter and that opened at a slant (60 degrees), which is very different from conventional models [7, 8, 9]. The straight pipeline was made up of stainless steel metal and was 1300 mm in length and 35 mm in diameter. In order to prevent external noise, the overall powder supply and tribocharge parts were shielded. The powder collection part was made up of a cyclone-type Faraday cup, unlike conventional models [7, 8, 9]. The cyclone-type Faraday cup had a separate metal container (30 mm in diameter) for easy measurement of the amount of collected powders. The other devices included two electrometers (Kethley,

6517A), a data logger (Graphtec, GL900), an electrostatic ionizer (Kasuga Denki Inc., KD-510W), an electrical balance (Shimazu, BL220H), and a timer.

The typical operating procedure for the tribocharge test is summarized as follows:
(1) Adjust the air pressure using the air pressure valve.
(2) Load 1 g of the sample powder into the powder port.
(3) Eliminate the initial charge of the sample powders using an electrostatic ionizer.
(4) Reset the electrometers (Q mode) connected to the pipeline and cyclone.
(5) Switch the air flow on, and run the sample powder through the pipeline where friction occurs.
(6) Measure the indication value when it stabilizes.
(7) Measure the amount of powder collected in the Faraday cup.
(8) Clean out the pipeline and cyclone part using an air spray in preparation for the next test.

The charge to mass ratio, q (nC/g), of the sample powder was obtained by dividing the charge, Q (nC), by mass, m (g), of the sample powder loaded into the Faraday cup:

$$q = \frac{Q}{m} \quad \text{(1)}$$

This experiment is mainly to determine how much the charge to mass ratio, q (nC/g), of the sample powders is affected by the size of the powder and the amount of air pressure. The cyclone-type Faraday cup and the pipeline were suspended electrically with insulators. All experiments were conducted in ambient conditions (air temperature: 22 ℃; relative humidity: 55 %).

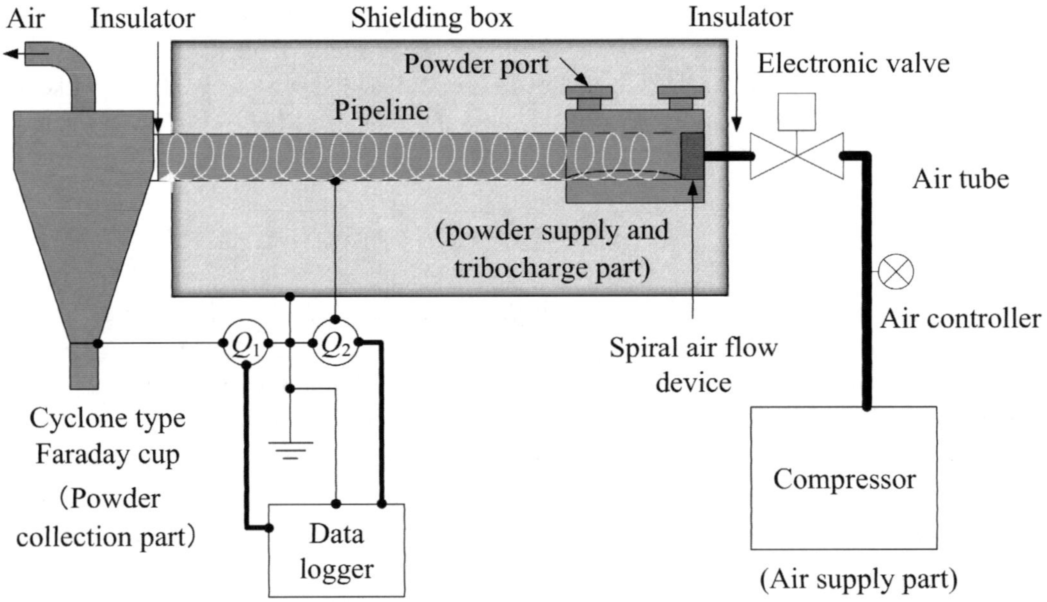

Fig. 1 Schematic of the spiral air tribocharge apparatus used in this study.

Photo 1 Structure of the powder port and spiral air flow device.

Sample powders

Glass beads are extremely good light reflectors and thus are widely used as the main component of the coating material for light-reflecting adhesive tape, background paint for traffic control sign boards, and paint for road signs. Four different grades of commercial glass beads were used in this experiment, as shown in Table 1. The particle size was determined using a laser diffraction system (LDA Win 1.21, dry type). Hereafter, they will be called by their abbreviation shown in Table 1. The analysis of the images from microscopy (Neoscope, JCM-500) revealed that the G-4 powder particles were spherical, as shown in photo 2. The other glass bead particles were the same shape as G-4.

Table 1. List of powder specimens used in experiments.

Specimen	Abbreviation	Diameter [μm]		
		D_{10}	D_{50}	D_{90}
Glass beads 1	G-1	22	62	85
Glass beads 2	G-2	93	131	175
Glass beads 3	G-3	130	180	241
Glass beads 4	G-4	162	225	294

D_X: X% volume average.

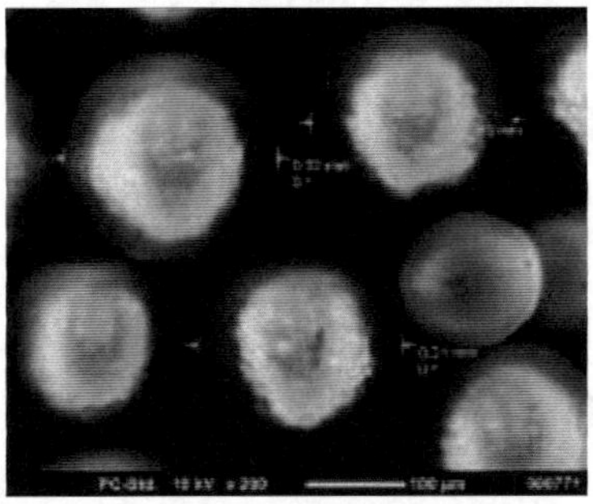

Photo 2 Shape of G-4 powder used in this experiment.

Experimental results and discussion

Figure 2 shows the typical change of charge detected over time when the G-3 particle was used and the air pressure was 0.15 MPa. It should be noted here that when no particle was used and only air was passed through the pipeline, the charge was detected as almost zero. The charge Q1 of the powder was found through the cyclone-type Faraday cup and then was confirmed through the charge Q2 obtained from the pipeline. As the powder particles passed through the pipeline for about 0.82 seconds, as seen in the Fig. 2, the charge clearly increased. Q1 and Q2 had almost the same trend and absolute values. This result confirmed the reliability of the test and showed that there was no disturbance by external noise.

In this study, all of the glass beads (G-1 to G-4) were negatively charged when stainless steel metal was used as the contact surface. The charge polarity of the materials relates to the work function (or ionization potential) of each material. The contact potential difference between two substances is defined by the difference between their fermi levels, which correlate with the work function. In general, materials with higher work functions tend to appropriate electrons from materials with lower work functions. This point will be carried out in our further study.

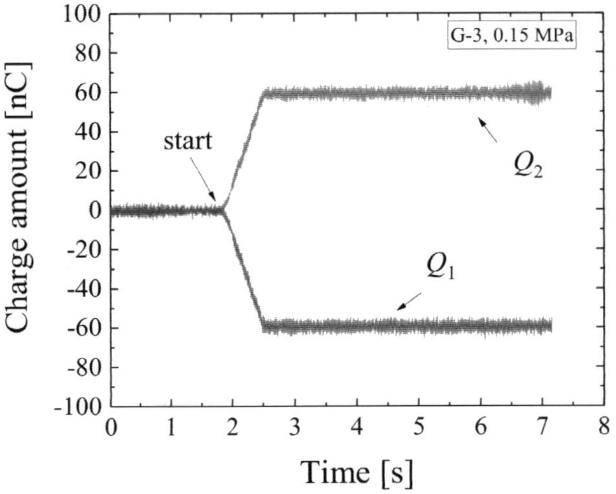

Fig. 2 Change of charge detected over time.

Figure 3 shows the charge to mass ratio of glass beads as a function of applied air. The range of applied air was between 0.1 MPa and 0.3 MPa in this study. When the applied air pressure was below 0.1 MPa, it was most likely not enough to send all of the powder through the pipeline. Each experiment was repeated five times, and the average values were reported. This graph does not contain an error bar, but these were about 10%. The results showed that the charge to mass ratio of both powders increased as the applied air increased. Although not shown in Figs. 2 and 3, the increase in air pressure also causes the slope of the graph to increase significantly. This can be explained by the fact that with increased pressure of applied air, the particles collide against the surface of the pipeline with more strength [9]. In order to investigate in more detail, the velocity of air at approximately 10 mm from the inside of the pipeline wall was measured using an anemometer (Kanomax, Anemomaster-6115). The approximate velocity of air flow was 9.5 m/s at 0.1 MPa, 12 m/s at 0.15 MPa, 15 m/s at 0.2 MPa, 20 m/s at 0.25 MPa, and 22 m/s at 0.3 MPa. It could be clearly seen that as the air pressure increased, the air velocity also increased. Another reason for this may be that because the air was applied in a spiral form, the particles may have collided more frequently against the pipeline when the air flow was increased. This exception will be investigated in detail in the future.

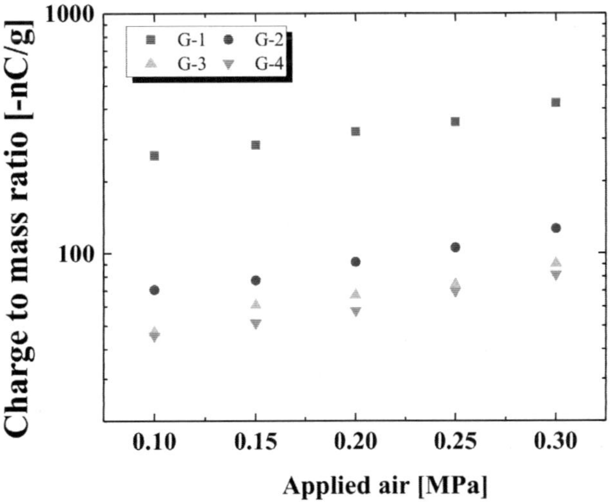

Fig. 3 Charge to mass ratio of glass beads as a function of applied air.

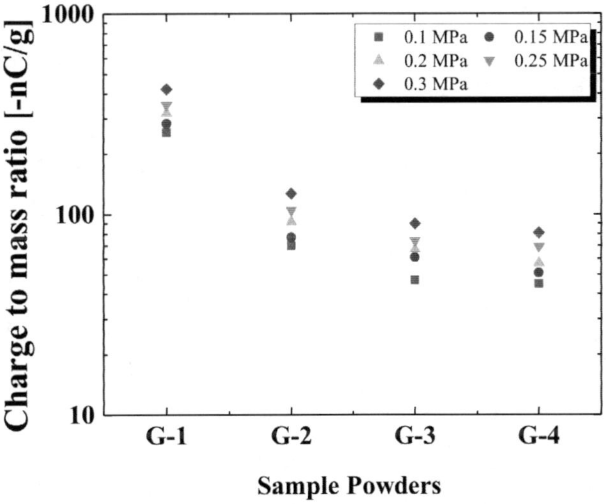

Fig. 4 Charge to mass ratio as a function of different glass bead sizes.

Figure 4 shows the charge to mass ratio as a function of different sizes of glass beads. The size distribution of each glass bead in this study was described in Table 1. The results showed that the charge to mass ratio of the glass beads increased as the powder size decreased. That is to say, smaller particles had a larger q than the larger ones. The total surface area of all of the smaller beads combined is greater than that of the larger ones, which results in a high charge to mass ratio development in the tribocharge. The value range of the q of the glass beads obtained in this study was about 40 nC/g to 100 nC/g. These values agreed with those obtained from the micronizing and pneumatic transfer process [10]. These values were high enough to generate incendiary electrostatic discharges during handling in industrial processes. This suggests that if glass beads are used with solvents in the cleaning process, they must be handled carefully, as incendiary discharges can ignite the solvents even though the glass beads themselves are nonflammable material.

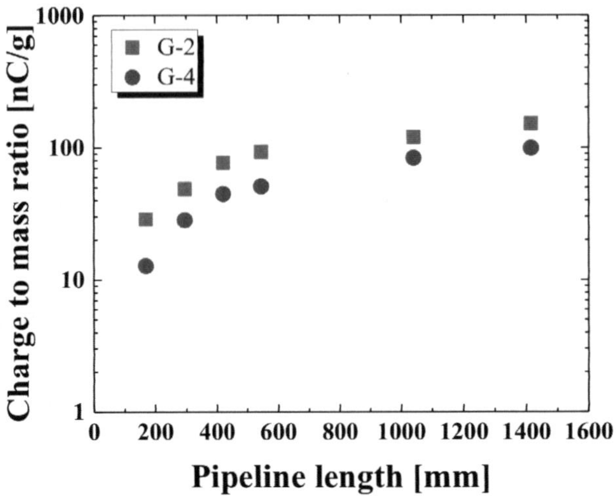

Fig. 5 Charge to mass ratio as a function of pipeline length.

Figure 5 shows the charge to mass ratio as a function of pipeline length. Six different pipeline lengths, *d*, were used: 170, 295, 420, 545, 1040, and 1415 mm. The glass beads used in this experiment were G-2 and G-4, and the air pressure was 0.35 MPa. Since the pipeline length changed with each test, it was difficult to attach the Faraday cup to the end of the pipeline. Therefore, in this study, only the value of q_2 was measured. As seen in Fig. 2, q_2 had the almost same value as q_1 but had an opposite sign. The length of the pipeline was measured starting from beginning of the powder port because the inside structure of the powder port was the same as that of the pipeline. Each experiment was repeated five times, and the average values were reported. The results showed that q_2 increased as the length of the pipeline increased for both sizes of glass beads. This can be explained by the fact that as the length increases, the time the glass beads come in contact with the pipeline also increases. The increase rate at the beginning section of the pipeline was especially large. It can also be seen through Fig. 5 that overall, G-2 had a larger q_2 than G-4 because of the total surface area. This confirms the reliability of the data shown in Fig. 4.

In addition, both beads seemed to reach saturation when the pipeline length was 1415 mm. It was compared to the maximum charge to mass ratio, q_c [nC/g] calculated from the well-known equation shown below [11]:

$$q_c = |\sigma_{max}|/(D \cdot \rho), \quad \text{-------------------------(2)}$$

where σ_{max} [kg/m^3] is the maximum surface charge density (27 μC/m^2); D [μm] is the particle diameter; and ρ is the particle density (1.5×10^3 kg/m^3).

In this study, the particle sizes of G-2 and G-4 were 131 μm and 225 μm, respectively, as shown in Table 1. The calculated maximum values q_c and experimental maximum values q_t are compared as shown in Table 2. The calculated and experimental values for both G-2 and G-4 agreed well with each other. Therefore, it can be seen that both G-2 and G-4 reached saturation.

Table 2. Calculated maximum values and experimental maximum values of the charge to mass ratio.

	q_c [nC/g]	q_t [nC/g]
G-2	137	151
G-4	80	99

Conclusion

In this paper, a novel tribocharging system that utilizes the Faraday cup was presented and the electrostatic charge behavior of glass beads was experimentally investigated. The results were as follows:
1) For all of the glass beads, the charge clearly increased as the glass bead particles passed through the pipeline. The charge amount of the glass beads and the charge amount obtained from the pipeline had almost the same trend and absolute values. This result confirmed the reliability of the test and showed that there was no disturbance by external noise.
2) All of the glass beads were negatively charged when stainless steel metal was used as the contact surface. The value range of the charge to mass ratio of the glass beads obtained in this study was about 40 nC/g to 100 nC/g.
3) The charge amount of the glass beads increased as the applied air increased. It could also be seen that as the air pressure increased, the air velocity also increased.
4) The charge amount of the glass beads increased as the powder size decreased due to the total surface area of the beads.
5) The charge amount of the glass beads (G-2, G-4) increased as the length of the pipeline. The increase rate at the beginning section of the pipeline was especially large.

Acknowledgments
The authors gratefully acknowledge Mr. Naoto Nogera for his assistance during the test.

References

1. F. Riedewald (2012): "A fire in a secondary pharmaceutical powder transfer operation," *Process safety progress*, 31, 4, pp. 390-392.
2. W. Kaialy W, "A review of factors affecting electrostatic charging of pharmaceuticals and adhesive mixtures for inhalation," *Int.J.Pharm*. Vol.503, pp.262-276.
3. N. Gibson (1997), "Static electricity — an industrial hazard under control," *J. Electrostatics,* Vol. 40-41, pp. 21-30
4. K. Choi, K. Choi, K. Nishimura (2015), "Experimental study on the influence of the nitrogen concentration in the air on the minimum ignition energies of combustible powders due to electrostatic discharges," *J. of Loss Prevention in the Process Industries.* Vol.34.pp.163-166.
5. K.S. Choi, H.Sakasai, and K. Nishimura (2015), "Experimental study on ignitability of pure aluminum powders due to electrostatic discharges and Nitrogen's effect," *J. of Loss Prevention in the Process Industries.* Vol.35.pp.232-235.

6. K.S. Choi, H. Sakasai, and K. Nishimura (2015), "Minimum ignition energies of pure magnesium powders due to electrostatic discharges and Nitrogen's effect," *J. of Loss Prevention in the Process Industries.* Vol.41.pp.144-146.
7. N. Masui, and Y. Murata (1981), "charging tendency of the flowing powder through a metal tube (II), " *The five Annual meeting of Institute of Electrostatic Japan,* pp. 244-247.
8. S. Watano, S. Saito, and T. Suzuki (2002), "Feedback control of powder electrostatic charge in a pneumatic conveying process," *J. Soc. Powder,Japan*, Vol. 39, pp. 496-502.
9. P. Boschung, M. Glor (1980), "Methods for investigating the electrostatic behavior of powders, J. of elelctrostatic," Vol. 8, pp. 205-219.
10. IEC50404(2004), "Electrostatics- Code of practice for the avoidance of hazards due to static electricity," p. 66.
11. M.Glor (1988), "Electrostatic hazards in powder handling," *Research studies press LTD.*, p.26.

509266 Spray-Drying of a Layered Double Hydroxide Nanosuspension

Abstract:

Over the past few years, layered double hydroxide (LDH) has attracted particular interests in both academia and industry. LDH is 2D anionic clay of brucite-like layered structure with various organic or inorganic anions located between layers. The diverse composition of LDH, and its associated physical and chemical properties open up new possibilities in various applications. LDHs are successfully applied as precursors in synthesis of supported catalysts, sorbents, carriers delivering bioactive molecules, matrices containing active pharmaceutical ingredients for control release, electrochemical biosensors, etc. LDHs are mainly synthesized in a form of nanosuspensions. The application of nano-sized sorbents or catalysts in the liquid-phase processes requires a special technique to separate them from the final product. An uncontrollable agglomeration of solid nanoparticles makes their industrial application impossible in a dry phase. Moreover, the toxic effects of nanoparticles on human health and ecosystem have been discussed in the literature.

Utilization of larger micron-size agglomerates of nanoparticles is one of the alternative ways to safely handle nanoparticles in the dry phase on the industrial scale. Spray drying is one of the well-known techniques that have been widely applied for production of microspheres from nano-sized particles. Using this technique, it is possible to control the size distributions as well as the morphology of spray-dried agglomerates. However, there are only a limited number of publications related to spray drying of LDH nanosuspensions. Moreover, spray-dried LDH agglomerates of rather wide size distribution, low surface area and irregular shape were reported in the literature.

In this study, we explore the effect of process parameters, such as spraying air flow rate, drying air temperature, slurry concentration and slurry feed rate on the size distribution, morphology, specific surface area and pore volume of LDH agglomerates fabricated by spray drying of nanosuspensions.

The round-shaped agglomerates with a diameter of more than 2 and less than 15 micrometers were obtained by spray drying of suspensions of nanoparticles with median diameter of 171 nm without addition of dispersant and binder. Our results confirmed that the spraying air flow rate has the dominant effect on the agglomerate median size and the width of agglomerate size distribution as well as on the agglomerate specific surface area. The small agglomerates with narrow size distribution were collected during drying at the high flow rate of spraying air. Our results also indicate that lowering the temperature of drying air and using the dilute nanosuspensions promote generation of small agglomerates.

The specific surface area of spray-dried LDH agglomerates is in the range of 49 to 101 m^2/g. The agglomerates of high specific surface area were obtained at low flow rate of spraying air, high slurry feed rate of concentrated slurry and using low temperature drying air. These results also confirm that drying at low drying rate yields to formation of agglomerates of high surface area. Similar trends were observed in terms of the variation of total pore volume with drying process parameters.

Thus, by adjusting the spray drying conditions, it is possible to produce agglomerates of LDH nanoparticles satisfying the product requirements.

513829 Impact of Spheroidization of UO_2 Powders on the Filling of Press Molds

Abstract:

In industry, the use of powders is very common. This underlines the interest in understanding the flow of powders in process lines (hopper, silos, transportâ€¦) and finding the link between powder physical characteristics and its rheological properties. This study deals with the flowing properties of a UO2 powder in a powder metallurgy process for nuclear fuel fabrication. A reproducible filling of the press molds is required to avoid mechanical grinding of the sintered pellets.

A powder is characterized by its particle size distribution but also by the morphology of particles. Morphology has a major impact on the cohesion and then on the flowability of powders [1]. UO2 powders are generally made up agglomerates which are composed of aggregates, themselves formed by crystallites (Figure 1). Agglomerates are primarily governing the powder flowing behavior. Their characteristics sometimes do not allow them to meet the expectations of the process, especially for the filling of the press molds.

For this reason, in a powder manufacturing process, a spheroidization operation (Figure 2) could precede the shaping of pellets to improve the filling properties.

In this study we observe the influence of particles morphology on the rheological properties of UO2 powders. Whereas there is no standard and universal method to characterize the flow of powders [2], there is direct (shear cells) and indirect methods (dynamic flowability measurements like permeability). We use both methods: the indirect methods based on qualitative measures of powder cohesion, the direct methods to determine the powder cohesion for a specific consolidation state through the Freeman shear cell [3]. While a shear cell measurement permits to determine the powder flow in quasi-static system, a permeability measurement shows the flow index in dynamic system. We identify the link between theses measurements and the quality of the filling.

References:
[1] Xiaowei Fu and others, «Effect of Particle Shape and Size on Flow Properties of Lactose Powders», Particuology, Advances in Characterization and Modeling of Particulate Processes, Apr. 2012.
[2] P. Juliano, B. Muhunthan, and G. V. Barbosa-Cánovas, « Flow and shear descriptors of preconsolidated food powders », J. Food Eng., vol. 72, no 2, p. 157-166, Jan. 2006.
[3] J. Schwedes and D. Schulze, « Measurement of flow properties of bulk solids », Powder Technol., vol. 61, no 1, p. 59-68, Apr. 1990.

508203 Modelling a Twin Screw Granulator Using the Discrete Element Method

Abstract:

Wet granulation is a process used to create larger stable agglomerates (granules) from fine powders. This has many desirable outcomes such as improving flowability, compactibility, and homogeneity. Granulation is commonly employed in the food, pharmaceutical, detergent, and fertilizer industries, but despite its wide adoption it is often inefficiently operated, with high recycle ratios in continuous processes and high rejection rates in batch processes. Wet granulation is the most common type of granulation and, in pharmaceutical applications, it is a critical step in tablet manufacturing that affects the uniformity and compactibility of the final dosage form.

Traditionally batch granulation was the favoured granulation process method in the pharmaceutical industry due to the challenges faced from adopting a continuous approach, such as cost or inability to monitor product quality reliably. Since the introduction of Quality by Design (QbD) and Process Analytical Technology (PAT) by the FDA in 2003 there has been a move towards new manufacturing techniques such as continuous processing.

Although tremendous efforts have been made to gain scientific insight into the granulating process, a fundamental understanding of wet granulation is still lacking due to the complexity of the mechanisms involved. With twin screw granulation becoming a popularly employed method of wet granulation, an in-depth understanding of particle enlargement in the granulating process is necessary in order to improve the quality of the final product without the need for large-scale Design-of-Experiment studies.

Despite the extensive experimental research carried out for twin screw granulators in recent years, there has been little computational work carried out. This paper employs the Discrete Element Method (DEM) to study a 25 mm diameter, GEA ConsiGmaâ„¢ 1 twin screw granulator with a typical 60Â° forward configuration for kneading elements. The DEM simulations were conducted using the commercial code EDEM with a DEM contact model developed for cohesive solids implemented through an API. The contact model is based on an elasto-plastic contact with adhesion and uses hysteretic loading and unloading paths to model the elastic-plastic contact deformation. In these simulations, the adhesion is used to capture the effect of the binder liquid without the complication of modelling the liquid directly. The adhesion parameter is a function of the plastic contact overlap. The model has previously been shown to be able to predict the stress-history-dependent behaviour depicted by a flow function of the material. In this study the effect of particle size, gap tolerance and cohesion on the residence time distribution and local solid fraction at different elements in the granulator is assessed. The level of cohesion used is related back to uniaxial test results which can be used to measure the flowability of a cohesive granular solid.

508216 Hybrid Multiscale Modelling of a Twin Screw Granulator

Abstract:

Wet granulation is a process used to create larger stable agglomerates (granules) from fine powders. This has many desirable outcomes such as improving flowability, compactibility, and homogeneity. Granulation is commonly employed in the food, pharmaceutical, detergent, and fertilizer industries, but despite its wide adoption it is often inefficiently operated, with high recycle ratios in continuous processes and high rejection rates in batch processes. Wet granulation is the most common type of granulation and, in pharmaceutical applications, it is a critical step in tablet manufacturing that affects the uniformity and compactibility of the final dosage form.

Modelling processes such as wet granulation is not straightforward, but may be achieved using techniques such as Population Balance Modelling (PBM) and Discrete Element Method (DEM). PBM groups particles into different classes based on their properties and uses empirically derived differential equations to describe key processes such as aggregation or breakage. However, although PBM tracks the size and number of particles over time, it cannot account for the distribution of the particles and the effect of this local variability.

By contrast, DEM simulations track individual particles, and as such, phenomena such as segregation or liquid migration can be studied. DEM simulations also consider the effect that the geometry and its dynamics have on the particles' behaviour, which is something that PBM cannot account for directly. However, DEM simulations of fine materials can be computationally expensive and this often results in reduced numbers of particles being used to make the simulation computationally tractable. This is usually achieved through increased, unrealistic particle sizes or the simulation of isolated local zones without considering adjacent zones which may affect the behaviour.

Traditionally, large-scale Design-of-Experiment (DoE) studies would be needed to estimate key rate parameters for use in PBM modelling of granulation and this can be both costly and time-consuming. DEM simulations provide an alternative method to estimate these parameters. By coupling the two simulation methods, one can utilise the key benefits of each method to better simulate complex processes.

This paper presents a hybrid approach to modelling twin screw granulation via DEM–PBM coupled simulations to study a 25 mm diameter, GEA ConsiGma™ 1 twin screw granulator with a typical 60° forward configuration for kneading elements. The DEM simulation was conducted using the commercial code EDEM with a DEM contact model developed for cohesive solids implemented through an API. The PBM simulation is carried out with the commercial code gFORMULATE which is built on top on the advanced modelling platform gPROMS. A bi-directional coupling is implemented with information shared between the two simulation tools at specified intervals. Preliminary results for this coupled approach are presented and discussed.

509577 Blocking Rules for Discharging Granular Materials from a Flat Bottom Hopper

Abstract:

Hoppers and silos are widely used in storing granular materials in various industries, such as agricultural, chemical, food and pharmaceutical industries. Three key aspects need to be considered when hoppers are used to store powders and grains: 1) stress analysis; 2) hopper blockage ; 3) mass flow rate. A hopper must be designed strong enough to sustain the stress generated from the bulk materials during storage and discharge. On the other hand, a hopper must be designed properly so that the stored bulk materials can be discharged, i.e. without blockage and formation of ratholes and arches. Furthermore, the mass flow rate needs to be predicted and controlled. In this study, a combined experimental and numerical study was performed to determine the blocking rule for discharging granular materials from a flat bottom hopper. Granular materials of various sizes were considered and the minimal orifice size through which the bulk materials can be discharged was determined experimentally using a Flodex tester and numerically using DEM. The dependency of the minimal

orifice size on particle size was explored. An semi-empirical model was also developed. Using this model, the critical orifice size at which the flow will stop, i.e. blockage takes place, can be determined from the mass flow rate data. It was found that excellent agreement between the numerical results and experimental measurements was obtained. Both experimental and numerical results showed that, even for spherical particles, layered dead zones could be formed during the discharge from the flat bottom hopper. In addition, mono-layered dead zones were developed when large particles were used. Furthermore, the developed blocking rules were also verified experimentally and numerically.

508310 Characterization of Mesoscopic Structure in Cohesive Powder, Neat or Blended, By X-Ray Computed Tomography and Prediction By the Discrete Element Method

Abstract:

Understanding and predicting structural characteristics of cohesive powder systems is a technical challenge in many powder technology processes and applications. The discrete element method (DEM) [1] is a promising approach to model the structure and flow behavior of cohesive granular systems. Advances in experimental techniques such as X-ray computed tomography (XRCT) have now enabled characterization of fine cohesive powder systems with sub-micron resolution. Combined with standard bulk powder characterization methods (e.g. density), this technique allows for mesoscopic structural characteristics of cohesive powder systems to be compared with DEM simulation. In this study, DEM and XRCT are used to characterize the mesoscale structure, under controlled consolidation conditions, of two moderately cohesive 'model' (roughly spherical) powders with distinct material properties. The DEM simulations are based on a constitutive elasto-plastic model [2] that makes use of the JKR theory [3] to account for cohesive interparticle forces. Density and void structure obtained from DEM simulation are compared to characterization results obtained from XRCT image analysis. XRCT and DEM simulation approach is applied to examine structures of binary mixtures of the two model powders. The ability of DEM simulation and the contact model to accurately predict structural characteristics of single component and binary granular systems is discussed.

References

[1] P. A. Cundall and Otto D.L. Strack. A discrete numerical model for granular assemblies. Geotechnique, 29(1):47–65, 1979.

[2] C. Thornton and Z. Ning. A theoretical model for the stick/bounce behavior of adhesive, elastic-plastic spheres. Powder technology, 99(2):154– 162, 1998.

[3] K.L. Johnson, K. Kendall, and A.D. Roberts. Surface energy and the contact of elastic solids. In Proceedings of the Royal Society of London A: Mathematical, Physical and Engineering Sciences, volume 324, pages 301–313. The Royal Society, 1971.

511638 Numerical Research of Hydrodynamics in Gas-Solid Micro Fluidized Beds

Abstract:

Due to the advantages of less energy consumption, more safety and good operability, the miniaturization of fluidized bed, or named micro fluidized bed, has been extensively studied by a lot of researchers. The Micro Fluidized Bed Reaction Analyzer (MFBRA) was adopted by the Institute of Process Engineering (IPE), Chinese Academy of Sciences (CAS), to study the gas-solid reaction kinetics under isothermal differential condition[1, 2]. In the MFBRA, the hydrodynamics of micro fluidized beds play a vital role on the heat and mass transfer and which is a key to the accurate measurement of gas-solid reaction kinetics. In order to optimize and design the MFBRA, numerous experiments have been conducted to study the hydrodynamics of micro fluidized beds[3, 4]. Unfortunately, those works mostly focused on the bed pressure drop and minimum fluidization velocity, nevertheless, the detail information, such as local solid volume fraction and particle velocity profile in micro fluidized beds are reported rarely. It is because the existing experimental techniques, such as, optical fiber and capacitance probes are hard to measure those local parameters without disturbing the flow field.

The objective of this work is to study the hydrodynamics of micro fluidized beds and to investigate the isothermal characteristics during kinetics testing of graphite powder combustion in MFBRA. In this study, Discrete Particle Model (DPM) and Two-Fluid Model (TFM) are selected to investigate micro fluidized beds, respectively. The DPM is carried out by the commercial software EDEM-Fluent Coupling simulations. And for TFM, we used the Eulerian multiphase model in the ANSYS Fluent. From our numerical simulations, the detailed hydrodynamic characteristics of micro fluidized beds are explored systematically by comparing different column sizes and superficial gas velocity. Meanwhile, it is found that both DPM and TFM have the ability to model the MFBRA under practical operation condition with high superficial gas velocity. In the term of isothermal characteristics in micro fluidized beds, the temperature field distribution as well as the heat transfer between heat-carrier and graphite powder are explored by the combination of simulations and experiments.

Reference:

1. Xu, G., *Micro fluidized bed reaction kinetics analyzer. Supported by instru-ment development project of Chinese Academy of Sciences, China, Project numberY2005014. (in Chinese).* 2005.
2. Yu, J., et al., *Isothermal differential characteristics of gasâ€"solid reaction in micro-fluidized bed reactor.* Fuel, 2013. **103**: p. 29-36.
3. Liu, X., G. Xu, and S. Gao, *Micro fluidized beds: Wall effect and operability.* Chemical Engineering Journal, 2008. **137**(2): p. 302-307.
4. Guo, Q.j., Y. Xu, and X. Yue, *Fluidization Characteristics in Micro-Fluidized Beds of Various Inner Diameters.* Chemical Engineering & Technology, 2009. **32**(12): p. 1992-1999.

507312 An Improved Bubble-Based Drag ModelÂ for Accurate Coarse-Grid Two-Fluid Modeling of Geldart a Powder Bubbling Fluidization

Abstract:

An improved bubble-based energy minimization multi-scale (EMMS) drag model was formulated in which the original bubble-based EMMS drag model at macroscale bed level was extended to microscale sub-grid level for accurate coarse-grid two-fluid modeling (TFM) of Geldart A powder bubbling fluidization. At both the macroscale and microscale levels, the physically heterogeneous structures were modeled by three pseudo phases, i.e., bubble phase, emulsion phase, and inter-phase, where conservation and stability condition equations were applied. Through such scale and phase decompositions, the improved bubble-based EMMS drag model was formulated through a two-step macro-to-micro scheme, where the obtained drag coefficient is related to phase volume fractions, velocities, and accelerations within each coarse grid. Modeling results show that TFM combined with the improved bubble-based EMMS drag model can accurately reproduce the heterogeneous structures for a laboratory-scale Geldart A powder bubbling fluidized bed. At relatively coarse grid, the predicted axial and radial distributions of solid concentration agree very well with experimental measurements. This indicates that our proposed improved bubble-based EMMS drag model incorporated TFM is promising for accurate and economical simulation of engineering-scale Geldart A powder bubbling fluidization.

507860 Numerical Simulation on Fine Particle Transport Behaviour in Electrostatic Precipitators

Abstract:

Electrostatic precipitators (ESPs) are a major air pollution control device to remove particles from treated gas. Extensive research has been done on several aspects such as electrostatics, fluid dynamics, charging mechanism, and particle dynamics. The overall particle-collection efficiency of a modern industrial wire-plate ESP exceeds 99%. However, the removal efficiency of submicron particles by conventional ESPs is as low as 70â€“80% owing to the low charge carried by fine particles and the increase in mobility with decreasing size. Submicron particles possess a high superficial area, thus favouring toxic heavy metals and other pollutants. Therefore, fine particles are collected with much lower efficiency and further research is required to improve this aspect.

In this work, a numerical study on the charging and transport behaviour of fine particles has been carried out based on wire-plate electrostatic precipitators (ESPs) with multiple wire electrodes. The effect of the applied wire voltage, inlet height, and wire distance on particle charging and transport behaviour, and the influence of the precipitator structure on particle trapping are analysed in detail. The 2-D numerical model involves electric field, space-charge density, gas flow, and particle trajectories. The fluid-flow velocity is higher than 1 m/s in this study and the EHD flow has a negligible. A Lagrangian approach for particle transport is utilized with the trajectories of individual particles tracked by solving their equation of motion. The electric field and gas flow are calculated using the finite-volume method. The flow simulation is done using the commercial software FLUENT with other parts coded as user-defined functions. The following main results can be drawn:

(1) It was observed that the Y-displacement of the particles exhibited a universal increase for all particle sizes, especially for larger particles, at low injection positions ($Y < 25$ mm). The larger the particle size, the higher the Y-displacement of particles because larger particle sizes improve both the particle charging level and electric field strength. No significant improvements were found for submicron particles.

(2) Increasing the spacing of the discharge electrode within a certain range and therefore increasing the electric field, it was observed that the particle-trapping effect improved and the outlet level exhibited a universal increase for all particle sizes, especially for large particles.

(3) The ESP models M1 and M2 were used to investigate their influence on the outlet height. Except for particles with a diameter of 2 μm, all other particle sizes showed an increased outlet height in M2 compared to that in M0. In addition, the different discharge-electrode arrangements, especially in M2, exhibited significant effects on the charging and transport behaviour of particles with a diameter of 1 μm.

(4) Particles with an inlet height of 5 and 10 mm were studied in M1 and M2 precipitators. The changed discharge-electrode arrangement improves particle trapping, especially in M2. This is because of the electron trajectories being closer to the discharge electrodes (M2), giving the particles more charge and leading to a greater electric field.

507879 Lattice Boltzmann Simulations of Porous Particulate Flows

Abstract:

The comprehensive understanding of the particle-fluid interaction is very important for fluid-particles system, and thus many studies have been conducted for the particle-fluid system[1-3] of solid impermeable particles. In the real industrial fields such as the catalyst process in chemical engineering[4], the particles usually possess the porous structure which will have effects on the fluid-particles interaction[5-6]. However, there has been little open literature discussing the particle-fluid system of porous permeable particles. Therefore, the objective of the present study is to investigate the effect of the porosity and permeability on the porous particulate flows which are formulated by the volume-averaged governing equations[7] in terms of intrinsic phase average velocity. The lattice Boltzmann equation model (LBE)[8] is used to solve the governing equations. Firstly, we validate the LBE model by the numerical results existing in the literature for the flow around and through one porous particle. Then, the porous particles settling against gravity in a quiescent fluid, i.e., the DKT process[9] for porous particles, is investigated. The effects of the porosity and Darcy number on the porous particulate flows are studied numerically in details. Also, we present the simulation results of the fluidization of 512 porous particles. Our results demonstrate the significance of the porous structure of particles for the particles dynamics behaviors.

Reference

[1] Ladd, A.J.C., 1994. Numerical simulations of particulate suspensions via a discretized Boltzmann equation Part I. Theoretical foundation. J. Fluid Mech., 271, 285-310.

[2] Hill, R.J., Koch, D.L., Ladd, A.J.C., 2001. The first effects of fluid inertia on flows in ordered and random arrays of spheres. J. Fluid Mech. 448, 213-241.

[3] Van der Hoef, M.A., Beetstra, R., Kuipers, J.A.M., 2005. Lattice Boltzmann simulations of low-Reynolds-number flow past mono- and bidisperse arrays of spheres: results for the permeability and drag force. J. Fluid Mech. 528, 233-254.

[4] Zhao, Y.F., Li, H., Ye, M., Liu, Z.M., 2013. 3D numerical simulation of a large scale MTO fluidized bed reactor. Ind. Eng. Chem. Res. 52, 11354-11364.

[5] Noymer, P.D., Glicksman, L.R., Devendran, A., 1998. Drag on a permeable cylinder in steady flow at moderate Reynolds numbers. Chem. Eng. Sci. 53, 2859-2869.

[6] Zhu, Q.Y., Chen, Y.Q., Yu, H.Z., 2014. Numerical simulation of the flow around and through a hygroscopic porous circular cylinder. Comput. Fluids 92, 188-198.

[7] Wang, L., Wang, L.-P., Guo, Z.L., Mi, J.C., 2015. Volume-averaged macroscopic equation for fluid flow in moving porous media. Int. J. Heat Mass Trans. 82, 357-368.

[8] Chen, S.Y., Doolen, G.D., 1998. Lattice Boltzmann method for fluid flows. Annu. Rev. Fluid Mech. 30, 329-364.

[9] Fortes, A.F., Joseph, D.D., Lundgren, T.S., 1987. Nonlinear mechanics of fluidization of beds of spherical particles. J. Fluid Mech. 177, 467-483.

508370 Modeling of a Novel Multi-Particle Collision Model for Gas-Solid FlowsÂ

Abstract:

Fluidized beds are widely applied in the chemical, petrochemical, metallurgical, environmental and energy industries due to their favourable mass and heat transfer characteristics. In past few decades computational fluid dynamics has proven to predict hydrodynamics of fluidized beds. DPM (Euler-Lagrange) [1] and TFM (Euler-Euler) [2] are two most common modelling approaches for gas-solid flow in fluidized beds. In TFM the solid phase itself is treated as a continuum phase interpenetrating with the gas phase. Since the concept of individual particles has disappeared, the effects of particle-particle interactions in the TFM has to be incorporated in a statistically averaged way, via the solid phase pressure and viscosity. TFM is preferred model to study large scale fluidized beds but model suffer from high numerical diffusion. And the KTGF theory in TFM assume elastic collisions and is limited only to spherical particles [2]. In DPM, the gas phase is considered as the continuous phase whereas the particle trajectories are obtained by integrating Newtonâ€™s equations of motion. Particle-particle and particle-wall interactions are explicitly taken into account using various physical models such as the soft sphere model. DPM proves to be accurate modelling of gas-particle flow. However, lagrangian models quickly become computationally too expensive because of the large number of particle interactions. Therefore coarse graining of discrete particles is an emerging model that can handle large number of particles effectively. The new coarse-grained model developments are of vital importance for the advancement of large-scale modelling of gas-particle flows that are flexible to extended for polydispersity, and non-spherical particles. In this study a novel lagrangian multi-particle collision (MPC) method is developed for predicting dense particulate-fluid flow.

The MPC solver is developed in OpenFoam [5], where the fluid phase is treated as a continuum and the solids are modelled by lagrangian particles. The MPC model grew out of the multiphase particle-in-cell (MP-PIC) model [3], that employs a fixed Eulerian grid, and particles are traced in a lagrangian fashion, including a parcel approach where each simulated particle represents a large collection of real particles. Each parcel follows Newtonâ€™s equations of motion, where inter-phase momentum transfer is accounted through mapping and interpolating quantities back and forth between parcel locations and the eulerian grid. Particle collisions in MP-PIC are not fully resolved but are derived from particles stresses that are theoretically calculated from Harris and Crighton [4] stress closure. In the MPC model, we implemented a functional form for the particle stress tensor where the coefficients are obtained computationally by means of DPM simulations. The idea is to extend the model from spherical particles to non-spherical particles, for which theoretical particles closures are currently not available in the literature. The viscous stress and random forces are typically ignored in MP-PIC model due to their insignificant effect [3]. Importantly, we incorporated viscous stress and random forces representing scattering of particles by collisions, which are obtained from DPM simulations by measuring the actual variance. Details of model development and validation is presented in this study.

Results from MPC model are compared with MPPICFoam and, with DPM simulations from DEM-LIGGGHT [6], and MFiX [7] code respectively. Pseuduo-2D fluidized beds of dimension 0.01x0.15x0.45 m is considered for comparison with experiments [8]. Comparison is made on the basic of pressure drop, void fraction, bubble dynamics and solids motion. We observed that MPC is in good agreement with DPM, and experiments, when compared to MPPICFoam. Unrealistic over-packing of solids is predicted in MPPIC due to limitation of Harris and Crighton [4] model and the velocity field updates, whereas in MPC these limitations overcomes and no over packing of particles is predicted. Solid circulations are in good agreement with DPM results, where solid moving above in the center and downward near to the wall. Time-average porosity plots revels the path of bubble motion and is in close agreement with results of DPM. In conclusion a novel multi-particle collision model is developed and validated to study gas-solid fluidized beds. The model is efficient to handle large number of particles. In future study, we plan to extend the model for dense flow of non-spherical particles.

Acknowledgments

The authors thank the European Research Council for its financial support under its consolidator grant scheme, contract no. 615906 (NonSphereFlow).

References

1. N.A. Patankar, D.D. Joseph, 2001. Modeling and numerical simulation of particulate flows by the Eulerian Lagrangian approach, International Journal of Multiphase Flow, Volume 27, Issue 10, October 2001, Pages 1659-1684.
2. Gidaspow, D., 1994. Multiphase flow and fluidization continuum and kinetic theory description, Academic Press, Boston.
3. Andrews, M.J., O'Rourke, P.J., 1996. The multiphase particle-in-cell (MP-PIC) method for dense particulate ows, Int. J. Multiphase Flow 22, 379-402.
4. Harris, S.E., Crighton, D.G., 1988. Solutions, solitary waves and voidage disturbances in gas-fluidized beds, J. Fluid Mech. 195, 437.
5. OpenFOAM, 2015. The Open Source CFD Toolbox, Ver. 3.0.0. ESI Group, Paris, France.
6. https://www.cfdem.com/liggghtsr-open-source-discrete-element-method-particle-simulation-code
7. https://mfix.netl.doe.gov/mfix/
8. M. Goldschmidt, 2001. PhD thesis, Hydrodynamic Modelling of Fluidised Bed Spray Granulation.

508490 Dpm Analysis of Large Fluidized Catalytic Cracking (FCC) Reactors

Abstract:

DPM Analysis of Large Fluidized Catalytic Cracking (FCC) Reactors

Abstract:

Computational Fluid Dynamic (CFD) modeling is a powerful tool for equipment design and development and is widely used at UOP, particularly for the reactor and regenerator design of FCC process unit. Through CFD modeling, a technical solution was developed for a large capacity FCC process unit purposing an elevated amount of propylene production.

A key challenge in FCC process design is how to optimize the size of the reactor vessel, which is a function of the number and size of reactor cyclones (catalyst separationdisengagers). UOP developed a new cyclone cluster configuration via CFD study of the reactor internals, including UOP's VSS™ riser termination device and reactor cyclones, which reduce the reactor size significantly compare to the conventional design, while maintaining effective catalyst / vapor separation and mechanical reliability. The objective of the current CFD study is to evaluate the catalyst separation efficiency and performance of the new riser termination device technology, which accommodate more cyclones in a smaller reactor vessel.

To better understand the flow pattern, flow distribution and the gas residence time inside the UOP VSS™ , the analysis was broken down into two sections. Part 1 focused on single gas phase only and in part 2 the particles behavior is studied using the Discrete Phase Model (DPM) in Fluent.

The single gas flow analysis results showed that the flow in the VSS chamber section has the same rotational pattern as in conventional FCC/VSS units, where the catalyst will separate from the gas products. Also the gas flow distribution at each chamber arm is uniform, which is critical to catalyst separation efficiency.

The DPM analysis confirmed an equal catalyst particle distribution through all the cyclones and a high efficiency separation in the riser termination device. Having this information from CFD allows for the new design to move forward.

509623 Catalytic Reactor Design Using Multiphase CFD

Abstract:

Catalytic Reactor Design Usng Multiphase Cfddimitri Gidaspow the 2011 Book , “Computational Gas-Solids Flows and Reacting Systems “, Edited By S. Pannala, M. Syamlal and Tom O'Brien ,Engineering Science Reference, Shows the Progress We Have Made in Using Multiphase CFD in Modeling Fluidized Bed Chemical Reactors and Other Multiphase Systems, Such As the Pyroplastic Flow Emitted from Volcanoes. in This Presentation the Design of Two Types of Catalytic Reactors Will be Described: A Fischer_ Tropsch Slurry Bubble Column Reactora Polymerization Reactor the CFD Design of a New Slurry Bubble Column Reactor Was Already Described By Gidaspow ,Et Al in Chemical Engineering Science 134,784-799 (2015).in This Presentation the Design Will be Extended to Obtain 100, 000 Barrels of Diesel per Day, Twice That of the Largest Commercial Reactors. the Production of Synthesis Gas Using Oxygen or Carbon Dioxide in Catalytic Fluidized Beds Needed for the Fischer-Tropsch Reactor Will Also be Discussed Briefly. for the Polymerization Reactor Our CFD Code Described in “ Computational Techniques “ By D. Gidaspow and V. Jiradilok , Nova Science, 2009 Was Modified for Polymerization Reactors. Such Reactors and Their CFD Modeling Are Described in a Chapter in Pannala's Et Al Book from Professor Rodney Fox's Group. They Point out That the Fine Particle (80 microns) Catalyst Radial Distribution, Is a Remaining Issue the Computations to be Presented Will Show That the Catalyst Will Have a High Concentration at the Reactor Walls for Various Operating Conditions. This Leads to a High Rate of Reaction at the Reactor Walls and May Result in Melting of the Polymers and Sheet Formation , Forcing Reactor Shut-Down. in Some of Our Simulations There Is Large Bubble and Slug Formation Similar to Those Mentioned in Fox's Chapter. Acknowledgement : I Thank Dr. Mayank Kashyap of SABIC for a Valuable Discussion and His Papers “ Application of Multiphase Flow CFD in the Gas Phase Polymerization Process “ Fluidization XV. May 24,2016 and a Related Paper Presented at Doe NETL, July 2016

509627 Simulation of Large Particle Turbulent Fluidization in Riser Reactors By Coarse Grain DEM-CFD

Abstract:

The complex two-phase flow in the riser is an important element of many industrial circulating fluidized bed reactors. In some cases, e.g. gas-solid polymerization, relatively large particles are transported upward in a turbulent regime, in which the high gas velocity results mostly from heat removal requirements. Strong solids acceleration, radial and axial velocity profiles for both the gas and solid, intense clustering and inhomogeneities make it impossible to reliably predict the riser head or local bed density profile. Detailed numerical simulations proved capable of showing some of the peculiarities of the flow, but limitations in the scale, lack of grid independence, need for special drag corrections and sub-grid models still hinder confident use of such tools in design and scale-up, particularly in industrial applications characterized by particle size and Reynolds numbers that are not well covered by the available scientific literature.

In the present contribution, coupled Discrete Element Method-Computational Fluid Dynamics (DEM-CFD) simulations are carried out using a coarse graining technique for the solid particles. According to this method, small number of particles are grouped into a computational particle (*parcel*) and represented by it as a spherical "lump," to reduce the computational cost of the simulations and allow lab-scale systems to be represented in full detail. The investigation focuses on the influence of the coars-graining degree, i.e. the number of real particles represented by a *parcel*, on the simulated flow hydrodynamics both in terms of its advantages (a quantification of the computational savings) and disadvantage, primarily the real capability of the method to resolve the behaviour observed and produce results in line with experimental findings. The gas and solid motion are studied, with the aim to characterize the clustering phenomena observed during the up-flow and its effect on the pressure drop. Assuming prescribed gas and solid inlet velocity fields, the potential tendency to show core-annulus flow, the intensity of solids clustering and the local bed density within the riser are examined at different number of particles per *parcel*.

509640 Numerical Study of Particles Shape Effects on Solid-Liquid Fluidizations

Abstract:

Numerical study of particles shape effects on solid-liquid fluidizations

Esmaeil Abbaszadeh Molaei, Zongyan Zhou*

Laboratory for Simulation and Modelling of Particulate Systems, Department of Chemical Engineering, Monash University, Clayton, VIC 3800, Australia

Solid-liquid fluidizations have been observed in many industries due to their high heat and mass transfer rate, high operation flexibilities, and reduced back mixing of fluid and solid phases. In most of these multiphase operations, particle properties such as size, density, and shape may experience significant changes during the process because of attrition, coalescence, comminution or chemical reactions. The difference in particle properties can result in significant segregation problem in liquid fluidized beds. To gain an understanding of solid-liquid fluidisation behaviour, most existing studies, either experimentally or numerically, have focused on solid flow behaviour with mono-sized particles or binary and ternary mixtures with different particle sizes or densities. In spite of substantial studies on the hydrodynamics of liquid fluidized beds, the effects of particles shape on the solid-liquid flow behaviour are still less reported. Except for two physical experiments, the previous studies considered particles as spheres, and few studies have been remarked on liquid fluidization of non-spherical particles. However, particles are generally irregular in most of the processes such as ore beneficiations using liquid-solid fluidized beds separator. Particle shape can cause different flow behaviour in liquid fluidizations rather than that observed for ideal spherical particles. Nonetheless, the answers to some fundamental questions such as how and why particle shape affects flow phenomena are not clear, and hence still poorly understood.

Therefore, in this work, a combination of computational fluid dynamics (CFD) and discrete element model (DEM) [CFD-DEM] is used to examine the effects of particle shape on the flow characteristics in liquid fluidizations. In the simulations, ellipsoidal particles are used as they can represent a wide range of particle shapes from oblate to prolate particles. Different aspect ratios are employed, and results are analysed mainly in terms of particle flow patterns, pressure drop, and bed expansion. In addition, force and particle orientation analysis is also conducted to examine the effect of particle shape and explain the occurrence of entrainment phenomenon.

The results show that ellipsoidal particles intend to enter to freeboard region and entrainment phenomenon occurs, and consequently, the interface between the bed and freeboard cannot be observed clearly. In addition, the pressure gradient variation with bed height is uniform for spheres; however, this uniformity is replaced with â€œdistorted parentheses-shapeâ€ profile for disc-shape and elongated particles. Finally, bed expansion analyses show that spherical particles have the lowest bed expansion. On the other hand, disc-shape particles intend to expand more significantly than spherical and elongated particles. This causes the entrainment phenomenon which has been found to be due to the effects of particles shape on particles terminal velocity.

* Corresponding author. Tel: 61 3 9905 0846; email: zongyan.zhou@monash.edu.

514318 CFD Investigation on Gas-Solids Flow and Heat Transfer in Two Fluidized Catalyst Cooler

Abstract:

Our previously experimental work showed that the internal circulation of solids in an annular catalyst cooler (ACC) had an ability to enhance the heat transfer between bed and wall, comparing with the base catalyst cooler (BCC). The current work simulated in detailed hydrodynamics of gas-solids flow and heat transfer to advance our understanding of the ACC by using the two-fluid model. The effective particle diameter $d_\mathrm{p}^{\text{âŽ}}$ was determined to 170 Î¼m and the specularity coefficient Ït was 0.3 in the 3D simulations. The catalyst cooler can be divided to four distinct regions including draft tube region, annulus region, bottom region and particle diffluence region according to the distributions of solids volume fraction and solids velocity. The effect of gas-solids flow on bed-to-wall heat transfer was compared in the different regions of the BCC and ACC. And the effects of operational gas velocity and bed structure were discussed. The results showed that the proper selection of the ratio of gas velocities and the position of the heat transfer tube were needed to form a stable internal solids circulation in the ACC. The ACC had a combined hydrodynamic feature of up-flow and down-flow catalyst coolers with bigger solids volume fraction and smaller particle resident time, which are beneficial for improving the heat transfer between solids and wall.

Hydrodynamics and mixing characteristics of a new-type particle mixer

Mengxi Liu, Chunxi Lu, Zhenliang Meng

State Key Laboratory of Heavy Oil Processing, China University of Petroleum, Beijing 102249, PR China

E-mail: mengxiliu@sina.com (M. Liu)

Abstract:

In order to further enhance the particle mixing in the internally circulating fluidized bed, especially along the radial direction, a new-type particle mixer with slots on the draft tube was proposed in the present work. The hydrodynamics of the mixer was investigated by using a multi-scale computational fluid dynamics with the EMMS drag. The prediction in terms of cross-sectionally averaged solid holdup and solid velocity were compared to experimental data for validation. The results show that in draft tube a low pressure region can be observed near the slots and in annulus the pressure in the vicinity of the slots seems larger than that in the region between the adjacent slots, which induces horizontal particle immigration and mixing. Nearly 25.4% to 29.7% of gas in the draft tube bypasses and flows through the slots into the annulus. The ratio of the mass flux of the tracer gas bypassing into the annulus to that of total tracer gas is approximately equal to 0. With increase of the superficial gas velocity in the draft tube from 0.3 m/s to 0.45 m/s, the total solid mass flux increases from 4.87 kg/s to 10.86 kg/s, while the mass flux through slots decreases from 2.69 kg/s to 0.52 kg/s.

1. Introduction

The characteristics of particle flow and mixing have a considerable influence on the mass and heat transfer, as well as the product quality. According to the previous study, the mixing degree of particles along the axial and radial direction is quite different [1-5]. Mostoufi and Chaouki [4] discovered that the diffusivities varied from 3.3×10^{-3} m^2/s to 5.6×10^{-2} m^2/s along the axial direction and 2.6×10^{-4} m^2/s to 1.5×10^{-3} m^2/s along the radial direction. Berruti et al. [6] also revealed that the particle mixing rate for radial direction was an order of magnitude smaller than for the axial direction. Therefore, the weak particle flow along the radial direction greatly limits the mixing ability of a fluidized bed. Internally circulating fluidized bed (ICFB) has been widely recognized as an efficient fluidized bed reactor, in which orderly and macroscopic particle circulation occurs. Compared with the conventional fluidized bed, ICFB offers multiple particle circulation between the draft tube and the annulus, which leads to relatively strong particle convection, diffusion and shear, and thereby improves the particle mixing performance along the radial direction [7]. In order to further enhance the particle mixing for some processes in which a uniform particle mixing is required, a new-type particle mixer with slots on the draft tube was proposed in the present work.

With increase of the computational power, the computational fluid dynamics (CFD) provides an additional tool to predict particle flow and mixing in the local region, which is difficult with the experimental methods. In particular, the gas-solid flow in a gas-solid fluidized bed can be accurately described by the energy minimization multi-scale (EMMS) model [8]. In this paper, the hydrodynamics of the mixer was investigated by using a multi-scale CFD based on the EMMS drag model [9, 10].

2. Experimental apparatus, Materials and Operating conditions

The structure of the particle mixer is shown in Fig.1. The mixer has a height of 5 m and an inner diameter of 286 mm. A draft tube with a height of 1 m and an inner diameter of 215 mm is coaxially mounted into the column. The four rectangular slots of 80×40 mm are symmetrically opened on the draft tube at a height of h=860 mm, in order to strengthen the particle radial flow. A gas distributor plate and a ring distributor are installed below the draft tube and the annulus respectively. The height of the gap between the gas distributor plate and the draft tube is 60 mm. For the sake of the convenience, the mixer is divided into six zones: the pre-mixing zone, feed influence zone, draft tube zone, gas-solid separation zone, annulus zone and the bottom zone. A perforated plate is mounted at the bottom of the pre-mixing zone, which is connected with a central downcomer, extending into the draft tube. Particles are first mixed in the pre-mixing zone and then the pre-mixed particles flow downward through the central downcomer and into the feed influence zone. In the feed influence zone, the pre-mixed particles are mixed with the particles moving upward in the draft tube, flow upward and circulate between the draft tube and the annulus. In the annulus, particles move downward to the bottom. Then a part of particles flow out of the mixer and the other part flow horizontally into the draft tube and recirculate.

The solid particles used in the experiment were FCC catalyst (ρ_p=1498 kg/m^3), with an average diameter of 76 μm. Air (ρ_g=1.225 kg/m^3) was introduced into the mixer to fluidize the particles. The superficial gas velocity ranged from 0.3 to 0.45 m/s in the draft tube (based on the cross-sectional area of the draft tube) and remained at 0.05m/s in the annulus (based on the cross-sectional area of the annulus). The mass flux of cold particles remained at 13.8 kg/m^{2}s. As we only investigated the influence of slots on the flow and mixing of particles in this paper, no hot particles were induced into the mixer.

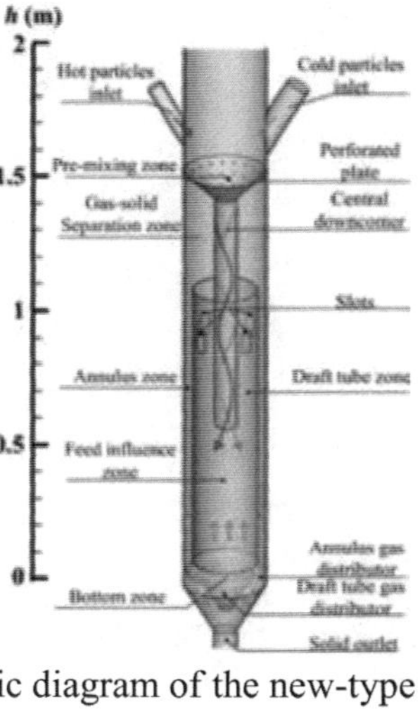

Fig. 1 Schematic diagram of the new-type particle mixer

3. CFD model and simulation method

The hydrodynamics of the mixer was performed using the Eulerian- Eulerian of Ansys Fluent

®13.0. The governing equations [11] include a set of mass and momentum equations for each phase, as shown in Table 1.

A 3-D mesh containing 580,000 cells was generated by using Gambit®2.4. The no-slip boundary was assigned for the gas phase and the partial-slip boundary condition developed by Johnson and Jackson [12] was used for the solid phase. In order to ensure a steady-state operation of the mixer, simulations were first performed for 30 s, and then the time-averaged data was analyzed for another period of 10 s. Other parameters settings in the model are shown in Table 2.

Table 1 Governing equations

Gas-phase Continuity equation $\frac{\partial}{\partial t}(\varepsilon_g\rho_g)+\nabla\cdot(\varepsilon_g\rho_g\boldsymbol{u}_g)=0$

Solid-phase Continuity equation $\frac{\partial}{\partial t}(\varepsilon_s\rho_s)+\nabla\cdot(\varepsilon_s\rho_s\boldsymbol{u}_s)=0$

Gas-phase momentum equation $\frac{\partial}{\partial t}(\varepsilon_g\rho_g\boldsymbol{u}_g)+\nabla\cdot(\varepsilon_g\rho_g\boldsymbol{u}_g\boldsymbol{u}_g)=-\varepsilon_g\nabla p+\nabla\cdot\boldsymbol{\tau}_g+\varepsilon_g\rho_g\boldsymbol{g}-\beta(\boldsymbol{u}_g-\boldsymbol{u}_s)$

Solid-phase momentum equation $\frac{\partial}{\partial t}(\varepsilon_s\rho_s\boldsymbol{u}_s)+\nabla\cdot(\varepsilon_s\rho_s\boldsymbol{u}_s\boldsymbol{u}_s)=-\varepsilon_s\nabla p-\nabla p_s+\nabla\cdot\boldsymbol{\tau}_s+\varepsilon_s\rho_s\boldsymbol{g}+\beta(\boldsymbol{u}_g-\boldsymbol{u}_s)$

Granular temperature equation

$$\frac{3}{2}[\frac{\partial}{\partial t}(\varepsilon_s\rho_s\Theta_s)+\nabla\cdot(\varepsilon_s\rho_s\Theta_s\boldsymbol{u}_s)]=(-p_s\boldsymbol{I}+\boldsymbol{\tau}_s):\nabla\boldsymbol{u}_s+\nabla\cdot(\kappa_s\nabla\Theta_s)-\gamma_s-3\beta\Theta_s$$

Gas-phase stress $\boldsymbol{\tau}_g=\varepsilon_g\mu_g[\nabla\boldsymbol{u}_g+(\nabla\boldsymbol{u}_g)^T]-\frac{2}{3}\varepsilon_g\mu_g(\nabla\cdot\boldsymbol{u}_g)\boldsymbol{I}$

Solid-phase stress $\boldsymbol{\tau}_s=\varepsilon_s\mu_s[\nabla\boldsymbol{u}_s+(\nabla\boldsymbol{u}_s)^T]+\varepsilon_s(\lambda_s-\frac{2}{3}\mu_s)(\nabla\cdot\boldsymbol{u}_k)\boldsymbol{I}$

Solid-phase pressure $p_s=\varepsilon_s\rho_s\Theta_s[1+2(1+e_s)\varepsilon_s g_0]$

Solid-phase shear viscosity $\mu_s=\mu_{s,col}+\mu_{s,kin}+\mu_{s,fr}$

Where $\mu_{s,col}=\frac{4}{5}\varepsilon_s\rho_s d_p g_0(1+e_s)\sqrt{\Theta_s/\pi}$, $\mu_{s,kin}=\frac{10\rho_s d_p\sqrt{\Theta_s/\pi}}{96\varepsilon_s(1+e_s)g_0}[1+\frac{4}{5}g_0\varepsilon_s(1+e_s)]^2$, $\mu_{s,fr}=\frac{p_s\sin\phi}{2\sqrt{I_{2D}}}$

Radial distribution function $g_0=[1-(\varepsilon_s/\varepsilon_{s,max})^{1/3}]^{-1}$

Diffusion coefficient $\kappa_s=\frac{150\rho_s d_p\sqrt{\Theta_s/\pi}}{384(1+e_s)g_0}[1+\frac{6}{5}g_0\varepsilon_s(1+e_s)]^2+2\rho_s\varepsilon_s^2 d_p(1+e_s)g_0\sqrt{\Theta_s/\pi}$

Collisional energy dissipation $\gamma_s=\frac{12(1-e_s^2)\varepsilon_s^2\rho_s g_0\Theta_s^{3/2}}{d_p\sqrt{\pi}}$

Heterogeneity index $H_d=\frac{\beta_e}{\beta}$, $\beta_e=\beta H_d$

Drag based homogeneous structure $\beta=\begin{cases}\frac{3}{4}C_d\frac{\varepsilon_s\varepsilon_g\rho_g|\boldsymbol{u}_g-\boldsymbol{u}_s|}{d_p}\varepsilon_g^{-2.65}, \varepsilon_g>0.8\\ 150\frac{\varepsilon_s^2}{\varepsilon_g}\frac{\mu_g}{d_p^2}+1.75\varepsilon_s\frac{\rho_p}{d_p}|\boldsymbol{u}_g-\boldsymbol{u}_s|, \varepsilon_g\le 0.8\end{cases}$

Where $C_d=\begin{cases}0.44 & Re_s\ge 1000\\ \frac{24}{Re_s}(1+0.15Re_s^{0.687}), & Re_s<1000\end{cases}$, $Re_s=\frac{\varepsilon_g\rho_g d_p|\boldsymbol{u}_g-\boldsymbol{u}_s|}{\mu_g}$

Table 2 Simulation Parameters

Gas-solid model	Eulerian-Eulerian
Solids Pressure	Lun et al.
Particle-wall specularity coefficient	0.5
Gas-wall friction coefficient	No-slip
Pressure-Velocity Coupling	Simple
Gradient discretization	Least squares cell based
Momentum discretization	Second-order upwind
Volume fraction discretization	Quick
Max. number of iterations per time step	40
Pressure outlet (atm)	1
Inventory (kg)	56
Incipient solid holdup	0.6

4. Results and discussion

4.1 Model validation

Fig. 2 gives the distribution of cross-sectionally averaged solid holdup and particle velocity in the mixer. As shown in Fig. 2(a), the solid holdup is larger in the bottom zone (h=0.04 m) and due to the accelerated upward motion of particles, the solid holdup decreases with increase of the axial height (when $h \leq 0.16$ m). The cross-sectionally averaged solid holdup is nearly constant in the draft tube and gas-solid separation zone of $0.4 \leq h \leq 1.25$ m. The cross-sectionally averaged particle velocity is shown in Fig. 2(b). The particle velocity is smaller in the bottom zone and keeps almost invariable in the draft tube and the gas-solid separation zone. Compared with the experimental data, the particle velocity seems slightly higher, while the solid holdup in the simulation agrees reasonably with the experimental results.

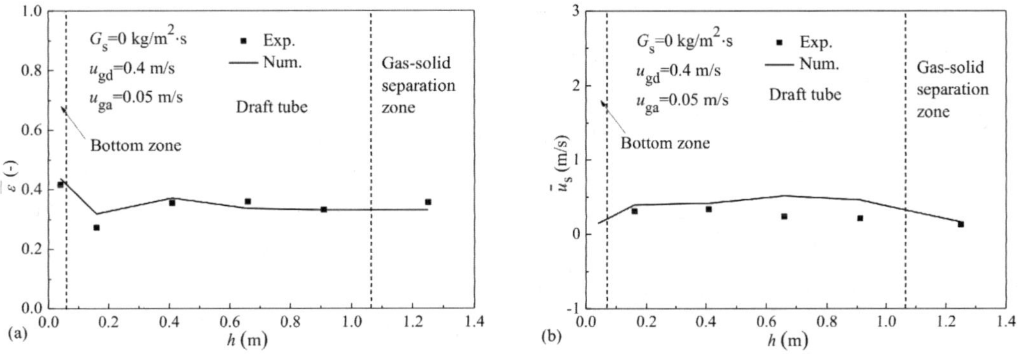

Fig. 2 Distribution of cross-sectionally averaged solid holdup and particle velocity as a function of axial heights

4.2 Pressure distribution near the slots

Fig. 3 shows the time-averaged contours of static pressure on the cross-section of h=0.9 m, the cross-section on the middle of the slots. It is seen that in the draft tube a low pressure region can be observed near the slots, smaller than the pressure in the region between the adjacent slots. This pressure drop forced particles inside the draft tube to move towards the slots and pass through them, which significantly facilitates the mixing of the particles along the radial direction. In the annulus the pressure in the vicinity of the slots seems larger than that in the region between the adjacent slots, inducing horizontal particle immigration and mixing.

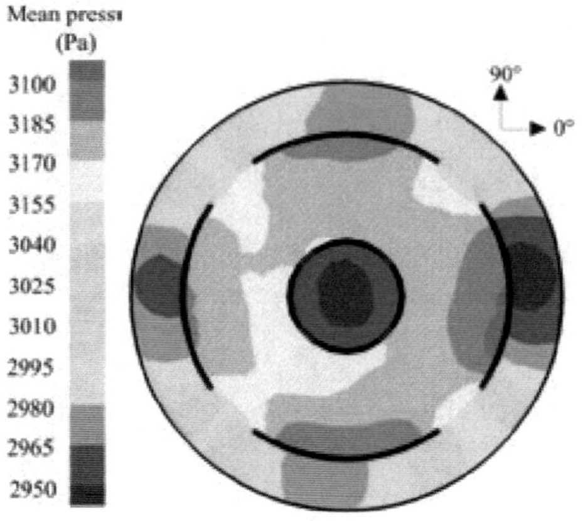

Fig. 3 Time-averaged contours of static pressure at the cross-section of the slots (u_{gd} =0.4 m/s, G_s=13.8 kg/m^{2}s, h=0.9 m)

4.3 Gas bypassing

In the present work, the gas bypassing behavior was investigated by CFD simulation, and a tracer gas having identical physical properties with air was defined. After the mixer running stably, the tracer gas was introduced into the mixer from the gas distributor plate to replace the original gas. A cross section (h=0.06 m) on the exit of the annulus was selected and the mass flux of the tracer gas through the cross section was registered. The results are shown in Fig. 4. f_B refers to the ratio of the mass flux of the tracer gas through the cross section (h=0.06 m) to that of total tracer gas. It is seen that f_B varies between -0.05 and 0.05 with a frequency of 10 Hz, indicating an unstable gas bypassing from the draft tube to annulus. This suggests that the gas bypassing into the annulus was finally brought out of the annulus by the circulating particles. The average f_B is approximately equal to 0, much smaller than that in a conventional ICFB (usually within 15%) [13], because massive particle circulation in the mixer makes it difficult to bypass from the draft tube to annulus.

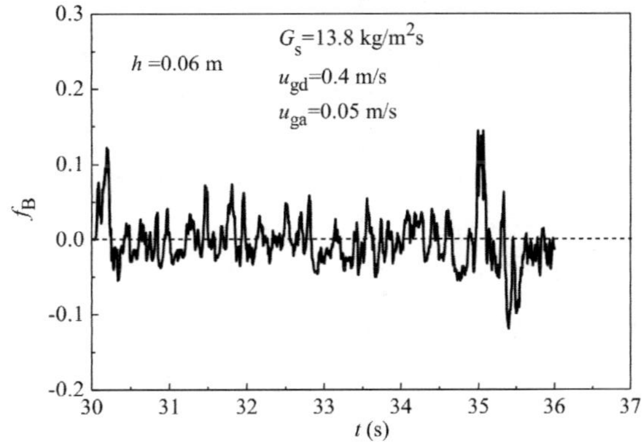

Fig. 4 Gas bypassing from draft tube to annulus in the bottom zone

As discussed previously, in the mixer there exists a pressure gradient from the draft tube to the vicinity of the slots, and then the annulus. Such a pressure difference leads to horizontal flows of particles and gas from the draft tube to the slots and then to the annulus. In the mixer the upside and downside of the slots is located at h=0.94 m and 0.86 m, respectively. Mass flux of the gas bypassing through the slots was obtained by monitoring the gas flux through the lower cross section (h=0.7 m) and the upper cross section (h=1 m) in the draft tube. Fig. 5 shows the gas mass flux under different operating conditions. The total gas mass flux in the draft tube increases with increasing gas velocity in the draft tube. With increase of the gas velocity from 0.3 to 0.45 m/s, the mass flux of the gas bypassing through the slots increases from 0.157 to 0.221 kg/m^{2}s. But the mass flux of bypassing gas does not increase linearly with the gas velocity. It is also seen that the mass ratio of the gas through the slots to that in the draft tube drops from 29.7% to 25.4%.

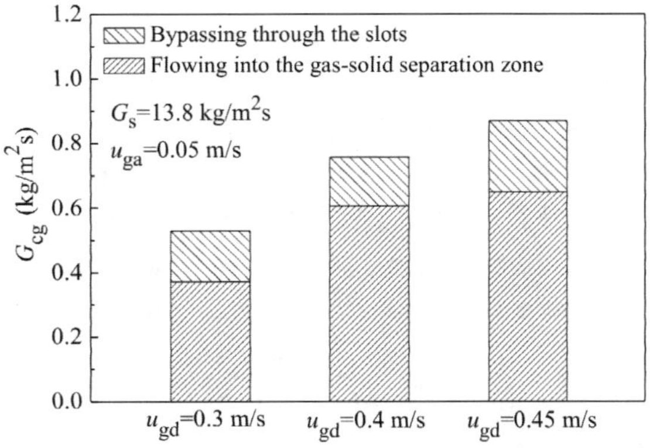

Fig. 5 Gas mass flux through the slots and the gas-solid separation zone as a function of superficial gas velocity in the draft tube

4.4 Particle mass flux

Fig. 6 shows the particle mass flux in the mixer. It is seen that the mass flux increases with increase of the superficial gas velocity in the draft tube. The mass flux through the slots accounts for about 50% of the total mass flux when u_{gd}=0.3 m/s. Because the axial particle velocity rose with increase of the superficial gas velocity from 0.3 m/s to 0.45 m/s in the draft tube, making it difficult to change the direction of the particles. Meanwhile, the radial particle velocity above the draft tube increases. Consequently, with the increase of the superficial gas velocity in the draft tube , the mass flux through the slots considerably decreased, from about 55.3% of the total mass flux (about 2.69 kg/s) at u_{gd}=0.3 m/s to only 4.8% (about 0.52 kg/s) at u_{gd}=0.45 m/s. Meanwhile, the total particle mass flux in the mixer increased from 4.87 to 10.86 kg/s as u_{gd} increased from 0.3 to 0.45 m/s.

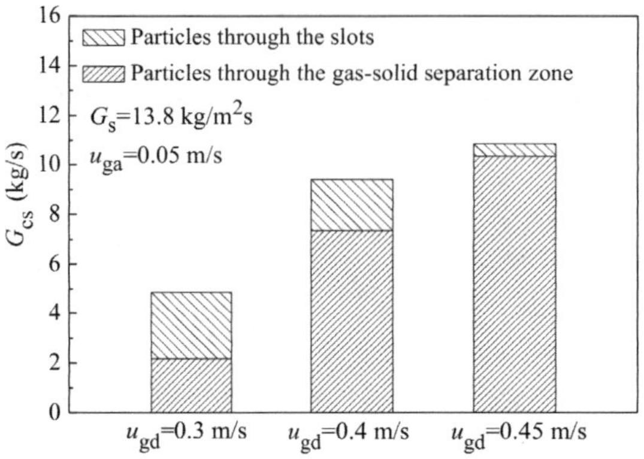

Fig. 6 Particle mass flux through the slots and the gas-solid separation zone as a function of superficial gas velocity in the draft tube

5. Conclusions

The characteristics of particle flow and mixing have a considerable influence on the mass and heat transfer, as well as the product quality. In order to further enhance the particle mixing in the ICFB, especially along the radial direction, a new-type particle mixer with slots on the draft tube was proposed in the present work. The hydrodynamics of the mixer was investigated by using a multi-scale computational fluid dynamics with the EMMS drag. The prediction in terms of cross-sectionally averaged solid holdup and solid velocity were compared to experimental data for validation. The results show that in draft tube a low pressure region can be observed near the slots and in annulus the pressure in the vicinity of the slots seems larger than that in the region between the adjacent slots, which induces horizontal particle immigration and mixing. Nearly 25.4% to 29.7% of gas in the draft tube bypasses and flows through the slots into the annulus. The ratio of the mass flux of the tracer gas bypassing into the annulus to that of total tracer gas is approximately equal to 0. With increasing the superficial gas velocity in the draft tube from 0.3 m/s to 0.45 m/s, the total solid mass flux increases from 4.87 kg/s to 10.86 kg/s, while the mass flux through slots decreases from 2.69 kg/s to 0.52 kg/s.

Notation

G_s	solid mass flux (kg/m^2 s)	G_{cg}	gas mass flux in the mixer (kg/m^2 s)
G_{cs}	solid mass flux in the mixer (kg/s)	h	height (m)

p	pressure (Pa)	R	radius of the reactor (mm)
r	radial position of measurement points (mm)	t	time (s)
u_{ga}	superficial gas velocity in the annulus (m/s)	u_{gd}	superficial gas velocity in the draft tube (m/s)
$\bar{u}$	cross-sectionally averaged axial particle velocity (m/s)		

Greek

ε_g	voidage	ε_s	solid holdup
$\bar{\varepsilon}_s$	cross-sectionally averaged solid holdup	ρ	density (kg/m^3)

References

[1] F. Niklasson, H. Thunman, F. Johnsson, B. Leckner, Estimation of solids mixing in a fluidized-bed combustor, Ind Eng Chem Res, 41 (2002) 4663-4673.

[2] L.H. Shen, M.Y. Zhang, Effect of particle size on solids mixing in bubbling fluidized beds, Powder Technol, 97 (1998) 170-177.

[3] L.H. Shen, M.Y. Zhang, Y.Q. Xu, Solids Mixing in Fluidized-Beds, Powder Technol, 84 (1995) 207-212.

[4] N. Mostoufi, J. Chaouki, Local solid mixing in gas-solid fluidized beds, Powder Technol, 114 (2001) 23-31.

[5] D.Y. Liu, X.P. Chen, Lateral solids dispersion coefficient in large-scale fluidized beds, Combust Flame, 157 (2010) 2116-2124.

[6] F. Berruti, D.S. Scott, E. Rhodes, Measuring and Modeling Lateral Solid Mixing in a 3-Dimensional Batch Gas Solid Fluidized-Bed Reactor, Can J Chem Eng, 64 (1986) 48-56.

[7] M.M. Fang, K. Luo, S.L. Yang, K. Zhang, J.R. Fan, Computational Fluid Dynamics-Discrete Element Method Investigation of Solid Mixing Characteristics in an Internally Circulating Fluidized Bed, Ind Eng Chem Res, 52 (2013) 7556-7568.

[8] J. Li, M. Kwauk, Particle-fluid Two-phase Flow: the Energy-minimization Multi-scale Method, Metallurgical Industry Press, Beijing, 1994.

[9] Z.S. Shi, W. Wang, J.H. Li, A bubble-based EMMS model for gas-solid bubbling fluidization, Chem Eng Sci, 66 (2011) 5541-5555.

[10] K. Hong, Z.S. Shi, W. Wang, J.H. Li, A structure-dependent multi-fluid model (SFM) for heterogeneous gas-solid flow, Chem Eng Sci, 99 (2013) 191-202.

[11] D. Gidaspow, Multiphase flow and fluidization: Continuum and Kinetic Theory Descriptions, Academic Press, Boston, 1994.

[12] P.C. Johnson, R. Jackson, Frictional–collisional constitutive relations for granular materials, with application to plane shearing, J. Fluid Mech. 176 (1987) 67–93.

[13] B.H. Song, Y.T. Kim, S.D. Kim, Circulation of solids and gas bypassing in an internally circulating fluidized bed with a draft tube, Chem Eng J, 68 (1997) 115-122.

514498 Power Spectral Density Analysis of Pressure Signal in 18 m Circulation Fluidized Bed Riser

Abstract:

Studies on the gas-solids flow characteristics using Power Spectral Density based on the pressure data from different axial locations in a riser of diameter 80 mm and height 18 m. Results show that the pressure signal fluctuation is mainly affected by gas-solids interaction, which included gas-solids friction, particle-particle collision and particle-particle aggregation. With the increasing solids circulating rate, frequency of particle interactions including particle-particle collision and particle-particle arrogation become strong and the pressure fluctuation also increased. It can be seen that the pressure signal fluctuation intensity decreased gradually with the increasing axial position of the riser. The low frequency but high power phenomenon nearly disappears in the fully developed zone near the top of the riser. Based on the results, it can be concluded that the gas-solids interaction is stronger at the bottom of riser. On the other hand, the gas-solid interaction is more homogeneous, and the pressure fluctuation is relatively weak in the fully developed zone.

514514 Characteristics of Pressure Fluctuations in Particle-Transport Inclined Pipe of a Circulating Fluidized Bed

Abstract:

In a large circulating fluidized bed device, the characteristics of pressure fluctuation in a particle-transport inclined pipe under different solid circulating flux were investigated based on measuring the dynamic pressure of gas-solid two-phase flow by using FCC equilibrium catalyst as the particles. The experimental results showed that with increase of particle mass flux, the flow patterns of FCC particles above and below the butterfly valve in the inclined pipe both successively changed, and the fluctuation amplitude of dynamic pressure increased. The standard deviation of the dynamic pressure below the butterfly valve possessed a linear relation with the particle mass flux, which could be used to characterize the real-time dynamic pressure in the inclined pipe. By wavelet analysis, it is indicated that there exists a main frequency caused by inclined pipe flow structures in the pressure fluctuations above and below the butterfly valve, besides, there is a secondary frequency in the pressure fluctuation below the butterfly valve caused by gas turbulence and gas-solid interactions. With the increase of particle mass flux, the wavelet energy ratio of the secondary frequency gradually decreased. The fluctuation pressure caused by discharge in the inclined pipe is a low frequency pressure and the excitation source of inclined pipe vibration.

508323 Dynamic Modeling of Attrition and Reactions in Circulating Fluidized Bed Reactors

Abstract:

Fluidized bed reactors are widely applied in chemical processes due to the excellent gas-solid contact. A problem that occurs in fluidized beds is the mechanical stress that particles are exposed to during the fluidization, which causes particle attrition. Additionally, fluidized beds are often operated in batch mode, which will cause ever-changing operating conditions with time. This makes fluidized bed modeling especially challenging, because the distributed particle properties, like size, wetness or reaction state, have to be tracked over the whole course of the simulation.

At the Institute of Solids Process Engineering at Hamburg University of Technology, novel process simulation software named DYSSOL is developed, particularly tailored for solids processes. The software is based on the flowsheeting concept. This means that chemical unit operations, like for example chemical reactors or separators, are put into process modules, which are interconnected into a process flowsheet and then mathematically solved. To make dynamic simulations available, new dynamic models had to be proposed for the involved equipment. The models themselves are based on empirical and semi-empirical correlations for hydrodynamics and kinetic particle reaction models are integrated.

The dynamic flowsheeting software was then used to simulate a system of interconnected fluidized bed reactors, which are used for Chemical Looping Combustion of solid fuels. This process is especially interesting, because heterogeneous reactions take place. Further, the process is a reactor-regenerator system causing the reactive solid to be oxidized and reduced and, hence changing its conversion state. Attrition plays a major role for the cost effective operation of the process.

In a small lab scale bubbling bed reactor, the jet and bubbling attrition rate constants for the solid oxygen carrier were experimentally determined under ambient conditions. In a small stand-alone cyclone the cyclone attrition was investigated. Furthermore, the reaction kinetics of the solid material with fuel gases was experimentally determined. All this information was included into the flowsheet simulation of a pilot scale circulating fluidized bed reactor, which is actually operated in our institute to deliver validation data.

The dynamic flowsheet simulation of the pilot plant, which was enhanced with kinetic and attrition information of smaller scale facilities, was conducted over 1000 seconds. A fuel load change was simulated and the chemical conversion was tracked. Over the whole simulation, the attrition and small particles leaving the reactor system was also covered. When comparing the simulation with experiments it could be shown that the gas conversion was sufficiently calculated and the conversion state of the oxygen carrier tracked correctly. The attrition modeling, however, with its parameters measured had to be adapted, so that it fitted the outlet particle flows measured under reaction conditions.

The financial support of DFG (Deutsche Forschungsgemeinschaft) within the priority program SPP 1679: â€œDynamic simulation of interconnected solids processesâ€ is gratefully acknowledged; grant number HA 6935/2-1.

514599 Analysis of FCC Cyclone Fault Diagnosis Technology Based on Particles Information

Abstract:

Cyclones are key equipment in fluid catalytic cracking unit (FCCU), which are used for separating the catalyst particles from oil gas or flue gas. Hence, it is very necessary to carry out condition monitoring and fault diagnosis for the running cyclone. The properties of particles vary at different positions in the particle circulation loop, such as size distribution, morphology and element composition, etc. In consequence, the particles contain much information about operation. Once the cyclone operates abnormally, the information of run-off catalyst particles appear abnormal too, and there exists a corresponding relationship between them. So, taking the catalyst particles as the carrier of fault information, this paper put forward cyclone fault diagnosis technology by extracting catalyst particle information, such as particle packing density, particle size distribution, particle micrographs etc, to build the relationship model between the cyclone fault and catalyst particle information.

514600 The Multi-Hole Throttling Effect and the Erosion Characteristics of the High Pressure Natural Gas

Abstract:

Based on Euler-Lagrange multi-phase flow numerical simulation method, the single hole and multi-hole throttling process of the high pressure natural gas was studied by using the RNG *k*-*Îµ* turbulent model, the Sutherland Viscosity Law compressible fluid viscosity model, and the advanced erosion model numerical simulation. The calculated mass flow variation law in the throttle choke is comparable with the theoretical calculations, which verified the adopted models and calculating methods. And then the throttling effect and the erosion characteristic of the high pressure natural gas was studied to check its specific law with different throttle holes by using the advanced erosion model. The results showed that, high pressure natural gas can reduce pressure and expansion after flowing through the single hole and multi-hole throttle choke, in which lead a high-speed jet region came up near its outlet. But the maximum jet velocity of multi-hole throttle choke was significantly lower than that of the single hole. At the same time the multi-hole uniform jet may reduce the vortex entrainment and the wall erosion damage, in which can be easier to mix the cold core and achieve stable pressure recovery at the role of the multi-jet mixing effect. Therefore, the multi-hole throttling can effectively restrain the erosion damage and the gas hydrate formation probability.

514601 Numerical Simulation of Flow Field In a Gas Pipe Distributor of the FCCU Regenerator

Abstract:

The flow field in the pipe distributor used in regenerator of FCCU (Fluid Catalytic Cracking Unit), was analyzed based on the numerical simulation. Firstly, the distributing gas performance of the pipe distributor was investigated. The result shows that the pressure increases along the branch pipe of the pipe distributor and the axial flow rate decreases in the branch pipe, which leads to the nozzle flow rate increasing gradually. Furthermore, an obvious bias flow was found in the inlet of branch pipe, which brings some nozzle flow rate less than the designed flow rate and causes an uneven gas distribution in the fluidized bed as well. Meanwhile, the momentum equation with variable mass flow was applied to analyze the flow in the pipe distributor. The flow type in the branch pipe belongs to the "momentum exchange controlling model". The pressure in the branch pipe always increases from the inlet to the end, and the pressure distribution is not consequentially uniform. The uneven gas distribution of the pipe distributor results in fluidized gas non-uniformity in the bed and erosion in the branch pipe and nozzle. Finally, according to the pressure characteristics in fluidized bed system, relationship between the pressure drop of branch pipe and pressure drop of nozzles, as well as of the critical nozzle pressure drop and the designed nozzle pressure drop, were discussed. The gas distributor performance could be optimized by means of the structural improvement of branch pipe.

514602 Full-Loop Simulation of Gas-solids Flow in Circulating Fluidized Bed of FCC System

Abstract:

Circulating fluidized beds are widely used in petrochemical, coal, metallurgy, and other industrial fields. A typical circulating fluidized bed (CFB) includes the riser, gas-solid separator, downer and solids flow control device. To obtain detailed information and get better understand of gas-solids flow in CFB, computational fluid dynamics (CFD) had been applied extensively in many systems such as circulating fluidized bed combustion (CFBC) systems and fluid catalytic cracking (FCC) systems. In most studies, the riser, the core part in CFB system, was isolated as a single simulation object by applying proper boundary conditions. However, there are some limitations of riser-only simulation, without consideration of dynamics in solids circulation rate and effects of other part of CFB system. Therefore, the full-loop simulation is a more promising way to describe complex gas-solids interaction in CFB by treating the whole CFB system as a simulation object. Nowadays, most of current full-loop simulations were focused on CFBC systems, there was little work about FCC system, with different type of particles and recycle mechanism from CFBC system.

In this work, both full-loop and riser-only simulations were carried out by TFM approach to investigate a pilot-scale gas-solid circulating fluidized bed of FCC system which has been set up in our group. The simulation condition was chosen according to experiments with different superficial gas velocity, valve opening and fluidizing air flow rate. Detailed comparison between full-loop and riser-only numerical simulations has been conducted with respect to the hydrodynamics inside the riser, which showed full-loop simulation had advantages on predicting dynamics of solids circulation rate and the gas-solids flow behavior. Distribution of pressure, solids holdup and velocity in system was investigated which was helpful to understand the hydrodynamics in CFB. The effects of operating condition were then studied, which revealed the system stability mechanism.

506758 Immobilization of Sulfur-Oxidizing Bacterium, *Thioalkalivibrio Sp.* D301 on Magnetic Nanoaparticles and Biodesulfurization

Abstract:

There is an urgent need to develop desulfurization methods for halo-alkaline wastewater. *Thialkalivibrio sp.* D301, an obligatory chemolithoautotrophic, haloalkaliphilic, sulfur-oxidizing bacterium, was used in the removal of sulfide in wastewater. We developed a process to immobilize D301 cells by using superparamagnetic Fe_3O_4 nanoparticles (NPs) under halo-alkaline conditions. Fe_3O_4 NPs were synthesized by coprecipitation of ferric and ferrous salts in the presence of N_2 gas. TEM image shows the spherical shape of Fe_3O_4 NPs with an average diameter of 10 to 15 nm. The particles had superparamagnetic properties. The saturation magnetization (Î´$_s$) of the Fe_3O_4 NPs was 73.985 emu g$^{\text{âˆ’1}}$ with zero remanence and coercivity. Analysis of zeta potentials and hydraulic diameters showed that the Fe_3O_4 NPs and bacterial cells interact mainly through the nano-size effect. The NPs-coated cells have the same sulfur oxidization activity as that of free cells. Experiment results showed that the immobilized cells still retained >90% of their sulfur oxidization activity after six times of reuse. The cells coated with Fe_3O_4 NPs did not interact through electrostatic attraction. The mechanism of NPs adsorption mainly involved the nano-size effect. Magnetic immobilization is a convenient technique for HA-SOB immobilization under halo-alkaline conditions and is a promising technology for large-scale industrial applications.

507944 Simulation of Bulk Solids and Granular Systems By Using Combined Discrete Element Models

Abstract:

Granular materials can be found in many industrial processes from chemical to energy, agriculture, mining, pharmacy and many other fields. They are almost always opaque, making the acquisition of experimental data below the surface difficult. DEM (Discrete Element Method) simulations are an effective way to study the physics of granular materials because they can simulate the micro-dynamic behavior of granular material on an individual particle scale and the motion of every particle can be traced. Since DEM was first considered, many modified discrete element models have been developed. However, most existing DEM simulation are just for spherical particles or a specific type of non-spherical particles. In this work, a combination of several types of discrete element models including spherical models, super-ellipsoids, multi-sphere models (with no interaction and deformation between sub-spheres), and bonded-spheres models (with interaction and deformation between sub-spheres) were used to simulate the flow of granular materials with many different shapes. Several simulation cases including the flow of particles in tumblers, blenders, and other devices were performed to validate the modeling approaches. The simulation results show that it is practical and efficient to combine different discrete element models together for simulating both spherical and non-spherical particles in bulk solids and granular systems.

515703 Material Optimization of Perovskite Films using High Throughput Synthesis and Multi-Dimensional Analysis

Abstract:

Over the past years, Perovskites received special attention in the field of renewable energy as they are one of the prime candidates for energy storage and conversion materials. Perovskites act as the active light harvesting layer of a solar cell and provide excellent performance.[1] A common synthesis strategy is through the liquid phase under inert atmosphere.[2,3] However, this comes with the challenge of a huge parameter space that in principle would have to be first synthesized, then characterized and finally optimized based on the therefrom deduced knowledge.

To overcome these limitations, we follow a fully automated approach, which allows automated high throughput (HT) synthesis and HT characterization of perovskites with high accuracy. As an efficient route to explore the parameter space, we use design of experiments (DoE) to find optimized settings for process parameters such as synthesis and annealing time and temperature, but also material aspects like the molar ratio of the reactants.

For all experiments, a Chemspeed© technologies (CST) swing XL platform with a liquid handling tool and a heatable shaking rack was used. The robotic platform is enclosed within an Mbraun© glovebox with inert atmosphere (Nitrogen). After precursor preparation, deposition was realized by a liquid handling tool on glass substrates inserted in a customized aluminum 24 well plate, which is then transferred onto a hot plate for film formation.

In a first step, after annealing at different temperatures, a characterization cascade was developed and applied that contains the subsequent elimination of samples that exhibit undesired properties and side products. Fast absorption analysis, photoluminescence (PL) measurements and finally X-ray diffraction (XRD) were performed to identify optimum chemistry and process parameters. Although this is a very reliable procedure, it is still comparatively demanding in terms of the time effort. Therefore a more efficient, purely optical approach by which the experimental parameter space is only explored in terms of rapid emission analysis was developed in a second step. Noteworthy, both strategies led to the same optimum conditions, however, at significantly reduced time effort.

In conclusion, our work shows how by the combination of automation, DoE and efficient characterization, optimum conditions for new materials like perovskites are rapidly identified. This opens the door towards faster material development which is important whenever new perovskite structures, e.g. lead free materials for industrial applications, are strived for.

Literature

1. Green, M.A.; Ho-Bailie, A.; Snaith, H.J., "The emergence of perovskite solar cells", Nature Photonics 2014, 8, (7), 506-514.
2. Baikie, T.; Fang, Y.; Kadro, J.M.; Schreyer, M.; Wei, F.; Mhaisalkar, S.G.; Graetzel, M.; White, T.J., "Synthesis and crystal chemistry of the hybrid perovskite (CH 3NH3)PbI3 for solid-state sensitised solar cell applications", J. Mater. Chem. A 2013,1 , 5628-5641.
3. Yan, K.; Long, M.; Zhang, T.; Wei, Z.; Chen, H.; Yang, S.; Xu, J., "Hybrid Halide Perovskite Solar Cell Precursors: Colloidal Chemistry and Coordination Engineering behind Device Processing for High Efficiency", J. Am. Chem. Soc. 2015, 137, (13), 4460–4468.

509509 Solid Liquid Separation Via Particle Flow Instability

Abstract:

A new separation strategy using novel particle-flow-instability physics is successfully developed for a *difficult-to-separate suspension* in which there is some combination of a small density difference between solid and liquid, high viscosity, and small-sized particles. The method we propose here requires no dilution for medium, high or extremely high viscosity slurries and can produce effective solidâ€“liquid separation with low capital and operational inputs. The results have relevance for many biotechnological and/or chemical processes.

509543 Milling and Grindability Assessment of Pharmaceutical Materials

Abstract:

Milling is one of the most important unit operations in the pharmaceutical industries. Most of the Active Pharmaceutical Ingredients (APIs) and excipients are produced by crystallisation, and grown to a sufficiently large particle size suitable for isolation and drying. Therefore most of the materials have to be milled to desirable particle sizes. Milling is influenced by a combined effect of milling dynamics (input power, mill geometry, etc) and material properties. Considerable research has been carried out on analysing the particle behaviour in various mills. Nevertheless, the prediction of the milling rate and milled product properties (e.g. product size distribution) is still very challenging.

In this work, the milling behaviour of several pharmaceutical materials in the PicoPlex mill (pin mill unit of the Picoline, Hosokawa Alpine) has been studied. A combination of experiments and Distinct Element Method (DEM) simulations is used to explore the dynamic behaviours of the materials in the mill. The results are then used to validate a novel approach of grindability assessment of pharmaceutical materials using the Scirocco disperser of the Malvern Mastersizer 2000 (Bonakdar et al. 2016). Based on this method the ease of grindability of the materials in the mill can be predicted.

Bonakdar, T., Ali, M., Dogbe, S., Ghadiri, M., and Tinke, A. (2016). A method for grindability testing using the scirocco disperser. *International Journal of Pharmaceutics,* 501: 65-74

507857 Bi-Directional Thermal Control of Twin Screw Granulation Process Via a Specialised Annular Heat Pipe

Abstract:

The continuous process of twin screw granulation forms a hotbed in pharmaceutical research as it is expected to succeed the current batch production of granules. In essence, the scale up of the material output is a function of production time, instead of equipment size (for batch granulators). This way, a â€œjust-in-time productionâ€ of the desired amount of material can be performed according to specific needs. In this work, a twin screw granulator was utilised in the granulation of both the placebo and active pharmaceutical formulations. Subsequently, the same equipment was used in granulating ceramic raw material powders for tile pressing. During processing, the spatial constraint in the barrel and screw compartments caused friction, resulting in excess heat generation. For the thermally-sensitive pharmaceutical formulations, this must be controlled to avoid denaturation of the granule products. A common solution is by using flowing cool water, but a good control has been proven to be difficult to achieve as different granulation zones generate different heat levels (kneading blocks > conveyer blocks). Heat pipe is a passive and highly effective thermal conductor and stabiliser. A carefully-designed annular heat pipe, filled with a small amount of methanol was successfully employed to stabilise the temperature in the barrel space, cooling the granulating material before it reached its critical temperature. For the ceramic powders, the methanol filling of the heat pipe was replaced by water and it was then heated up to provide an in-situ drying operation as the powders granulating. The in-situ drying in this instance, reduces energy consumption for the downstream drying processing as the product granules contain less moisture to be removed. The annular heat pipe, with its bi-directional operation of heating and cooling is a technically viable temperature controller for twin screw granulators. Compared to the established active cooling methods, the heat pipe provides a passive alternative, with only a discreet amount of fluid required to last its operational lifetime, thereby providing an economic thermal control solution.

Mixing Grains with Different Elongation in a Rotating Drums

Claudia Piacenza[1,2], *Marco Marconati*[1], *Colin Hare*[1], *Andrea C. Santomaso*[2], *Marco Ramaioli*[1,*]

[1]*Department of Chemical and Process Engineering, Faculty of Engineering and Physical Sciences, University of Surrey, Guildford GU2 7XH, UK*
[2] *APT Lab, Department of Industrial Engineering, University of Padova, Padova, Italy*

[*] *Email: m.ramaioli@surrey.ac.uk*

Abstract

Mixing grains with different properties is a remarkably challenging process, relevant to many industrial applications. Rotating drums have been used extensively as model systems to study granular media flow and mixing and segregation. Numerous studies considered the mixing of grains with different sizes, but only few studies considered shape and elongation, which have been already identified as important characteristics affecting packing and inducing segregation. In this contribution, the mixing of binary mixtures of grains having the same volume, but different elongations was studied experimentally. The mixing dynamics of a layered granular medium was characterized in a rotating drum, highlighting the impact of the grain shape, of the drum angular speed and of the drum filling ratio. A mixed or segregated state is reached very rapidly, but experiments are continued to verify that the state reached is a steady state. The experiments demonstrated that below a critical difference of elongation, the grains can be mixed effectively. Conversely, when grain elongations are very different, a central core is formed, rich in the more elongated grains. In this case only limited mixing can be achieved. These results can guide the formulation of mixtures of grains with different elongations.

1. Introduction

Mixing is a process operation aimed at obtaining a homogeneous system starting from two or more components. On the contrary, when there is a lack of homogeneity within the mixture segregation occurs. Segregation phenomena represent a relevant problem in the case of mixing solid particles having different physical properties such as particle size, density, shape or surface roughness. Differently from liquids and gases, solid particles are not thermal activated system [1] and so in order to obtain a certain degree of mixing they need to be mechanically activated. Typically three mixing mechanisms are largely recognized:

- dispersion, analogous to diffusion in fluids, it involves single particles that randomly move due to collision between each other;
- convection, it involves a portion of the bulk material, with a given direction and velocity;
- shear mixing: it is peculiar to granular solids and it is a combination of the two previous mechanisms.

Powder mixers are relatively simple machines of low capital cost. They are classified into two types: tumbling mixers and agitated mixers. Tumbling mixers are the simplest. A totally enclosed vessel is rotated about an axis causing the particles within the mixer to tumble over each other on the mixture surface. Tumbling vessels provide low-shear environments. Convective mixers instead are characterized by the action of impellers or paddles to move the powders around and generate a well-mixed product. Convective mixers provide more shear into the mixture and can be used with cohesive materials.

Rotating drums belong to the tumbling mixers. They are extensively used in the chemical and process industries. With respect to mixing processes, rotating drums are often used to study and understand mixing and segregation phenomena; as a result, many experimental and numerical studies were developed on granular behavior in rotating drums [1-3]. Different granular flow regimes can occur in the drum, each one with its specific flow behavior. In particular, different flow regimes can be assessed by using the dimensionless Froude number, $Fr = \omega^2 R/g$ [4], where ω [s^{-1}] is the drum rotation speed, R [m] the drum radius and g [m/s^2] the gravitational acceleration. With $10^{-4}<Fr<10^{-2}$ and filling level $\varphi>0.1$ rolling regime takes place. This motion is characterized by a uniform, steady flow of a particle layer on the surface (cascading or active layer) while the larger part of the bed (plug flow region) is moved upwards by solid rotation with the rotational speed of the wall. A uniform and good mixing is achievable with this type of motion, but when particle size distribution is very broad segregation can appear. The rolling bed is preferred, since is the simplest regime providing favorable conditions for heat transfer and high quality of the product even when mass flow rates are large.

When particles to be mixed have different properties such as different size, different density or different shape segregation typically occurs and particles tend to separate. [5, 6]. Difference in size is by far the most important cause of particle segregation. Particle size distribution influences also strongly bulk properties such as packing fraction and bulk modulus [7].

Different size segregation mechanisms can be identified:

- trajectory mechanism: according to the Stokes' law the velocity of fine particles is lower than that of big particles. This can cause segregation when small particles move through air;
- percolation mechanism: if a bulk material is disturbed, a rearrangement in the particle packing occurs and for fine particles it is easier to fall down in the inter-particle voids;
- inertial mechanism: in pouring and filling operations bigger particles tend to flow selectively to the bottom of the pile since they have a higher momentum;
- elutriation mechanism: in presence of air the very fine particles can remain in suspension while larger ones tends to settled. This can occur in charging storage vessels or hoppers.

A homogeneous binary granular mixture, with differences in size and density, in a two-dimensional rotating drum typically gives radial segregation. Considering a rotating drum operating in the rolling regime filled with particles of different sizes, bigger particles mainly tend to the edge of the system, while smaller particles accumulate to the centre of the drum. The phenomenon is fast and the rotational speed of the drum seems not to affect the phenomenon; the size ratio of the particles is, instead, the parameter that mainly affects size segregation phenomena. Most of the experimental investigations about segregation in the rotating drum involved spherical or nearly-spherical particles; the experimental studies of the flow of non-spherical particles are, in fact, rather limited [6,8]. The shape of the particles, however, strongly influences the motion in the system. In particular, the static and dynamic angles of repose of non-spherical particles are found to be very different from the ones measured for spherical particles suggesting that differences in particle mobility can affect the segregation process [9].

2. Materials and methods

2.1 Particles

Both spherical and non-spherical particles were used for the experiments. Their size (diameter d for spheres, major and minor axis for ellipsoids a and b respectively) , aspect ratio AR, projected area A_p, volume V, and true density ρ are reported in Table 1.

Table 1. Physical properties of the particles

Type	d [mm]	a [mm]	b [mm]	AR [-]	A_p [mm^2]	V [mm^3]	ρ [g/cm^3]
Spheres S	8	-	-	1	50.3	268	1.09
Ellipsoids E1	-	10.48	6.98	1.5	57.5	267	0.96
Ellipsoids E2	-	18.44	5.26	3.5	76.2	267	0.98

Spherical particles with size 8 mm were used for the experiments. The ellipsoids were 3D printed with *AR* equal to 1.5 and 3.5. They were printed with the Ultimaker 3D printer and made of polylactic acid (PLA). The software used to realize the 3D design of the particles used in this work is FreeCAD, which is an open-source parametric 3D modeller allowing to obtain STL files.

The true density of all the particles used for the experiments was measured by gas pycnometry (Multivolume Pycnometer 1305, Micromeritics). The measured density of the spheres was very close to the typical density of the nylon 66 (1.14 g/cm^3). The PLA ellipsoids had a density that was lower than the typical density of the PLA (1.24 g/cm^3). This is because the particles were printed with a set infill density of the 20%, so they where partially empty inside.

2.2 Tumbling mixer

The quasi-2D drum was made of two transparent polycarbonate disks with a diameter of 30 cm and a thickness of 6 mm; between them a wooden spacer was placed, with the same external diameter, an internal diameter of 25 cm and a thickness of about 9.1 mm. The spacer allowed to create an internal volume of finite thickness separating the two polycarbonate disks.

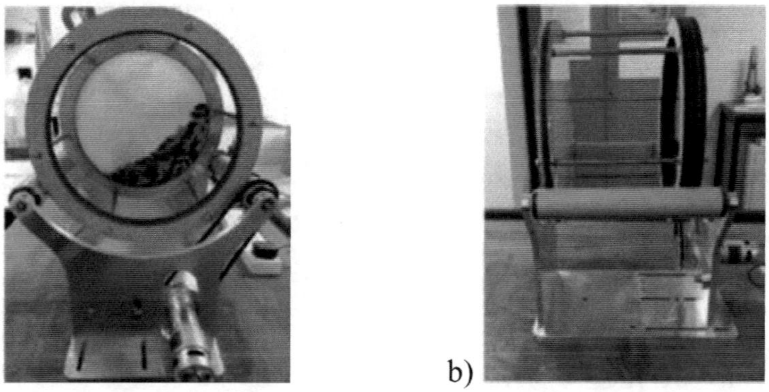

a) b)

Figure 1. a) Front view and b) side view of the rotating drum (the drum is on the left side of the rigid structure).

The internal lateral wall of the volume was covered with sandpaper in order to provide friction and avoid undesired slip of the particles. To keep the drum in vertical position and allow its rotation, three additional wooden spacers were used, at a distance of about 16.3 cm. This rigid structure could rotate at variable rotational speed thanks to a motor connected to a voltage supplier. Two rollers mounted on a rigid support made the rotation of the drum possible; a pulley transmitted the motion to the driving roller, while the second roller was passive. The rotational speed could be changed by varying the motor voltage. The structure could be removed from the rollers and by unscrewing the bolts, the drum could be easily opened so that to load/unload the particles.

2.3 Image acquisition

Frontal pictures of the drum were taken by a color camera and analyzed by the image processing software ImageJ [10]. To avoid the reflection of the camera and other objects on the front transparent disk, a squared LED panel was used which allowed to have a dominant diffused light illuminating the drum. The LED had a power of 8 W and a color temperature of 4000 K; it was placed behind the drum and opposite to the camera. A circular sheet of paper, which was attached to the drum between it and the LED, diffused the light properly. The camera was a Basler ace camera – model acA2000-165uc, with a maximum resolution of 2040 × 1086 pixels. The focus and the diaphragm of the camera could be adjusted manually.

2.4 Experimental plan

The attention was focused on mixtures of particles with different shapes, since segregation phenomena due to size had been widely investigated in the past. Three mixtures were considered, with two possible initial top-bottom segregated configurations. Filling level and rotational speed where changed also as detailed in Table 2.

Table 2. Experimental tests performed with differently shaped particles.

AR 1 [-]	*AR* 2 [-]	ω [rpm]	φ [-]	Init. Config.
1	1.5	4.22	25%	E1/S
1	1.5	4.22	25%	S/E1
1	1.5	4.22	47%	E1/S
1	1.5	4.22	47%	S/E1
1	1.5	11.51	25%	E1/S
1	1.5	11.51	47%	E1/S
1	3.5	4.22	25%	E2/S
1	3.5	4.22	25%	S/E2
1	3.5	4.22	47%	E2/S
1	3.5	11.51	25%	E2/S
1	3.5	11.51	25%	S/E2
1	3.5	11.51	47%	E2/S
1.5	3.5	4.22	25%	E2/E1
1.5	3.5	4.22	25%	E1/E2
1.5	3.5	4.22	47%	E2/E1
1.5	3.5	4.22	47%	E1/E2

*AR*1 and *AR*2 indicate the aspect ratios of the particles in the binary mixture. It is reminded that the spheres (*AR*=1) and the two types of ellipsoids (*AR*=1.5 and *AR*=3.5) had the same volume. The E1/S initial configuration indicates a segregated configuration with short ellipsoids above spheres (this was assumed as basis configuration); the S/E1 configuration is inverted with spheres above short ellipsoids. A similar convention is applied to the experiments considering mixtures of ellipsoids. All the experiments were performed with binary mixtures with a composition of 50% and 50% on a number (and hence also volume) basis. The filling level φ of the 25% was obtained with a total number of particles equal to 200; with 320 total particles the filling level of the drum was equal to 47%.

2.5 Image analysis procedure

Image analysis techniques were used to identify the particles on the basis of their different shape. The procedure was automated using customized ImageJ macros. Contrast and brightness of the images were first adjusted. Color images were then set to gray scale and segmented by applying a threshold. The operation was carried out twice to separate light and dark particles with different thresholds. Watershed segmentation was then used to separate touching objects. This last operation was more difficult for the particles at the border of the drum; in some case, in fact, the ellipsoids were erroneously identified as spheres. Also, the presence of the border of the drum complicated the watershed operation.

The centre of mass of each particle ($x_{m,i}$ and $y_{m,i}$) was estimated and the radial distance (d_i) from the centre of the drum (x_c, y_c) calculated as:

$$d_i=[(x_c-x_{m,i})^2+(y_c-y_{m,i})^2]^{0.5}$$

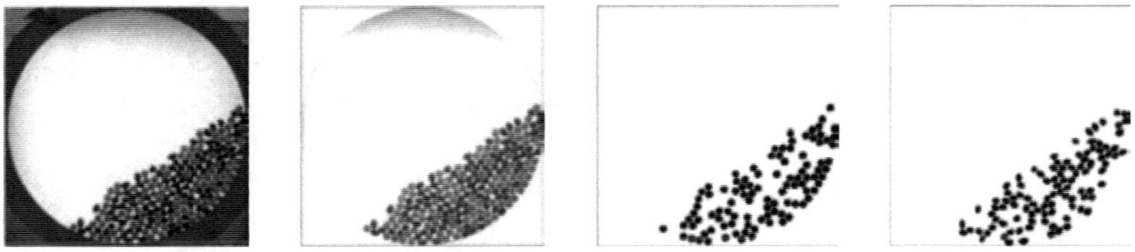

a) b) c) d)

Figure 2. Example of image processing steps. a) Original colour image; b) grayscale conversion and circle cut from the original picture; segmentation applied to c) light particles and d) dark particles.

The mean value of distance (d_m) for the spheres and for the ellipsoids was then calculated as:

$$d_m = \Sigma\, d_i/N$$

Only for very elongated ellipsoid (AR=3.5) the watershed operation failed resulting in over-segmentation errors. Single particles could not be identified unequivocally in this case; only one center of mass for all the elongated ellipsoid was estimated and the average distance (d_m) of the particles from the drum center calculated.

3. Results and discussion

Figure 3 shows two typical experiments, carried out at the lowest rotational speed and filling ratio, with ellipsoids (AR=1.5 at left and AR=3.5 at right) on top and spheres on bottom in the initial and final configuration. It is evident even by visual inspection that the final state presented some level of radial segregation and that a different final segregation intensity was reached in the two systems.

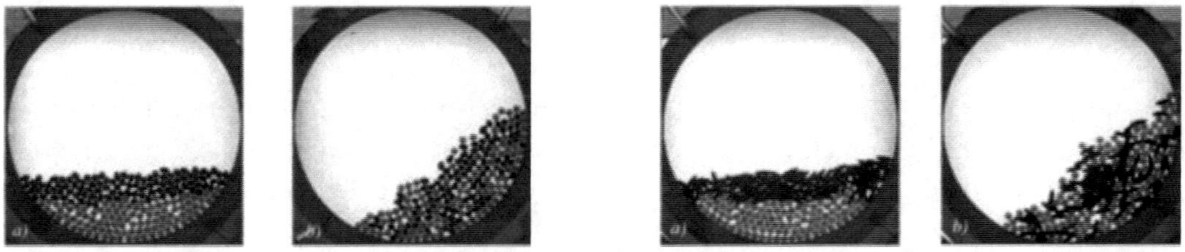

Figure 3. Drum at a) t=0 and b) t=30 min. On the left the initial configuration is E1/S (using ellipsoids with AR=1.5) and on the right the initial configuration is E2/S (AR=3.5). In both experiments ω=4.22 rpm and φ= 25%.

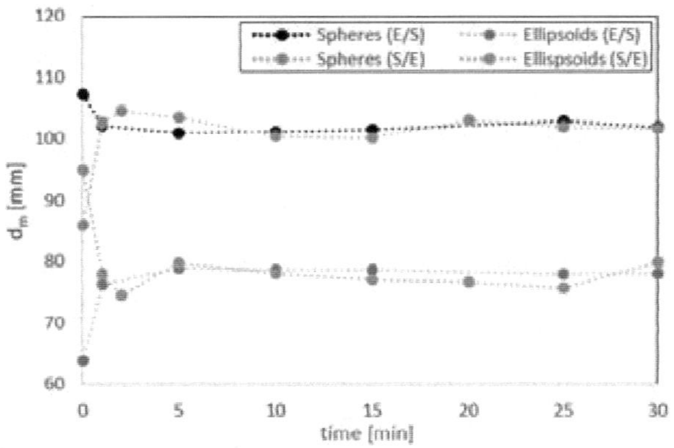

Figure 4. Evolution of the mean radial distance vs. time for the mixture of spheres and ellipsoids with AR=3.5. The initial configuration E2/S is compared to the initial configuration S/E2.

A more quantitative analysis was obtained by plotting the trend of the mean radial distance (d_m) of the two species at different time instants. As a typical example of the system evolution, the case of E2/S

is shown in Figure 4. The d_m reached a steady value after a short time. Dedicated experiments on a shorter time scale (2 minutes, not shown here) revealed that steady state value for d_m was reached after 30s for mixtures with ellipsoids E1 (AR=1.5) and after 60s for mixtures with E2 (AR=3.5).
At steady state, the radial distance was in both cases smaller for the ellipsoids, meaning that they mainly remained concentrated on top or at the center of the granular bed while the spheres mainly concentrated at the periphery of the drum (close to the drum wall). This tendency was much stronger for ellipsoids E2. So a slower segregation dynamics compensated by a larger segregation intensity was observed with the more elongated particles.
As Figure 4 shows, inverting the initial configuration from E/S to S/E did not affect the dynamics and the final result in all the experiments performed with these particles. Initial configuration resulted to have no effect on the segregation propensity of the mixtures analyzed.

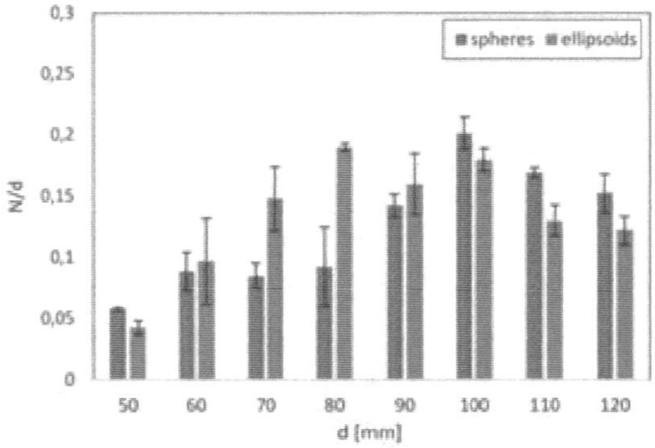

Figure 5. Distribution of the radial distance of spheres and ellipsoids with AR=1.5, obtained by averaging the results of three repetitions. Other conditions are t=30 min, ω=4.22 rpm, φ= 25%.

A more in-depth analysis can be carried out considering the distribution of the radial distance of the spheres and ellipsoids separately. The radial distance from the centre of the drum (d) was divided in classes of 10 mm and the number of each type of particles (N) in each class was computed. To obtain the distribution, N was normalized on the radial distance as N/d and plotted versus d. In the case of completely homogeneous mixtures and assuming that the medium takes the shape of a circular sector, N/d should be constant and equal for the two types of particles in each class. The peak of the distribution of the ellipsoids was found at lower distance than the peak of the distribution of the spheres. Moreover, at lower distance, the number of ellipsoids was higher than the number of spheres, especially for distances between 70 and 90 mm; at higher distance (closer to the drum wall), instead, the number of spheres was higher than the number of ellipsoids, thus confirming the previous analysis based on d_m.

Also, binary mixtures made of the two types of ellipsoids E1 and E2 (AR=1.5 and AR=3.5 respectively) were taken into account at the same filling levels of the previous experimental tests and at the lower rotational speed. Stable segregation was again observed with intermediate results with respect to the two previous cases (S+E1 and S+E2). So as a general comment on these experiments, stable segregation was always observed when mixing particles having same volume but different AR. The intensity of segregation increased with increasing the difference in AR between the particles. In particular particles with larger AR tended to accumulate at the center of the granular bed, while particle with lower AR accumulated at the periphery. Segregation occurred in the surface active layer where the more elongated particles, because of their reduced propensity to flow (particle rolling was prevented by their elongated shape), were systematically overtaken by the less elongated (and faster) particles. In this way a spontaneous separation of particles occurred in the active layer, with those having a lower AR arriving first at the drum wall, while those having an higher AR remaining trapped in the core.

3.1 Effect of filling level

The comparison of the behavior of the mixtures at the two filling levels can be carried out by normalizing the mean radial distance d_m of the two types of particle on the mean distance of all the particles d_c, calculated as $(d_{m,\ spheres}+d_{m,\ ellips.})/2$. The trends of d_m/d_c are shown in Figure 6

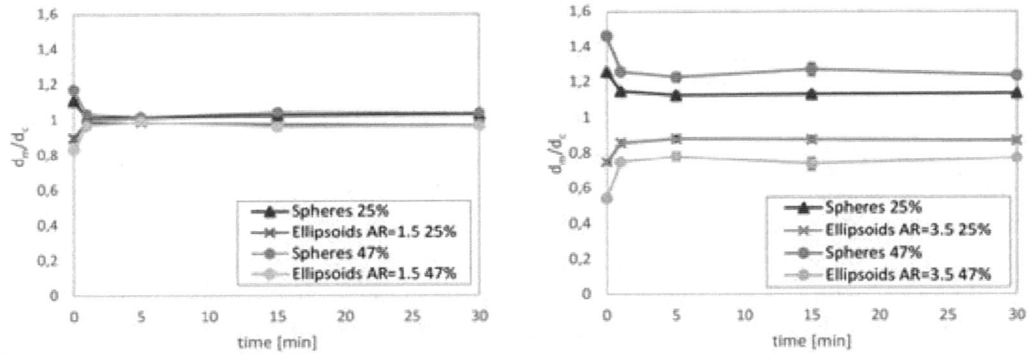

Figure 6. Trend of d_m/d_c vs. time for the mixture of spheres and ellipsoids with *AR*=1.5 (left) and AR=3.5 (right) at two different filling levels.

In the case of mixture of spheres and ellipsoids with *AR*=1.5, the curves at different filling levels overlap, showing that the behaviour of the mixture was not affected by the filling level of the drum. In the case of the other two mixtures (S+E2 and E1+E2, not shown), the curves did not overlap , especially for the case S+E2 shown in Figure 6 (right). An explanation can be found in terms of bed geometry. When increasing the filling level (below 50%) the length of the surface active layer increased as well. This gave more chances for particles to separate, in particular when the difference in *AR* was large enough (S+E2 and E1+E2). However for the system S+E1 where the difference in *AR* was the smallest, the difference in active layer length in the two cases was not sufficient to create observable differences in the segregation pattern.

3.2 Effect of rotational speed

In addition to experiments carried out at 4.2 rpm, tests were repeated at 11.5 rpm with spheres and ellipsoids (25% filling level). Figure 7 shows that rotational speed did not affect the segregation patterns, at least in the range of the analyzed speeds.

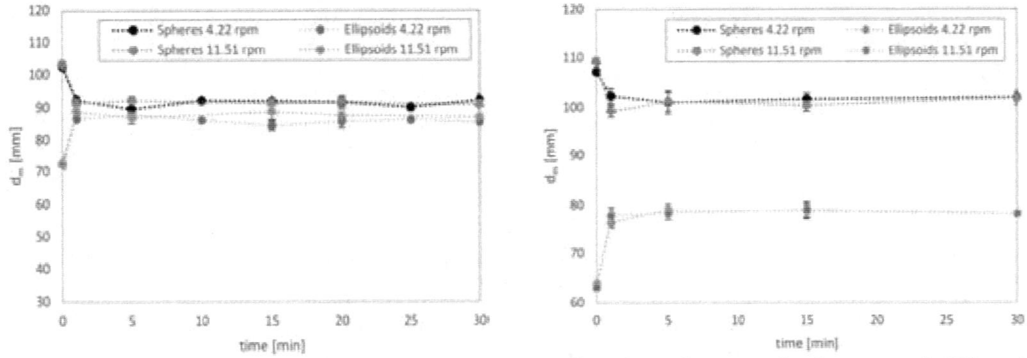

Figure 7. Trend of the mean radial distance vs. time for the mixture of spheres and ellipsoids with *AR*=1.5 (left) and *AR*=3.5 (right) at two different rotational speeds.

Conclusions

From the experimental tests described in the previous paragraphs, it was found that binary mixtures of particles with different shape but with the same volume do not mix uniformly and segregation was always observed. In particular, the particles with lower aspect ratio tended to stay closer to the drum walls, while the more elongated particles were found preferentially at the centre of the granular bed. The most significant case was represented by the mixture of spheres and ellipsoids with AR=3.5; for this mixture, in fact, the segregation was more evident and the non-spherical particles clearly accumulated in the centre of the granular bed. The same final state was obtained regardless of whether the longer particles were initially located to the top or the the bottom of the mixture. A proportionality between AR difference and segregation intensity was observed. Some variables such as rotational speed and initial bed configuration were found to not affect the segregation dynamics and intensity. Only filling level (in addition to AR) was found to affect segregation intensity. A mechanistic explanation was given in terms of differences in particle flowability occurring in the surface active layer (confirmed also by an analysis of the angles of repose in the drum). In this specific case therefore percolation and inertial effects (typical of size segregation) were not the main driving mechanisms for the observed segregation (the particle indeed had the same volume and similar mass so that sterical and inertial phenomena were minimized). Segregation mainly occurred because of differences in particle velocity in the active layer due to a reduced mobility of particles with larger AR.

References

1. GDR MiDi (2004), "On dense granular flows", The European Physical Journal E, 14, 4, 341–365.
2. Jarray A., Magnanimo V., Ramaioli M., Luding S. (2017) Scaling of wet granular flows in a rotating drum, EPJ Web of Conferences, 140,03078.
3. Volpato S., Canu, P., Santomaso A.C. (2017), "Simulation of free surface granular flows in tumblers", Advanced Powder Technology 28(3), 1028-1037.
4. Mellmann J. (2001), "The transverse motion of solids in rotating cylinders – forms of motion and transition behaviour", Powder Technology, 118, pp 251-270.
5. Windows-Yule C.R.K., Scheper B.J., van der Horn A.J., Hainsworth N., Saunders J., Parker D.J., Thornton A.R. (2016), "Understanding and exploiting competing segregation mechanisms in horizontally rotated granular media", New Journal of Physics, 18, 2, 023013, 2016.
6. Santomaso, A.C., Olivi, M., Canu, P. (2004), "Mechanisms of mixing of granular materials in drum mixers under rolling regime", Chemical Engineering Science, 59 (16), 3269-3280.
7. Kumar N., Magnanimo V., Ramaioli M., Luding S. (2015), "Tuning the bulk properties of bidisperse granular mixtures by small amount of fines", Powder Technology, 293, 94-112.
8. Ramaioli M., Pournin L., Liebling Th. M., (2005), "Numerical and experimental investigation of alignment and segregation of vibrated granular media composed of rods and spheres", Proceedings of the 5th International Conference on Micromechanics of Granular Media, 2, 1359-1363.
9. Santomaso, A.C., Canu, P. (2007), "Single particles properties vs. bulk flowability", Proceedings of 5th International Conference for Conveying and Handling of Particulate Solids.
10. Schneider, C. A., Rasband, W. S., Eliceiri, K. W. (2012), "NIH Image to ImageJ: 25 years of image analysis", Nature methods 9(7): 671-675.

507441 Generation of Particles with a Special Morphology By Desublimation of Copper Phthalocyanine

Abstract:

The morphology of submicron particles becomes more and more important concerning new functional materials. For example the special electric properties of particles can be used in electronic devices [1]. Moreover the electric, optical or catalytic properties of particles can be optimised by adjusting the morphology of the particles. This may imply a risk for the health but makes it possible to realize specific product characteristics. The morphology of the particles should therefore be considered e.g. for filter tests and for the developing and testing of sensor systems.

To explore the characterization of the particle morphology the DFG-AiF cluster project "multi-parametric characterization of particle-based functional materials using innovative online measurement systems" (MPaC) includes the development of online sensor systems for the size and structure of the particles as well as the development of generation methods for particles with a special morphology. The production of spherical particles can be achieved with different commercial aerosol generators. Several generation techniques for special particles are described in literature [2] [3] [4]. To test the sensor systems with particles of different shapes a new type of aerosol generator was developed. The aim for the new generator is to produce the particles in the gaseous phase and provide the particles directly for sensor or filter testing.

The developed generator uses the desublimation of copper phthalocyanine (organic pigment) to structure the surface of titanium dioxide. The bulk material copper phthalocyanine is dispersed into a nitrogen gas stream and transported to a tube furnace. Here the bulk material sublimates and the gas stream carries the material to a laval nozzle where the adiabatic expansion takes place. This leads to a rapid decrease of the temperature behind the laval nozzle and an oversaturation in the gaseous phase. In addition to the decrease of the temperature core particles of titanium dioxide were initiated behind the laval nozzle. The copper phthalocyanine desublimates on the core particles and causes a structured surface. The morphology of the generated particles is characterized by SEM. The particles show a punctual coating of the titanium dioxide particles with copper phthalocyanine. The size of the titanium dioxide particles is about 500 nm and the punctual coating of copper phthalocyanine measures a few nanometres.

The generation method and the design of the generator will be presented and the morphology of the generated particles will be described.

[1] K. Deppert, M. H. Magnusson, S. B. Carlsson, B. Gustafson, T. Junno, . L. Montelius, B. J. Ohlsson, L. Samuelson und L. E. Wernersson, „Future electronic devices by aerosol techniques," *Journal of Aerosol Science,* Nr. 31, 2000.

[2] J. Raula, A. Kuivanen, A. Lähde, H. Jiang, M. Antopolsky, J. Kansikas und E. . I. Kauppinen, „Synthesis of image-leucine nanoparticles via physical vapor deposition at varying saturation conditions," Journal of Aerosol Science, Nr. 38, 2007.

[3] M. Heurlin, M. H. Martin, D. Lindgren, M. Ek, R. L. Wallenberg, K. Deppert und L. Samuelson, „Continuous gas-phase synthesis of nanowires with tunable properties," Nature, Nr. 492, 2012.

[4] K. Nielsch, K. Deppert, M. H. Magnusson, G. Dumpich, E. Kruis und H. Fissan, „Formation of ultrafine particles from powders of compounds with incongruent sublimation behavior," Journal of Aerosol Science, Nr. 28, 1997.

516184 Rubbery Milling of Seed Endosperms for Improved Sustainability by Natural Functionality Preservation

Abstract:

Increasing demand for functionalised goods addressing individual needs requires a profound understanding of process-induced alterations. Development of models and process-structure-property relationships are essential to control the resulting quality and properties of processed food. A wide range of food systems contain solid or semi-solid particulates as major structure building blocks. Generation and tailoring of such particulates strongly depends on their size distribution, shape, morphology, phase behavior (amorphous/crystalline), interfacial properties and related inter-particle force interactions.

A novel processing technology was developed widening the frame of possible alterations, aiming towards tailoring such particle properties generating novel food structures or preserving techno-functional characteristics of the raw material. The technology employs targeted adjustments of the initial moisture content in combination with milling temperature. The resulting transition from the glassy to the rubbery state generates different material properties and processing behaviour. The technology is highly versatile, owing to the universality of moisture induced state transitions.

Seeds from locus bean gum or guar gum were wetted with excess water in order to adjust a rubbery state of the seed endosperm material. In this state milling within a novel type of impact mill with superimposed drying and sifting steps causes more gentle material disintegration compared to conventional dry milling. This preserves the high molecular weight fraction of e.g. galactomannans (micro-scale structure) from degrada-tion and allows at the same time the adjustment of a micro-porous morphology (macro-scale structure) improving wetting, water uptake and dispersing characteristics of the resulting thickener powders. Such structure preservation /modification increased the thickening characteristics extraordinarily as will be demonstrated.

From this prominent rubbery milling example we have entered in related systematic work addressing a wider range of seed endosperms with various â€œnature-synthesizedâ€ func-tionnalities to be further addressed.

/1/ Pollard, M., Eder, B., Fischer, P., and Windhab, E. J. (2010) "Characterization of galacto-mannans isolated from legume endosperms of Caesalpinioideae and Faboideae subfamilies by multidetection aqueous SEC." *Carbohydrate Polymers*, 79(1), 70-84.

/2/ Illmann, S., Stoob, M., Sieber, C., Konz, A.K. and Windhab, E.J. (2012) "New Thickening Quality of Galactomannan Polysaccharides by Tailored Impact Milling" *XVIth Int. Congress on Rheology*, Lisbon, Portugal, August 5-10

503422 Particle Technology Education at Purdue University

Abstract:

Degree programs and coursework at Purdue University focused on particle technology will be presented. Additional information may be found at: https://engineering.purdue.edu/CP3

506453 Teaching Particle Technology in Portugal â€“ University of Coimbra

Abstract:

Education in Particle Technology started in Portugal in the 80â€™s (last years of the decade). The Chemical Engineering degree of the University of Coimbra was the first engineering degree, in the country, to include this topic in the curriculum. Particle Technology was recognized by us, at that time, essential for the education of chemical engineers, in parallel to education in the traditional areas of Chemical Engineering which were still focused on monophasic systems.

Since then, Particle Technology has always been present in the Chemical Engineering curriculum of the University of Coimbra, in different configurations. During the first years particle characterization was combined with design of operations involving particulate systems (grinding and pneumatic and hydraulic transport). From 2005 to 2010, with the curriculum reformulation in relation with the Bologna Process, the two topics started to be addressed in different courses: one course on Materials Characterization and the process topic in a course entitled Structuring and Stabilization Processes, dealing with colloids and emulsions, grinding, agglomeration and spray drying and, for the first time, an integrated modelling strategy for particulate processes based on the population balance formulation was introduced in this course. The two courses were compulsory, at this time.

From the school year 2011/2012 till present, the configuration of this area in the Chemical Engineering degree of the University of Coimbra changed again, even if not deeply (see https://apps.uc.pt/courses/EN/course/361). The course on materials characterization, which includes general notions on particle characterization is kept as an autonomous course included in the 1st cycle of the Chemical Engineering degree of the University of Coimbra. A second course on design of processes involving particulate systems now exists in the 4th year of the MSc degree (1st year of the 2nd cycle), entitled Disperse Systems Technologies, which is an optional course, and which deals with all types of disperse systems, i.e gas, liquid or solids as disperse phase. This course starts with a review of the mechanisms/interactions occurring at the interface with the disperse phase, goes on, again, with a chapter on population balances modelling, and proceeds with strategies for disperse systems stabilization, linking this subject with formulation concepts, ending with the design of some processes involving particulate systems, such as colloids and emulsions, crystallization, grinding, agglomeration and spray drying. This course is also offered to other engineering degrees, such as Materials Engineering and Biomedical Engineering.

Also at the PhD level a course on Disperse Systems Technologies can also be offered during the formal lectures period of the PhD program on Chemical Engineering (or others), depending on the students background and interests.

Since the introduction of the Particle Technology topic in the curriculum of the Chemical Engineering degree of the University of Coimbra, the feedback from employers has been very positive, and the competences in this field acquired by our students have been quite valued by the job market. Till today, the Chemical Engineering degree of the University of Coimbra is still the one, in Portugal, with a stronger emphasis on this topic in the curriculum. In our curriculum, the topic of Particle Technology interacts directly with another course on Product Design, also in the 4th year of the MSc degree and, of course, also with the Project Design course (5th year of the MSc degree), besides other more traditional courses in a Chemical Engineering degree such as in Reaction Engineering and with courses related to environmental technologies.

At present, we have started developing multimedia tools for the teaching/learning of this topic, especially in the field of particle characterization, to be made available on-line, as is the case of the interactive video on nanoparticles characterization (http://labvirtual.eq.uc.pt/siteJoomla/index.php?option=com_content&task=view&id=277&Itemid=470, in Portuguese). Also, the European Federation of Chemical Engineering (EFCE) is considering investing in this type of tools for the field of particle technology, as the result of a collaboration between the Working Groups of Characterization of Particulate Systems and Education.

Acknowledgement

The authors would like to acknowledge the financial support of Pest/C/EQB/UI0102/2013 from FCT/MCTES (PIDDAC), co-financed by the European Regional Development Fund (ERDF) through the program COMPETE (POFC).

506629 Highlights of Particle Technology Teaching in Singapore

Abstract:

Our approach to education in particle technology is to incorporate elements of our research in this field, relevant industrial case studies and apply a Technology Enhanced Learning (TEL) approach for implementing pedagogical interventions to enhance the learning experiences of students.

For supercritical CO_2 processing, case studies in extraction and drying are used to teach different chapters in the curriculum such as:

- Particle characterization
  - The bulk density, size, shape of real industry samples are given to students for analysis using software such as ImageJ
- Flow through packed bed
  - The pressure drop and scale-up of packed bed design for supercritical CO_2 flow through packed bed are used as a worked example and assignment problem to help students appreciate the role of density and viscosity in the pressure drop in a packed bed
- Drying
  - Supercritical CO_2 as an alternative drying media as compared to hot air or superheated steam. Case studies and examples provided

For gasification and chemical looping combustion applications, case studies in the treatment of wastes and carbon capture are similarly used as part of the syllabus:

- Particle characterization
  - The effect of polydispersity on various fluidization characteristics (entrainment, transport dis-engaging height, clusters, segregation, etc)
  - The effect of agglomeration
- Reliability of correlations and simulations
  - Correlations in literature give predictions orders-of-magnitude apart
- Scale-up
  - Rule-of-thumbs in scaling up the various regimes

A design software has been developed within the paradigm of Technology Enhanced Learning (TEL) to facilitate learning via a repeated practice approach. The software generates detailed solution steps to typical engineering design problems encountered within this module. Students can compare the complete solutions generated with their own hand calculations, and thereby apply a repeated practice approach towards their learning of engineering design calculations. The software has been successful in enhancing the performance of students in the final examination.

507116 Particle Technology Course at the University of Salerno

Abstract:

The Course of Particle technology at the University of Salerno is a 6 EC, that is a 60 hours course including lectures, as well as classroom and laboratory practical. The course dedicated to Master student in Chemical Engineering. It a compulsory course within the â€œinnovative process and nanotechnologies curriculumâ€ can be optioned by any mechanical or chemical engineering student at the master level.

The course includes basic particle size and shape characterization concepts, fundamentals on interparticle forces, fundamentals of powder mechanics and silo discharge, fundamentals on-fluid particle interactions and fluidization.

The main learning outcome is the ability of the student to understand the phenomena occurring in processes involving particulate solids and to carry out basic design operation on storage and fluidized bed units.

507382 Professional Master of Engineering Degree in Particle Technology

Abstract:

Particles are ubiquitous in the process industries yet nearly absent in the undergraduate education of chemical engineers. This poster descibes a professional Master of Engineering degree program at University of Delaware (the MEPT) that closes that gap by exclusively focusing on particulate matter, its processing, and its incorporation into products. The program can be completed in a single year although we welcome already-employed students who wish to take it part time. Now in our fourth year, we will describe our program and our experiences including guiding distance learners successfully through the program.

In its 1-year realization, a core curriculum of 18 credits is augmented by 6 credits of electives which the student may use to tailor the program to their specific interests. The program is capped by a 10-week, 6-credit summer internship. The 12 core credits taken in the fall teach the fundamentals of particle transport, particle properties & characterization, particle formation and transformation kinetics and the mathematics of particulate systems. The 6 core credits taken in the spring teach the student to apply the fundamentals to both (a) equipment design (in a unit operations course) and (b) product and process design (in a product engineering course).

The presenters of this poster are the program directors, both with long industrial careers preceding this venture into education. In addition, we have developed a strong network of industrial partners who share a wealth of additional experience and expertise with our students both as adjunct professors and as internship mentors. We have successfully placed international students in internships at US companies via the curriculum practical training program (CPT) who have gone on to practice in industry after graduation under the optional practical training program (OPT).

507780 Recent Developments in Particle Technology at the Universidad Tecnica Federico Santa Maria

Abstract:

The Department of Mechanical Engineering of the Universidad Técnica Federico Santa María formed in 2003 a Research Center for Transporting Granular Materials (CITRAM) to support educational activities and experimental research, as an extension for those ME students taking the regular course on Bulk Solids Handling and those developing a thesis in this area.

Over the years, many ME students have studied, investigated and modeled different topics including: pneumatic conveying of solid particles of different nature, size and density; identification of flow patterns developed in the pipeline, minimum conveying velocity and pressure drop; abrasive wear using a novel rubber-wheel tester; segregation of ores in stockpiles and silos; flow of granular materials in hoppers and chutes; heat transfer of granular materials; among others. The aim of this poster is to present and share the results of this work.

508520 Using Perusall to Enhance Student Learning of Particle Technology at Graz University of Technology

Abstract:

Perusall is a digital platform that allows students to collaboratively annotate textbooks, papers, and other similar classroom reading assignments. In order to enhance student learning, Perusall was introduced to help teach Particle Technology at Graz University of Technology (TU Graz, Austria). At Graz University of Technology, two Particle Technology courses exist in sequence; the first course is focused on fundamentals topics, whereas the second course is focused more on applications and separations topics. Approximately 60-75 students are registered for each course. For both courses, Perusall was used to assign reading materials, which were primarily journal articles that complemented the course topics. Although these assignments were non-compulsory due to national requirements, students were asked to use Perusall to read and annotate the course readings. Student participation and input regarding this digital approach to learning was recorded. This poster will highlight how Perusall was implemented in both Particle Technology courses as a tool to enhance student learning.

508524 A Graduate Course in Fluidization and Gas-Solid Flow Systems

Abstract:

The objective of this course is: a) to provide students with a fundamental understanding and design of Fluidization and Fluid- Particle flow systems, b) to provide students with an understanding of Computational Fluid Dynamics (CFD) and the kinetic theory approach to fluid-particle flows c) to provide students with the knowledge required to solve flow, heat and mass transfer problems in processes which involve particles and d) to provide students with sufficient knowledge to read and understand technical articles in areas related to fluid-particle processes. The course is designed for Master and PhD students in Chemical, Mechanical and aerospace engineering. The course includes: Fluidization concepts and regimes including bubbling, turbulent and circulating beds; Particle characterization, diagnostics and size reduction and agglomeration/granulation; Concept and design of Standpipes, non- mechanical valves, gas distributors, cyclones and hoppers; CFD approach to fluid-particle flow systems and fluidization including kinetic theory and dispersed phase modeling; Computational transport phenomena of fluid-particle systems including method of moments; and examples from energy, chemical and pharmaceutical industries. In addition the students will visit PSRI to see large-scale fluidized beds in operation.

The students are required to finish a team project. The teams consist of 3-4 students working on a project which will count as their final exam. The projects will primarily involve background research (about 80%) but some of the investigation (about 20%) should involve taking an additional step from what is currently known. This additional step can take the form of a research proposal, new theory, new insight based on current information, new calculations, new simulation and/or new data. The results of the projects will be communicated to the class and to the Course Instructors via a written report of 10-20 pages, and an oral presentation involving the professors and all of the group members.

508742 Education of Fluid-Solid Multiphase Flow at Department of Mechanical Engineering, Osaka University

Abstract:

In mechanical engineering field, education curriculum in undergraduate course concentrates on the basic subjects of mechanical engineering such as mathematics, mechanics, material mechanics, fluid mechanics, thermodynamics, control engineering, design, etc. Concerning the course relating to powder technology, we have a class of "Fluid-Solid Multiphase Flow" in the curriculum of graduate school of engineering.

The contents of "fluid-solid two-phase flow" is as follows;

1 Introduction: examples of multiphase flow

2 Definitions and fundamental concepts

3 Particle response time and Stokes number

4 Fluid force acting on particle

5 Fluid force acting on particle in dense system

6 Particle-wall and particle-particle interaction

7 Heat and mass transfer

8 Continuous phase flow equations

9 Modeling of fluid-solid multiphase flow

10 Discrete particle modeling and simulation

11 Numerical simulation method for continuous phase flow

This course includes basic concepts of powder technology, dynamics of particles in fluid flow, derivation of basic equations for locally-volume-averaged fluid flow, discrete particle modeling including DEM (Discrete Element Method), and basics of numerical analysis and modeling.

509420 Bulk Solids Handling Education at the KSU Bulk Solids Innovation Center

Abstract:

Kansas State University Bulk Solids Innovation Center (BSIC) was started in 2015 with university and industry partnership and is located in Salina, Kansas. This center is dedicated to bulk solids handling research, education and outreach and equipped with world class pneumatic conveying system (both dense and dilute) full scale test lab and other material testing equipment. The vision of this center is to become valued resources for companies who produce or handle bulk solids during their operations. Currently, this center offers short courses in various subject lines related to bulk solids handling such as storage and flow, pneumatic conveying systems, discrete element method, dust explosion and control and mixing and segregation. Duration for these short courses varies from two to three and half days and offered as live course with class room teaching and practical training. The fundamental course on bulk solids handling is offered as distance education online format. This center is also in the process of starting two year online (distance education) master degree program named professional master of technology specialized in bulk solids handling and one year graduate certificate program (online) in bulk solids handling.

EXPERIMENTAL INVESTIGATION AND MODELING OF MOVING BED TO MOVING DUNE TRANSITION

R. Dabirian, I. Gavrielatos, R.S. Mohan, and O. Shoham, The University of Tulsa, Tulsa, OK, USA

Abstract

Solid particles are encountered in oil and gas pipelines when producing from unconsolidated reservoirs. Solid accumulation can lead to increase in pressure loss, erosion, corrosion, production decline etc. The stratified flow regime is most vulnerable to particle deposition, owing to low phase velocities that occur in this regime. Various sand flow regimes may occur, whereby particles can stop moving and accumulate along pipe either as stationary dunes or a stationary bed. The objective of this paper is to investigate experimentally and theoretically the transition boundary between two of the regimes, namely, concentrate solid flow (moving bed) and moving dunes.

Experimental data are acquired in a 10 cm (4 in.) ID pipe, using air and water. Spherical glass beads with specific gravity of 2.65, 45-90 μm and 125-250 μm in diameter, at concentration up to 10,000 PPM, are utilized. The data include the effect of the superficial velocities, particle size and particle concentration on the different sand flow regimes, including the moving bed and moving dunes. The acquired data show that the transition between moving dunes and moving bed is strongly dependent on the above mentioned parameters.

A mechanistic model is developed for the prediction of the transition between moving dunes and moving bed, based on rolling mechanisms for particles located between two sand piles. When the particles roll up on a pile, transition to separated moving dunes occur, otherwise, moving bed is formed along the pipe. Comparison between the acquired data and model predictions show a discrepancy of 15%.

Introduction

Efficient sand transport in pipelines is important for flow assurance. The petroleum industry encounters sand production in reservoirs with low formation strength. Removal of particles can be accomplished by operating at high capacity; however, this solution can result in operational problems, such as erosion and equipment failure. On the other hand, particle transport under low phase velocities may result in particle deposition and correspondingly excessive pressure drop. Thus, it is desirable to operate at flow conditions that enable sweeping and transport of all particles along the pipe. Several studies are published on particle transport mechanisms in gas-liquid-solid multiphase flow. These include studies on the critical sand velocity, as well sand flow regimes, as summarized below.

Critical Velocity

Angelson et al. (1989) developed a multiphase model by extending Wicks's (1971) single phase flow model. The model predicts required gas and liquid velocities to avoid particle deposition. Sand transport experimental data were acquired by Oudeman (1993) in a 0.07 m ID pipe, utilizing two particle sizes of 150 μm and 300 μm and developed a correlation for prediction of efficient transport rates of sediments. Gillies et al. (1997) extended the single-phase sand transport correlation developed by Meyer-Peter (1948) for sand transport in multiphase flow. Salama (2000) developed a correlation to predict the minimum required velocity for full particle suspension in multiphase flow. Experimental studies on sand transport under intermittent flow were performed by Stevenson et al. (2001), whereby a correlation was developed for the particle velocity as function of the mixture velocity. Stevenson and Thorpe (2002) developed a model to predict minimum velocity to initiate motion of particles in intermittent flow. They also performed a series of experiments and developed a correlation for sand transport under stratified flow. The drift flux model has been used by Danielsion (2007) to extend his developed single-phase flow model to multiphase flow. Yan (2010) performed experiments for sand transport in horizontal and near horizontal pipes for single and multiphase flows, with particle diameter of 180 μm and liquid viscosities of 1 and 20 cp CMC water, as well as 105, 200 and 340 cp mineral oils. The extension of Oroskar and Turian (1980) model to multiphase flow has been proposed by Hill (2011) to predict minimum flow rate required to move particles continuously in stratified flow and intermittent flow. Ibarra et al. (2012) aquired experimental data for particle size of 211-297 μm , and particle concentration up to 0.01 m/m. Similarly, an extension of Oroskar and Turian (1980) model was proposed for stratified flow to predict the minimum mixture velocity leading to sand deposition. More recently, Dabirian et al. (2016) and (2017) acquired experimental data for three particle sizes of 45-90, 125-250, and 425-600 μm with concentration up to 0.01 m/m, and developed a model to predict the critical sand deposition velocity under stratified flow.

Sand Flow Regimes

Flow regime studies for solid-liquid flow were published by Cheremisinoff (1986), Doron and Barnea (1996) and Ayazi Shamlou (1970). Later studies on sand flow regimes in gas-liquid-solid multiphase flow were presented, such as Oudeman (1993) who classified sand flow regimes for multiphase as suspension, moving bed and stationary bed. Hill (2011) defined six sand flow regimes under stratified and intermittent flows, including homogeneous, heterogeneous, scouring, moving dunes, moving bed/dunes and stationary bed. Ibara et al (2012) applied the same sand flow regimes definitions as Oudeman (1993) for stratified flow. Recently, Dabirian et al. (2016) defined a comprehensive sand flow regime under stratified flow, as shown in Figure 1. According to this study, the flow regimes include fully dispersed solid flow (homogeneous), fully dispersed solid flow (heterogeneous), dilute solids at wall, concentrated solids at wall, moving dunes, and stationary dunes.

The literature review confirms that all experimental and theoretical studies have focused on the critical sand deposition velocity and the critical suspension velocity, considering stationary bed and fully dispersed solid flow. Also, the previous studies did not take into account the mechanism causing the sand flow regime transitions boundaries. The objective of this study is to acquire data for the critical velocity leading to the transition between the concentrated solids at wall (moving solid bed) and moving dunes, considering the mechanisms occuring when particles stop moving for these flow regimes. A mechanistic model is proposed to predict the above transition boundary under different experimental conditions.

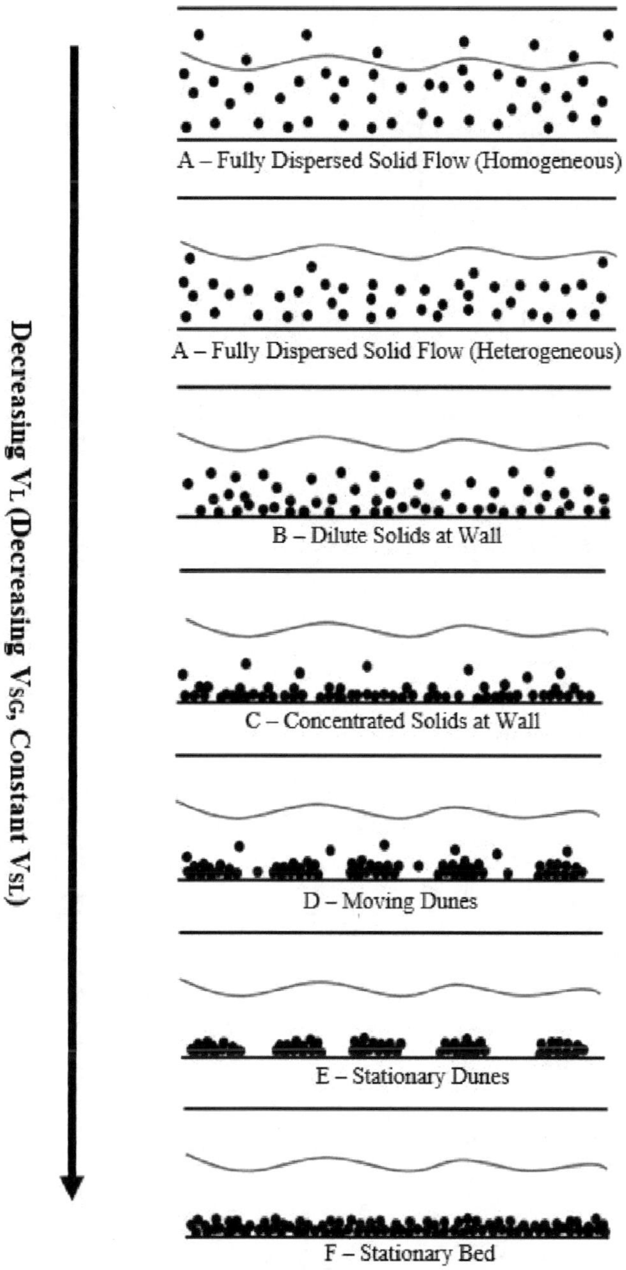

Figure 1. Sand Flow Regimes (Dabirian et al. 2016)

Experimental Program

This section presents a brief description of the test facility and the test matrix, as given below.

Test Facility

A multiphase flow loop is designed and constructed, which is utilized to study particle transport under multiphase flow, as shown in Figure 2. Air is supplied by 60 hp electric compressor with 120 psig maximum operating pressure. The solid-liquid slurry is prepared in a 100 gallon tank which is equiped with two electronic batch mixers in order to prepare a homogeneous slurry. Additionally, a recirculation line is installed to inject the slurry back to the bottom of the slurry tank to promote mixing. The outline of the tank is connected to a centrifugal pump with 2 HP electric motor and a diaphragm pump with maximum 35 GPM flow capacity. The centrifugal pump is used to inject the slurry in to the test section, while a diaphragm pump is applied to remove all particles from the tank. The flow rate of the prepared slurry is measured by an Ultrasonic Doppler flow meter with measuring range of 0.05 to 9 m/s. The test facility consists of 10 cm ID PVC pipe 11 m. in length, which can be set up either horizontal or inclined slightly upward. The facility is equiped with two pressure transducers installed upstream and downstream of the test section in order to measure the pressure drop. To measure the liquid height in the pipe, an ultrasonic proximity sensor is installed with sensing range of 100 to 600 mm. Downstream of the test section, a Gas-Liquid-Cylindrical-Cyclone (GLCC) is installed to separate air from the slurry, whereby the slurry is recirculated back to the slurry tank and the air is vented. More details regarding the experimental facility can be found in Dabirian (2016).

Test Matrix

Table 1 shows the test matrix for this experimental study. The experiments are performed with air and water, and spherical glass beads as the solid phase. Two superficial liquid velocities, namely, V_{SL} = 0.05 and 0.1 m/s, and superficial gas velocities in the range of 3 to 12 m/s are chosen, underwhich stratified flow occurs. Two particle sizes of 45-90 μm and 125- 250 μm are used with particle concentrations up to 10,000 ppm.

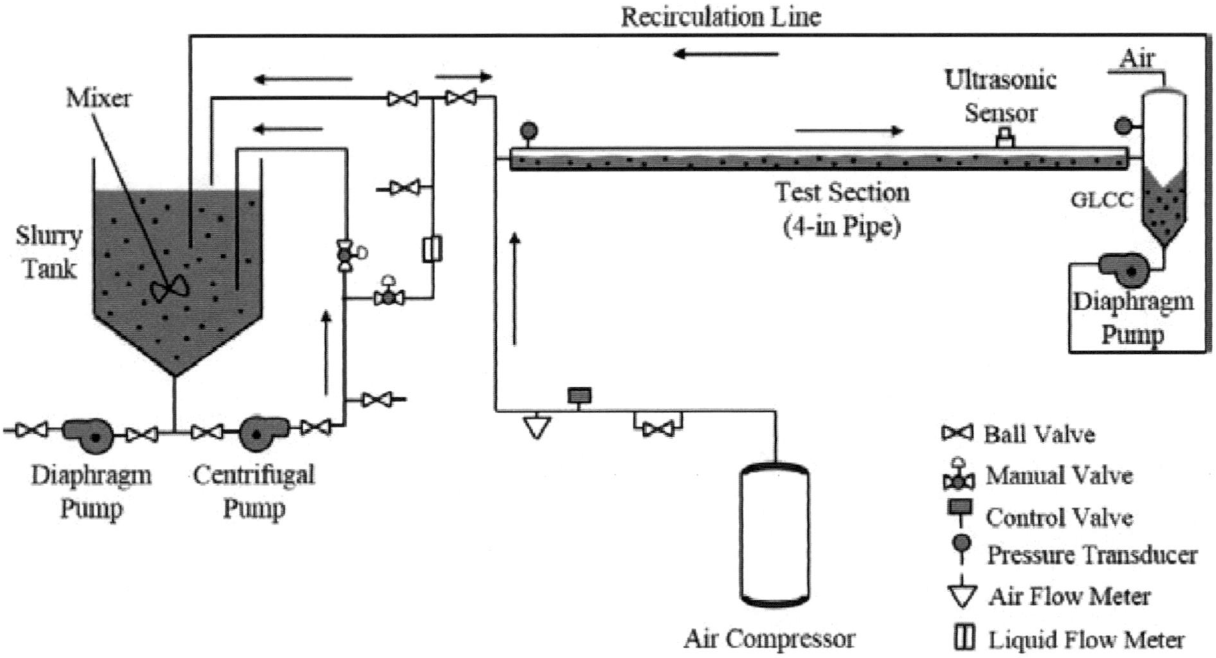

Figure 2. Schematic of Test Facility

Table 1. Experimental Test Matrix

Variable	Range	Units
Pipe Inside Diameter	0.1	m
Pipe Inclination Angle	0	degree
Superficial Gas Velocity	3 - 12	m/s
Superficial Liquid Velocity	0.05, 0.1	m/s
Liquid Viscosity	1	cp
Particle Specific Gravity	2.65	-
Particle Size	45 - 90, 125 - 250	µm
Particle Concentration	100 - 10000	PPM

Concentrated Solids Flow/ Moving Dunes Transition

Figure 3 shows a schematic of flow regime sequences, which are observed visually during the experimental data acquisition. As shown, the dilute solids at wall regime is observed at relatively high liquid velocity where the particles move on the pipe wall by rolling, sliding and saltation mechanisms (Figure 3.A). With decreasing liquid velocity, a continuous narrow sand bed with a flat profile is created on the wall, whereby the top particles move under the same 3 mechanisms (Figure 3B). With reducing further the liquid velocity, a wavy sand bed profile is seen owing to accumulation of small particles behind bigger ones, which tends to form sand piles on the bed, whereby the rolling mechanism is dominant. Once the particles between the sand piles are rolled up to the top of the sand piles, gaps are generated between the piles, and sand dunes are formed; otherwise, a continuous bed with a wavy profile is obseved along pipeline.

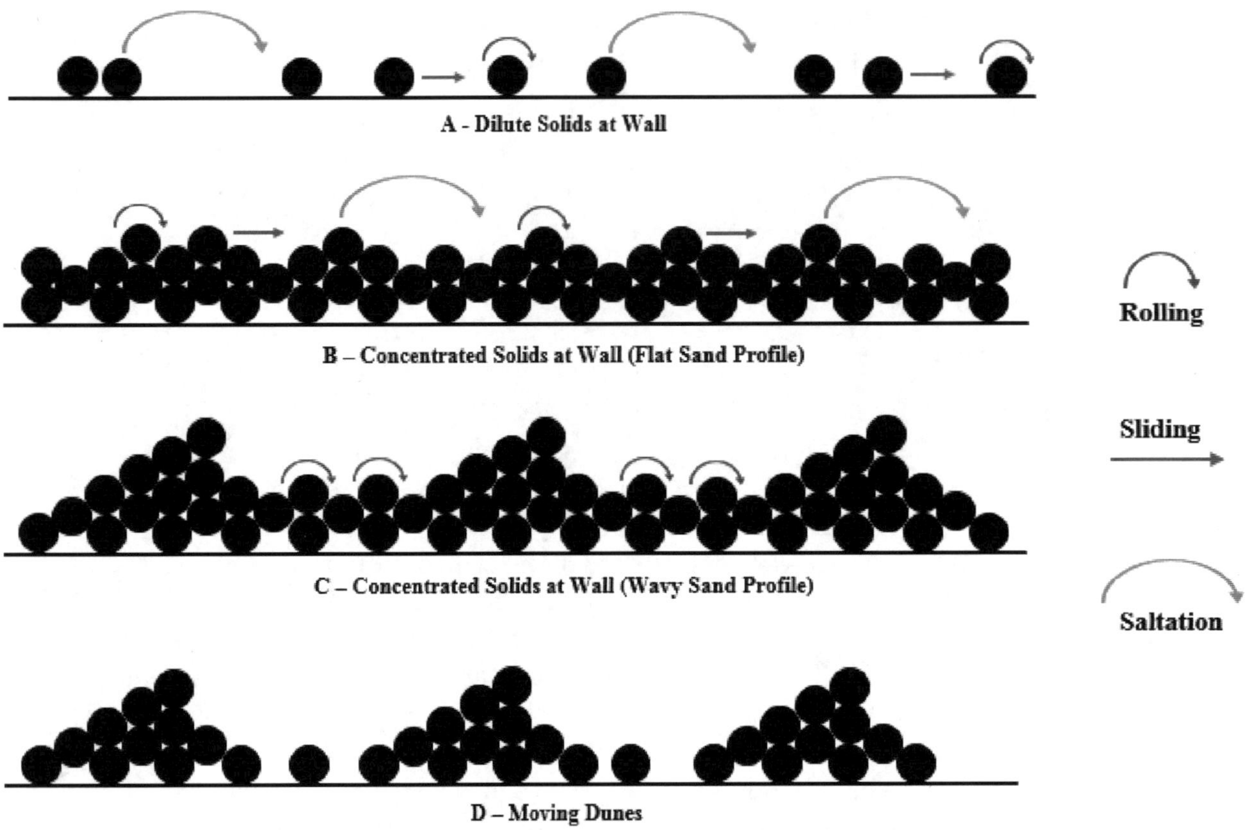

Figure 3. Schematic of Flow Regime Sequences

Modeling

A mechanistic model for prediction of critical velocity leading to the transition between the concentrated solids flow (moving bed) and moving dunes is presented. Figure 4 shows a schematic of transition between the concentrated solids flow and moving dunes. As mentioned, sand piles are observed at relatively low liquid velocities where particle rolling between the piles is dominant. The model for the critical velocity is developed based on rolling of particles trapped between two sand piles.

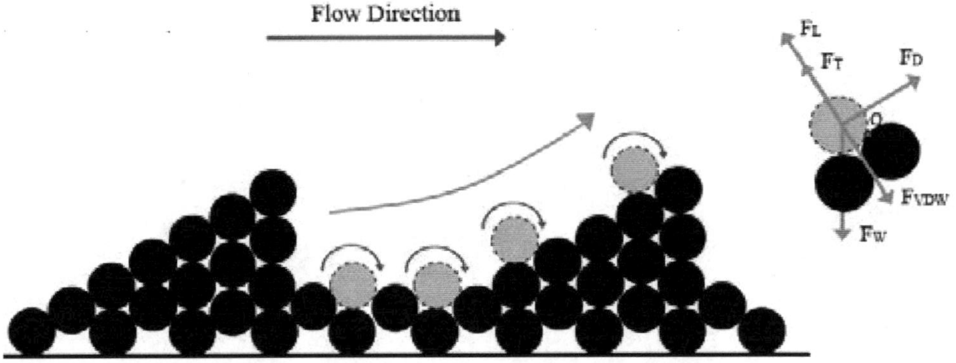

Figure 4. Schematic of Transition between the Concentrated Solids Flow and Moving Dunes

The tendency of particles to roll up to the top of the sand piles can be determined by net torque balance around point O, as given by Eq. (1). Once the net torque is greater than zero, sand dunes are created in pipe, otherwise a continuous stationary bed is observed on the bottom wall. In this study, the liquid velocity, V_L, is accounted as the critical velocity when the net torque is zero.

$$T_{net} = L_D\left(F_D\right) - L_W\left(F_W\right) + L_{T,L,VDW}\left(F_T + F_L - F_{VDW}\right) \quad \text{.........................} \quad (1)$$

In the above equation, the variables of F_D, F_L, F_T, F_W and F_{VWD} are respectively drag, lift, turbulent, apparent weight gravity and Van der Waals forces. Table 2 shows the summary of the forces used for the Eq. (1).

The levers for the drag force (L_D), lift force (L_L), turbulent force (L_T), apparent weight force (L_W), and Van der Waals force (L_{VWD}) are calculated based on the equations provided in Table 3. According to Colver (1998), the angle of the pile front is 45°; and β is the repose angle of particles, which is assumed to be 60°.

Table 2. Summary of the Forces Acting on a Single Particle

Force	Equation	Reference
Drag Force	$F_D = \frac{1}{2} C_D \rho_L A_D V_{LL}^2$	Cimbala and Cengel (2008) Soepyan et al. (2016) Dabirian et al. (2016)
	$C_D = 0.2\left(\frac{\text{Re}_P}{10}\right)^{-1.72}$	Soepyan et al. (2016) Dabirian et al. (2016)
	$\text{Re}_P = \frac{\rho_P V_{LL} d_P}{\mu_L}$	
	$A_D = \left(\frac{\pi d_P^2}{4}\right)\left[0.5 + \frac{\beta}{180} + \frac{\sin(2\beta)}{2\pi}\right]$	Doron and Barnea (1993) Dabirian et al. (2016)
	$y^+ = V_{LL}^+ + a\left[e^{kV_{LL}^+} - 1 - kV_{LL}^+ - \frac{1}{2}\left(kV_{LL}^+\right)^2 - \frac{1}{6}\left(kV_{LL}^+\right)^3 - \frac{1}{24}\left(kV_{LL}^+\right)^4\right]$	Leif (1972) Dabirian et al. (2016)
	$y^+ = \frac{yV_\tau}{\upsilon_L}$	
	$V_{LL}^+ = \frac{V_{LL}}{V_\tau}$	
	$V_\tau = V_{LL}\sqrt{\frac{f}{8}}$	
Turbulent Force	$F_T = \begin{cases} 0, & \text{if} \quad \text{Re} < 2100 \ or \ d_p < \delta \\ \frac{1}{4}\rho_L V_\tau^2 A_P w, & otherwise \end{cases}$	Oroskar and Turian (1980) Soepyan et al. (2016) Dabirian et al. (2016)

	$\mathrm{Re} = \dfrac{\rho_L V_L d_L}{\mu_L}$	
	$\delta = \dfrac{5\mu}{\rho_L V_\tau}$	
	$w = \dfrac{2}{\sqrt{\pi}}\left[\dfrac{2}{\sqrt{\pi}}\gamma \exp\left(-\dfrac{4\gamma^2}{\pi}\right) + \int_{\gamma}^{\infty} \exp\left(-\dfrac{4\gamma^2}{\pi}\right) d\gamma\right]$	
	$\gamma = \dfrac{V_S(1-C_V)^2}{\upsilon_L}$	
	$V_S = \dfrac{g d_P^2(\rho_P - \rho_L)}{18\mu_L}$	
Lift Force	$F_L = 1.615 \dfrac{\mu_L V_{LL} d_P^2}{\upsilon_L^{0.5}}\left(\dfrac{dV_{LL}}{dy}\right)^{0.5} F(r^+)$	Saffman (1965) Wang et al. (1997) Ramadan et al. (2001) Dabirian et al. (2016)
	$\dfrac{dV_{LL}}{dy} = \dfrac{V_\tau^2}{\upsilon_L}\dfrac{dV_{LL}^+}{dy^+}$	Ramadan et al. (2001) Dabirian et al. (2016)
	$\dfrac{dV_{LL}^+}{dy^+} = \dfrac{1}{1 + a\left[k e^{kV_{LL}^+} - k - k^2 V_{LL}^+ - \dfrac{1}{2}k^3 V_{LL}^{+2} - \dfrac{1}{6}k^4 V_{LL}^{+3}\right]}$	Leif (1972) Dabirian et al. (2016)
	$F(r^+) = \dfrac{20.9(r^+)^{2.31}}{10^{1.962\log r^+ + 1.412}}$	Wang et al. (1997) Dabirian et al. (2016)
	$r^+ = \dfrac{d_P}{2}\left(\dfrac{\upsilon_L}{\dfrac{dV_{LL}}{dy}}\right)^{-0.5}$	
Van der Waals Force	$F_{VDW} = \left(\dfrac{A_H}{12}\right)\left(\dfrac{d_P}{S^2}\right)$	Cabrejos (1991), Soepyan et al. (2016) Dabirian et al. (2016)
Apparent Weight Force	$F_W = \dfrac{\pi d_P^3}{6}(\rho_P - \rho_L)$	

Table 3. Force Levers

Lever	Equation
Drag Force	$L_D = \frac{d_P}{2}\sin(\beta)$
Lift, Turbulent and Van der Waals Forces	$L_L = L_T = L_{VWD} = \frac{d_P}{2}\cos(\beta)$
Apparent Weight Force	$L_W = \frac{d_P}{2}\cos(\theta - \beta)$

Comparison Study

This section presents a comparison between the acquired experimental data and the predictions of the proposed model. Figures 5 and 6 present the collected experimental data and model predictions for particle size of 45-90 µm with V_{SL} = 0.05 m/s and 0.1 m/s, respectively. The symbols of purple diamond and yellow triangle represent the concentrated solids flow and moving dunes, respectively. As mentioned, the in-situ liquid velocity required for transition between the concentrated solid and moving dunes flow regimes is designated as the critical velocity.

As shown in both figures, for concentrations less than 2000 ppm, the experimental data exhibit a slight increase for the critical velocity with concentration, while the critical velocity doesn't change significantly for higher concentrations. The model predictions confirm that the predicted critical velocity is not affected significantly by the particle concentration, and constant values are predicted for the critical velocity over wide ranges of concentrations. This phenomenon is due to the turbulent force, which is the only force taking into account the effect of concentration. Oroskar and Turian (1980) derived the equation for the turbulent force under high particle concentrations, namely, > 20,000 ppm, whereby the force does not vary considerably at low particle concentrations, i.e., <10,000 ppm.

Comparison between model predictions and the acquired data show than the model over-predicts the critical velocity over the entire particle concentration range. Figure 7 and 8 show the comparison for particle size of 125-250 µm with V_{SL} = 0.05 m/s and 0.1 m/s, respectively. The data demonstrate that sand dunes occur at particle concentrations less than or equal to 2,000 ppm, while a continuous bed is observed for higher concentrations. Although the data show a slightly increasing critical velocity trend with concentrations, the model shows a flat trend for the critical velocity with particle concentrations. In general, a good agreement between the model predictions and acquired data is observed for particle size of 125-250 µm. The comparison between the model and experimental data confirms an averaged discrepancy less than 20% for all conditions.

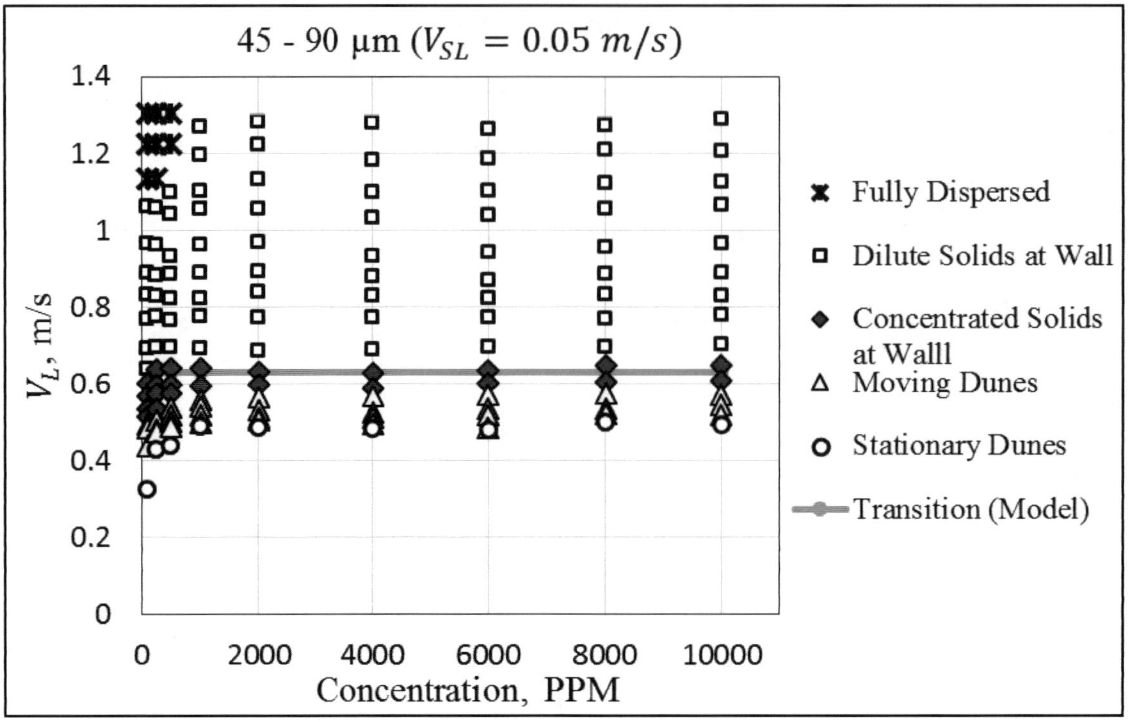

Figure 5. Experimental Data for V_{SL} = 0.05 m/s and Particle Size of 45 – 90 µm

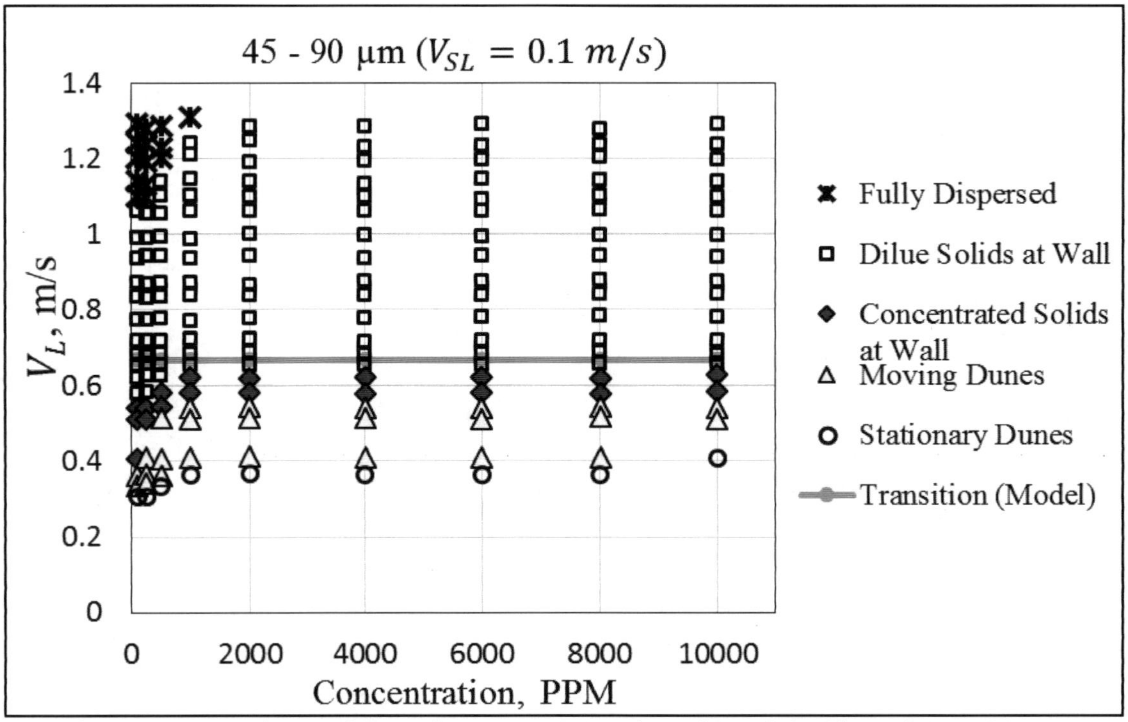

Figure 6. Experimental Data for V_{SL} = 0.1 m/s and Particle Size of 45 – 90 µm

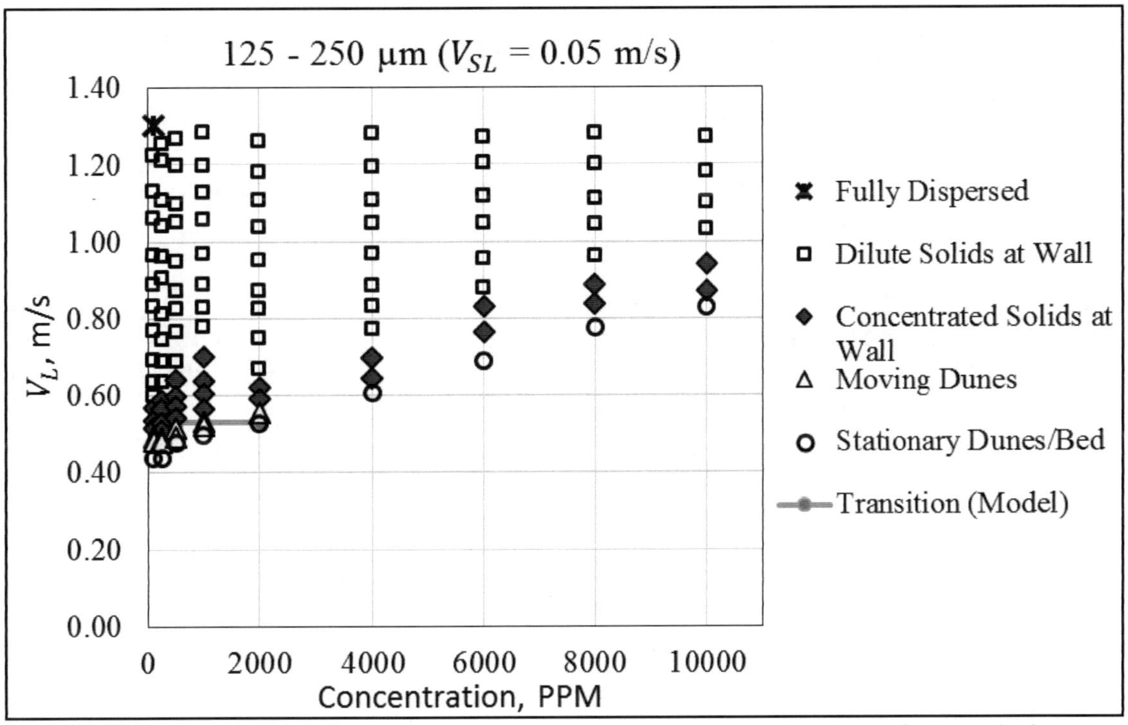

Figure 7. Experimental Data for V_{SL} = 0.05 m/s and Particle Size of 125 – 250 μm

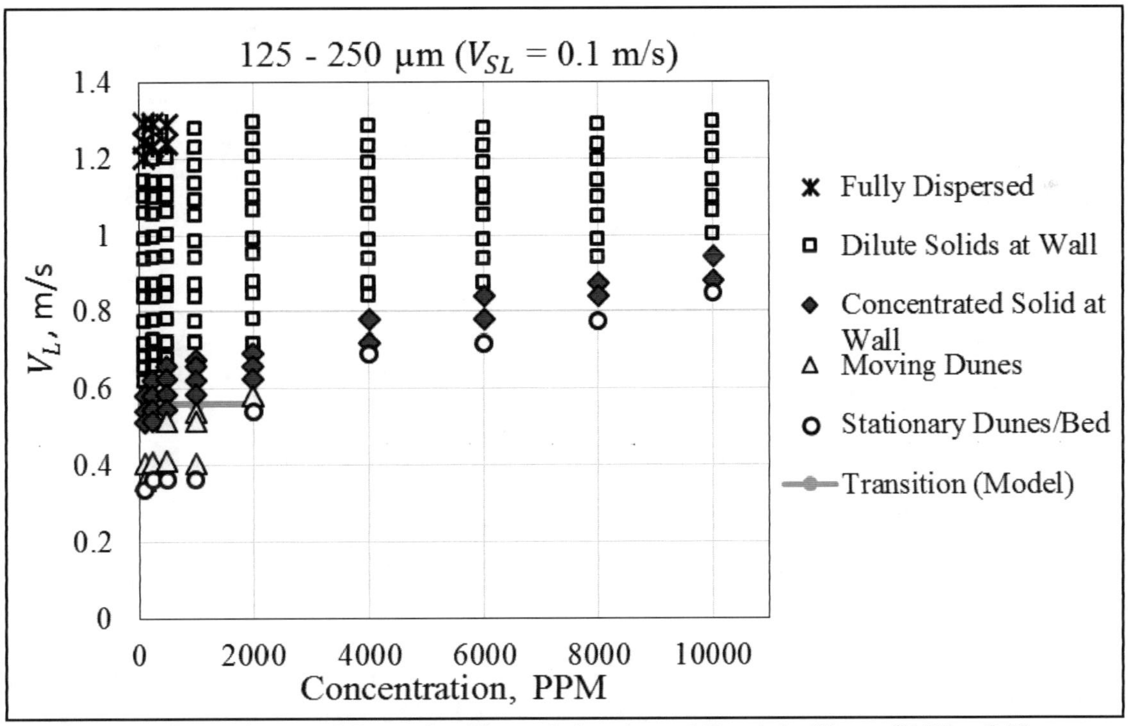

Figure 8. Experimental Data for V_{SL} = 0.1 m/s and Particle Size of 125 – 250 μm

Conclusions

Experimental data for sand transport in stratified flow under low particle concentrations, namely, <10,000 ppm, are acquired. The effect of various parameters, such as gas and liquid velocities, particle size and particle concentration, on the sand flow regimes is considered. The data show that sand dunes occur for particle size of 45-90 μm under various particle concentrations, while the sand dune regime is observed at low particle concentrations < 2000 ppm for particle size of 125-250 μm.

A mechanistic model based on rolling mechanism is developed to predict the transition boundary between the concentrated solid flow and moving dunes regimes. The model is derived based on a net torque balance on a particle located between sand piles. The results show that the model predictions are a weak function of particle concentration. The model predicts the critical velocity for 45-90 μm at particle concentrations greater than 2000 ppm with high accuracy. There is also a good match between the model and the experimental data for the 125-250 μm particle. In general, good agreement between the model predictions and experimental data confirm an averaged discrepancy less than 20% for all conditions.

Nomenclature

a	=	Leif Model Coefficient
A_D	=	Upper Area of Particle Projected on a Plane Normal, [m^2]]
A_H	=	Van Dar Waals Force Coefficient, [J]
A_P	=	Pipe Cross Sectional Area, [m^2]
C_D	=	Drag Coefficient, [-]
C_V	=	Particle Volume Concentration, [m^3/m^3]
D	=	Inside Pipe Diameter, [m]
d_P	=	Particle Diameter, [μm]
f	=	Friction Factor, [-]
F_D	=	Drag Force, [kg.m/$\boldsymbol{s^2}$]
F_f	=	Friction Force, [kg.m/$\boldsymbol{s^2}$]
F_L	=	Lift Force, [kg.m/$\boldsymbol{s^2}$]
F_T	=	Turbulent Force, [kg.m/$\boldsymbol{s^2}$]
F_{VDW}	=	van der Waals Force, [kg.m/$\boldsymbol{s^2}$]
F_W	=	Apparent Weight Force, [kg.m/$\boldsymbol{s^2}$]
$F(r^+)$	=	Factor for Modified Saffman's Equation, [-]
k	=	Leif Model Coefficient
L_D	=	Drag Force Lever, [$\boldsymbol{m}$]
L_L	=	Lift Force Lever, [$\boldsymbol{m}$]
L_T	=	Turbulent Force Lever, [$\boldsymbol{m}$]
L_{VDW}	=	van der Waals Force Lever, [$\boldsymbol{m}$]
L_W	=	Apparent Weight Force Lever, [$\boldsymbol{m}$]
N_{Re}	=	Modified Reynolds Number, [-]
r^+	=	Factor for Modified Saffman's Equation, [-]
Re_L	=	Liquid Phase Reynolds Number, [-]
Re_P	=	Particle Reynolds Number, [-]
S	=	van der Waals Force Coefficient, [m]
T_{net}	=	Net Torque, [**N. m**]
V_τ	=	Friction Velocity, [m/s]
V_L	=	Liquid Velocity, [m/s]
$V_{L,L}$	=	Local Liquid Velocity, [m/s]

$V^{+}_{L,L}$ = Dimensionless Local Liquid Velocity, [-]

V_{SG} = Superficial Gas Velocity, [m/s]

V_{SL} = Superficial Liquid Velocity, [m/s]

w = Factor assigned for Conversion of Turbulent Energy to Heat

y = Distance from Dune, [m]

y^{+} = Dimensionless Distance, [-]

Greek Letters

β = Particle Repose Angle, [degree]

θ = Angle of front Dune, [degree]

δ = Viscous Sublayer Thickness, [m]

μ_L = Liquid Viscosity, [kg/m.s]

ρ_L = Liquid Density, [kg/$\boldsymbol{m^3}$]

ρ_P = Sand Particle Density, [kg/$\boldsymbol{m^3}$]

υ_L = Liquid Kinematic Viscosity, [$\boldsymbol{m^2}$/s]

References

1. Angelsen, S., Kvernvold, O., Lingelem, M., and Olsen, S. (1989), "Long-Distance Transport of Unprocess HC Sand Settling in Multiphase Pipelines". In Proceedings of the Fourth International Conference on Multiphase Flow, 19-21.
2. Ayazi Shamlou, P. (1970), "Hydraulic Transport of Particulate Solids", Chem. Eng. Commun., 62, pp. 233-249.
3. Cabrejos, F.J. (1991), "Incipient Motion of Solid Particles in Horizontal Pneumatic Conveying". Pittsburgh, PA.
4. Cheremisinoff, N.P. (1986), "Encyclopedia of Fluid Fechanicas". Golf Pub Co.
5. Cimbala, J.M., Çengel, Y.A. (2008), Essentials of Fluid Mechanics: "Fundamentals and Applications". The McGraw-Hill Companies, Inc., New York, NY, USA.
6. Colver, T.J. (1989), Pocket Reference. Sequoia Publishing, Inc., Littleton, Colorado.
7. Dabirian, R., Mohan, R., Shoham, O., Kouba, G. (2016), "Critical Sand Deposition Velocity for Gas Liquid Stratified Flow in Horizontal Pipes". J. Nat. Gas Sci. Eng. 33.
8. Dabirian, R., Mohan, R. and Shoham, Ovadia, (2017), "Mechanistic Modeling of Critical Sand Deposition Velocity in Gas-Liquid Stratified Flow", Journal of Petroleum Science and Engineering, 156, 721-731.
9. Dabirian, R., Mohan, R., Shoham, Ovadia, and Kouba, Gene. (2016), "Solid Particles Flow Regimes in Air-Water Stratified Flow in a Horizontal Pipeline". Journal of SPE Oil and Gas Facilities, SPE-174960.
10. Dabirian, R. (2016), "Modeling and Experimental Investigation of Sand Transport in Gas-Liquid Stratified Flow". Ph.D. Dissertation. The University of Tulsa
11. Danielson, TJ. (2007), "Sand Transport Modeling in Multiphase Pipelines". In Offshore Technology Conference, Offshore Technology Conference, Paper No.18691.

12. Doron, P., Barnea, D. (1993), "A Three Layer Model for Solid Liquid Flow in Horizontal Pipes". Int. J. Multiph. Flow 19 (6), 1029–1043.
13. Gillies RG., McKibben, MJ., and Shook, CA. (1997), "Pipeline Flow of Gas, Liquid and Sand Mixtures at Low Velocities". Journal of Canadian Petroleum Technology, 9, 36, 36-42.
14. Hill, A.L. (2011), "Determining the Critical Flow Rates for Low Concentration Sand Transport in Two-Phase Pipe Flow by Experimentation and Modeling". M.Sc. Thesis, The University of Tulsa, Tulsa, OK.
15. Ibarra, R., Mohan, R. S., and Shoham O. (2014), "Critical Sand Deposition Velocity in Horizontal Stratified Flow". SPE International symposium and exhibition on formation damage control, Louisiana, USA.
16. Meyer-Peter, E. and Muller, R. (1948), "Formulas for Bed-Load Transport". In The 2nd Congress of the International Association for Hydraulic Research, 39- 64.
17. Oroskar, A.R., Turian, R.M. (1980), "The critical velocity in pipelines flow of slurries." AIChE J. 26 (4), 550–558.
18. Oudeman, P. (1993), "Sand Transport and Deposition in Horizontal Multiphase Trunklines of Subsea Satellite Developments". SPE Production & Facilities, 4, 8, 237-241.
19. Ramadan, A., Skalle, P., Johansen, S.T., Svein, J., Saasen, A. (2011), "Mechanistic Model for Cutting Removal from Solid bed in Inclined Channels". J. Pet. Sci. Eng. 30, 129–141.
20. Saffman, P.G. (1965), "The Lift on a Small Sphere in a Slow Shear Flow". Journal of Fluid Mechanics, 22, 2, 385- 400.
21. Salama, MM. (2000), "Sand Production Management". Journal of Energy Resources Technology, 122, 1, 29-33.
22. Stevenson, P., Thorpe, R.B., and Davidson, JF. (2000), "Incipient Motion of a Small Particle in the Viscous Boundary Layer at a Pipe Wall". Chemical Engineering Science, 57, 21, 4505-4520.
23. Stevenson, P., Thorpe, R. B., Kennedy, JE., and McDermott, C. (2001), "The Transport of Particles at Low Loading in Near-horizontal Pipes by Intermittent Flow". Chem. Eng. Sci, 56, 2149-2159.
24. Stevenson, P. and Thorpe, R. B. (2002), "Method Calculates Sand Velocity". Oil and Gas Journal, 47-50.
25. Stevenson, P. and Thorpe, R. B. (2002), "Velocity of Isolated Particles along a Pipe in Smooth Stratified Gas Liquid Flow". AIChE J., 5, 48, 963-969.
26. Stevenson, P., Thorpe, R. B., Kennedy, JE., and McDermott, C. (2001), "The Transport of Particles at Low Loading in Near-horizontal Pipes by Intermittent Flow". Chem. Eng. Sci, 56, 2149-2159.
27. Soepyan, F.B., Cremaschi, S., McLaury, B.S., Sarica, C., Subramani, H.J., Kouba, G.E., Gao, H. (2016), "Threshold Velocity to Initiate Particle Motion in Horizontal and Near Horizontal Conduits". Power Technol. 292, 272–289.
28. Wang, Q., Squires, K.D., Chen, M., McLaughlin, J.B. (1997), "On the Role of Lift Force in Turbulence Simulations of Particle Deposition". Int. J. Multiph. Flow 23 (4), 749–763.
29. Wicks, M. (1971), "Transport of Solids at Low Concentration in Horizontal Pipes". In Zandi, I., ed., Advances in Solid-Liquid Flow in Pipes and its Application. Pergamon Press, Pennsylvania.
30. Yan, W. (2010), "Sand Transport in Multiphase Pipelines". Ph.D. Dissertation, Cranfield University.

508449 A Meso-Scale Flow Model of Gas-Liquid-Solid Fluidized Beds with Acceleration Items

Abstract:

Mechanism modeling is an important method to quantify the flow heterogeneous phenomena of gas-liquid-solid fluidization systems. In this study, a meso-scale flow model with bubble and solid particle acceleration items was developed based on the energy minimum multi-scale theory (EMMS) to predict the global flow parameters of gas-liquid-solid fluidized beds. The consideration of acceleration item makes the model be capable of accurately predicting both the nominal-steady and the unsteady hydrodynamics of the gas-liquid-solid fluidization. Compared with the previously reported EMMS models of gas-liquid-solid fluidization in the literature, the prediction results of this new meso-scale flow model are closer to the experiment data. Furthermore, the predictions indicate that both bubble acceleration and particle acceleration have an influence on the flow behavior of gas-liquid-solid fluidized bed, such as bubble diameter and relative slip velocity, which seriously affect the momentum and mass transfer between phases. In addition, the acceleration behaviors of the unsteady-state give a reasonable explanation for bubble coalescence.

508396 Flow Regimes in Gas-Liquid-Solid Mini-Fluidized Beds with Single Gas Orifice

Abstract:

Flow regimes and their transitions in the gas-liquid-solid mini-fluidized beds (MFBs) with single gas orifice were studied experimentally in this paper. The diameter of the fluidized beds varied from 3mm to 5 mm and the bed height was 50 mm. The ranges of superficial gas and liquid velocity were 1.96×10^{-4} - 4.73×10^{-3} and 5×10^{-6}-4.2×10^{-2} m/s, respectively. Half-fluidization, slug, dispersed bubble and transport flow regimes in the three-phase MFBs were identified by analyzing minimum fluidization velocity, pressure drop, and entrainment velocity of solid particles based on experimental data of pressure drop and fluidization velocity as well as flow observations. The effects of solid particle and liquid properties, bed and gas orifice sizes, and static bed height on the flow regimes and transitions were investigated. Results show that the behavior of gas bubbles and wall effect affect the flow regimes and transitions. Obvious wall effect increases minimum fluidization liquid velocity in the liquid-solid mini-fluidized beds, which increases the transition liquid velocity between the half-fluidization and slug flow regimes. Moreover, wall effect makes Taylor bubble bigger at lower superficial liquid velocities due to the bubble coalescence and the transition boundaries expand. Solid particles aggregation, wall effect and bubble wake behavior are responsible for reduction of the minimum entrainment velocity of solid particles. Flow regime maps of the gas-liquid-solid MFBs were presented. Correlations were suggested for the flow regime boundaries.

509634 Systematic Investigation of Segregated Layer Inversion in Liquid-Fluidized Binary Beds

Abstract:

Liquid-fluidized beds of two or more components are encountered in a variety of industrial processes, such as mineral ore dressing, sorting/classification, coal beneficiation, separation in plastic waste recycling and in some biochemical reactors. Depending on the application, segregation may be a desirable or undesirable feature in the process. In some cases, size-driven or density-driven segregation occurs separately or one of them strongly prevails over the other, making predictions obtained using simple models rather reliable. When the solids to separate exhibit contrasting differences in size and density (like in many examples cited above), even the *direction of segregation* is not easily predictable, i.e. which one component tends to separate towards the surface and which one to sink to the bottom, giving rise for some systems to the so-called â €œlayer inversion phenomenon.â€

In the present contribution, the reliability of a recently introduced model for predicting the segregation direction of fluidized binary beds (the Particle Segregation Model, PSM) is examined by systematic experimental investigation of binary beds with different size and density ratios and at different bed composition. The experimentally obtained surface concentration, expansion properties and the layer inversion voidage are compared with the PSM predictions, offering a direct means of model validation.

506169 Intrusive Probes in Riser Applications

Abstract:

Many of the probes used to understand hydrodynamics in circulating fluidized bed risers intrude into the environment they are measuring, although assumptions are typically asserted that the intrusive probes do not affect the data collected. This could be a poor assumption in some cases and conditions. We found that intrusive fiber optic probe measurements consistently mis-predicted the solids concentration compared to the non-intrusive pressure drop measurements outside the fully developed flow region of a riser containing fluid catalytic cracking catalyst or glass bead particles. The discrepancy was sensitive to superficial gas velocity, solid circulation rate, probe position and flow direction. Barracuda VRâ„¢ computational fluid dynamics simulations confirmed this, and indicated that particle momentum was lost at the leading edge of the probe and particles were spilling over to the probe tip. Accordingly, new probe designs were proposed to mitigate the intrusiveness of a fiber optical probe for more accurate characterization.

508342 Detection of Fluidization with Real-Time Magnetic Resonance Imaging

Abstract:

Unfluidized areas (dead zones) within gas-solid fluidized beds can deteriorate the mixing and heat transfer between the gas phase and the solid phase, and thus lower the performance, homogeneity and efficiency of the desired process. A thorough understanding of the conditions that lead to the formation of dead zones is therefore key in order to design, operate and scale fluidized bed reactors reliably. The experimental detection of dead zones inside three-dimensional (3D) fluidized beds can provide valuable information to validate theories and numerical models. However, due to the opacity of most granular materials, such measurements are challenging and require the use of intrusive probes[1,2] or non-intrusive tomographic techniques, e.g. x-ray tomography[3], electric capacitance tomography (ECT), positron emission particle tracking (PEPT)[4] or magnetic resonance imaging (MRI)[5]. While intrusive probes perturb the granular dynamics, the tomographic techniques suffer from limitations in either their ability to measure particle motion (x-ray CT, ECT, PEPT) or from poor temporal resolution (MRI)[6].

Here we present a MRI methodology that can discriminate between fluidized and unfluidized areas in 3D beds of considerable size (cylindrical bed; diameter 190 mm and height 300 mm) at a frame rate of 20 Hz. The high temporal resolution of this technique allows studying not only the presence of dead zones but also their formation and disintegration. The presented measurement technique bases on a recent development in real-time probing of granular dynamics with MRI which uses array detection of the magnetic resonance and engineered granular signal sources[7]. It exploits both spin saturation effects and diffusion attenuation to create a strong dependency of the signal on particle motion.

We anticipate the presented technique to be a starting point for more sophisticated studies on the formation of dead zones in fluidized beds, such as determining the influence of internals, the distributor or bed geometry. Moreover, we envisage that the technique is suited to provide fundamental physical insights on the nature of fluidization, its relations to jamming and the glass transition[8].

References

1. Geldart, D. & Kelsey, J. R. The use of capacitance probes in gas fluidised beds. *Powder Technol.* **6,** 45–50 (1972).

2. Werther, J. & Molerus, O. The local structure of gas fluidized beds -II. The spatial distribution of bubbles. *Int. J. Multiph. Flow* **1,** 123–138 (1973).

3. Mudde, R. F. Double X-ray Tomography of a Bubbling Fluidized Bed. *Ind. Eng. Chem. Res.* **49,** 5061–5065 (2010).

4. Wildman, R. D., Huntley, J. M., Hansen, J. P., Parker, D. J. & Allen, D. A. Single-particle motion in three-dimensional vibrofluidized granular beds. *Phys. Rev. E* **62,** 3826–3835 (2000).

5. Dennis, J. S. Magnetic Resonance Studies of Dead-Zones in Gas-Solid Fluidised Beds. (2013).

6. Pore, M. *et al.* Magnetic resonance studies of a gas-solids fluidised bed: Jet-jet and jet-wall interactions. *Particuology* **8,** 617–622 (2010).

7. Penn, A. *et al.* Real-time probing of granular dynamics with magnetic resonance. *Sci. Adv.* Accepted, (2017).

8. Goldman, D. I. & Swinney, H. L. Signatures of glass formation in a fluidized bed of hard spheres. *Phys. Rev. Lett.* **96,** 145702 (2006).

507536 Solids Mixing Studies in Fluidized Beds Using Fluorescent Tracer Technique

Abstract:

A fluorescent solid tracer technique has been used to study various properties such as solids mixing time, solids residence distribution time and the effects of internals on mixing in fluidized beds. Studying these parameters is of great importance in controlling the back-mixing of solids or verifying that solids segregate in a fluidized bed. The technique may also be used to compare the flow patterns of different solids, understand the hydrodynamics associated with different internals at different operating conditions, and to optimize the size and the shape of a reactor. In these sets of experiments, FCC catalyst was coated with a suitable UV dye in a 12-in (30-cm)-diameter spray column using a suitable solvent to dissolve the dye. The dyed catalyst was then injected into a 6-in (15-cm)-diameter Plexiglas^{TM} column fluidized at superficial gas velocities of 1 ft/s (0.31 m/s) and 0.25 ft/s (0.08 m/s) to analyze mixing of FCC equilibrium catalyst. Fluorescent probes, which were inserted at different axial and radial locations in the bed, captured the mixing phenomena of the catalyst in the bed. A photodiode converted an optical signal to an electrical signal and the data were recorded using a data acquisition system and was analyzed to study the mixing behavior of particulate solids in a small scale fluidized bed.

508410 *Stereoscopic PIV Study on Swirl Phenomena in Uniflow Cyclones*

Abstract:

Understanding the complex flow field in a uniflow cyclone is important for further design development. The uniflow cyclone is advantageous in certain applications due to the compactness e.g. in space-limited environments in engineering equipment and will be therefore investigated. This separator type uses centrifugal forces induced by a swirl vane inlet to separate particles from a continuous phase. The resulting flow field is characterized by a strong circumferential velocity component, which forces the fluid to rotate in downstream direction.

Stereo-PIV is used to evaluate the flow field along the cross section and to gain insights into the flow field characteristics [1]. 3C/2D-PIV is is applied to evaluate all three velocity components simultaneously in the measurement plane. Therefore the cameras and the laser optics are mounted on an 1D traversing mechanism for precise positioning of the optical PIV system in the measurement volume along the axis of the main flow. This study mainly focuses attention on detailed measurements on the flow field in the separation chamber of the uniflow cyclone to find the best geometric parameters by means of aerodynamic aspects [2,3]. For validation a set of dust-laden experiments are carried out to get information about separation efficiency and pressure drop (ISO 5011) [4]. As a result the effect on cyclone performance has been quantified with a consistent set of experimental data which shows the influence of vane angle and core diameter ratio variations.

[1] Pillei, M., Goller, R., Kofler, T., Wierschem, A., and Kraxner, M., “Experimental Investigation of the Velocity Distribution near the Swirl Generator of a Uniflow Cyclone for Performance Data Prediction,” In *18th International Symposium on the Application of Laser and Imaging Techniques to Fluid Mechanics*.

[2] Chang, F. and Dhir, V. K., “Turbulent flow field in tangentially injected swirl flows in tubes,” International Journal of Heat and Fluid Flow, vol. 15, no. 5, pp. 346–356, 1994

[3] Pillei, M.; Kofler, T.; Kraxner, M.; Fluid flow investigation in the separation chamber of a uniflow cyclone with 3C-PIV and CFD; Fachtagung “Lasermethoden in der Strömungsmesstechnik”; GALA Karlsruhe, 2014.

[4] ISO 5011. Inlet air cleaning equipment for internal combustion engines and compressors - Performance testing.

507396 Riser Dynamics â€“Â a Comparison of Scale

Abstract:

Risers in gas-solids circulating systems are used extensively in the petrochemical and energy industries as well as being proposed for advanced technologies from chemical looping combustion/gasification. The performance of these units has been linked to the meso-scale and the dynamic behavior of the flow field. New experimental data has been obtained in a 0.1 m diameter riser with a 5.3 m height. The particle is an 850 μm high density polyethylene with a density of 860 kg/m^2. Experiments were conducted primarily in the core-annular regime. In addition to standard differential pressure measurements along the riser, three-dimensional measurements of the solids volume fraction were obtained using a Tech4Image Electrical Capacitance Volume Tomography (ECVT) system. The dynamic structure of the riser solids fraction is obtained by analyzing the time series data for each voxel in the flow field. Specifically, the data were analyzed using higher level statistics (skewness and kurtosis) as well as chaotic correlation dimension and the entropy. These values are compared with published ECVT data using the same material from a 0.3 m diameter riser with a 16 m height. The effect of system scale is analyzed and discussed.

507361 Detailed Numerical Simulation of a Full-Loop Circulating Fluidized Bed Under Different Operating Conditions

Abstract:

Detailed numerical simulation of a full-loop circulating fluidized bed under different operating conditions

Yupeng Xu[1,*], Jordan Musser[1], Tingwen Li[1,2], William A. Rogers[1]

1 National Energy Technology Laboratory, Morgantown, WV 26505, USA

2 AECOM, Morgantown, WV 26505, USA

Abstract

Circulating fluidized beds (CFB) are widely employed in industry for a wide variety of gas-solid contacting operations. Over the past few decades, CFB hydrodynamics have been studied using various experimental techniques and by numerical simulations. Recently, several efforts have modeled full-loop CFBs using the two-fluid model (TFM) whereby gas and solids are treated as interpenetrating continua. However, this technique only resolves solid motion at the computational cell level. With the advancement of high-performance computing, higher fidelity models, like Computational Fluid Dynamics-Discrete Element Method (CFD-DEM) which tracks the trajectory of individual particles, have become increasingly popular as research tools. This is because particle-scale information including residence time, collision forces, and dispersion intensities are readily available for detailed analysis of complex flow phenomena.

In this first of its kind work, both experimental and computational studies of a benchtop, full-loop circulating fluidized bed are conducted. Experimental data include pressure drops across different components, solids circulation rate measured using Particle Image Velocimetry (PIV), and standpipe inventory height. Numerical simulations were performed with MFIX-DEM, a coupled Computational Fluid Dynamics-Discrete Element Method (CFD-DEM) solver. Comparison of experimental and simulation data shows good agreement with respect to system component pressure drops and standpipe inventory height. Additionally, the effects of drag model selection and particle representation (mono-sized particles versus a particle size distribution) are also reported.

Figure 1. Experimental setup and operation conditions (SLPM)

Figure 2. Direct comparison between experiments and simulations under different operating conditions (Left: experiments, right: simulation)

508046 Generating Macroscopic Quantities of Particle-Fluid Flows of Coarse Particles By an Averaging Method: Selection of Sample Sizes and Test of Solid Pressure Correlation

Abstract:

Generating macroscopic quantities of particle-fluid flows of coarse particles by an averaging method: Selection of sample sizes and test of solid pressure correlation

Q.F. Hou[1], Z. Y. Zhou[1], J. S. Curtis[2], and A.B. Yu[1,3]

[1]ARC Research Hub for Computational Particle Technology, Department of Chemical Engineering, Monash University, Clayton, VIC 3800, Australia

[2]College of Engineering, University of California, Davis, One Shields Ave, Davis, CA 95616, USA

[3]Centre for Simulation and Modelling of Particulate Systems, Southeast University - Monash University Joint Research Institute, Suzhou 215123, PR China

Abstract

Macroscopic quantities of particle-fluid flow such as solid density and stress tensor can be obtained from microscopic variables such as particle positions and interaction forces through averaging methods. In using average methods, spatial and temporal sample sizes play an important role. However, how to properly select sample sizes is not clearly understood. Furthermore, the validity of the obtained correlations using the obtained macroscopic quantities has not been well examined. By using the combined approach of computational fluid dynamics and discrete element method, this work is attempted to give answers to these two concerns. First, the effect of spatial and temporal sample sizes on averaged macroscopic quantities is discussed. Guidance for the selection of sample sizes is provided to generate macroscopic quantities which are compatible with continuum mechanics. Then the correlation between solid pressure and solid fraction is established and compared to the correlations available in the literature. It agrees quantitatively with experimental observations and semi-empirical correlations but agrees only qualitatively with kinetic theory predictions. The evidence proves that the averaging method is an effective approach in linking microscopic and macroscopic properties of particle-fluid flows. If the sample sizes are selected properly valid correlations can be formulated for continuum modeling.

Key words: fluidized beds, granular media, particle-fluid flows

508263 Structured Flow in Gas-Solid Fluidized Beds: Particle Clustering and Bubble Self-Organisation

Abstract:

Fluidized beds are widely used to put in contact solid and gas phases in a variety of industrial processes. The interaction force between the gas and the suspended solids and the energy dissipated in particle collisions originate complex flow features that are dependent on the solids properties e.g. spouting, channelling, jetting or bubbling. Despite the advance in design and control, multi-phase flow models still struggle to describe many of these structures and scale up requires from a significant amount of historical practice and experimentation. The performance of a bubbling fluidized bed for instance is heavily reliant on a good level of mixing, associated to the observed solid circulation and the bubble characteristics e.g. size, velocity. From a fundamental perspective, the formation of bubbles can be explained by the instability of the coupled momentum equations or any similar phenomenological model used to study the bed response and the pressure fluctuations associated to nucleation, coalescence and rupture. While this type of analysis allows the formulation of a theoretical framework [1], important effects such as particle polydispersity, segregation, particle inertia and shape remain poorly understood. Furthermore, hydrodynamic models leave open questions regarding the behaviour of dense granular systems where multiple particle contacts and local anisotropy become important in the transport of momentum [2]. This constitutes the general pitfall of numerical frameworks based in the kinetic theory of granular flow. While they remain the only alternative to support the design of full scale units, most kinetic models rely on a microscale description of particle interactions that is inconsistent with the complex granular rheology of granular media in the dense and quasi-static regimes [3].

Allowing for more degrees of freedom at the design stage is a promising way to introduce a higher level of control in the operation of bubbling fluidized beds. In recent years, several alternatives have been put forward to manipulate the time and spatial scales of particle and particle-fluid interactions at a micro- and mesoscale in order to render a more reproducible flow structure in the macroscopic system. In addition to other methods based in modifying the spatial distribution of forces in the bed e.g. particle-particle applying electric fields or particle-fluid using fractal injectors [4], the pulsation of the inlet gas has been identified as a promising candidate to structure the flow in bubbling beds. Pulsating the gas flow introduces a temporal dynamic in the drag force that can supress the natural instability of the gas-solid flow system. Ripples in sandy beaches or dunes in deserts are good examples of dynamic granular structures that emerge as a response to a periodic perturbation in the flow around them, e.g., tides and waves, wind gusts and eddies. Transferring the same principle into the operation of quasi-2D fluidised bed, we have demonstrated that under certain conditions it is possible for gas bubbles to reorganise into a regular pattern that propagates throughout the bed into an stable large scale flow structure [5]. Under these operating conditions one has a much tighter control over the resulting bubble size, velocity and separation, thus opening the way for an entirely new concept in unit control and design.

This contribution investigates how dynamic bubble flows emerge in the fluidization of Geldart B particles under a pulsed airflow. At certain conditions, bubbles rearrange into a triangular tessellation with a narrow bubble size and wavelength distribution, independent of the bed width. We discuss the role that the dissipation of energy through friction plays in the bubble nucleation mechanism and how the local changes in rheology can stabilise the bubble pattern. We report a series of experimental data in quasi-2D pulsating beds along numerical studies under different airflow conditions, inter-particle friction factors and bed dimensions. The transition from the formation of a surface waves in a shallow layer to the nucleation of bubbles in thicker beds is related to the analysis of large clustered regions appearing in the wake and around pairs of rising bubbles. A comparison between continuous and discrete numerical formulations given by Eulerian-Eulerian (TFM) and Eulerian-Lagrangian (CFD-DEM) frames is used to showcase the limitations of classical kinetic and rheology models in predicting the behaviour of the granular flow in a dense and plastic regime. The inherent difficulty of a kinetic model to deal with correlated particle velocities and the resulting anisotropic distribution of contacts can explain why the common closures for the solid stress fail to capture how bubble self-organise under dense conditions[5]. The explicit solution of the rheology through an Eulerian-Lagrangian frame however succeeds in tracking the local changes in granular rheology and is shown to reproduce the observed bubble dynamics [6].

References

[1] Batchelor G.K. (1993). Secondary instability of a gas-fluidized bed, J. Fluid Mech. 251, 359-311.

[2] Chialvo S., Sun J., Sundaresan S. (2012). Bridging the rheology of granular flows in three regimes Physical Review E 85, 021305 1-8.

[3] Chialvo S., Sundaresan S. (2013) A modified kinetic theory for frictional granular flows in dense and dilute regimes. Physics of Fluids 25, 070603.

[4] Coppens M-O, Ommen J.R, (2003). Structuring chaotic fluidized beds, Chem. Eng. J. 96 117-124.

[5] Wu K, De Martin L., Mazzei L., Coppens M-O. (2016). Pattern formation in fluidized beds as a tool for model validation: A two-fluid model based study. Powder Tech., 295, 35-42

[6] Wu K, De Martin L., Coppens M-O. Pattern formation in pulsed gas-solid fluidized beds â€" The role of granular solid mechanics. Chemical Engineering Journal (2017). In Press. https://doi.org/10.1016/j.cej.2017.05.152

509348 Studies on the Mechanism and Regulation-Control Method of a High Density Gas-Solids Circulating Fluidized Bed Downer

Abstract:

The efficient conversion and optimal utilization of heavy oil has great practical significance to the development of national economy. Fluid catalytic cracking (FCC) is a critical approach to deal with heavy oil. Riser, as the key part of the FCC unit, is characterized by the non-uniform â€œcore-annualsâ€ flow structure with relatively significant gas and solids backmixing which may result in low reactant conversion and product selectivity. Downer reactor, in which gas and solids move downward co-currently, has unique features such as the plug-flow reactor performance and relatively uniform gas-solids distribution compared to the riser reactor. Downer is therefore acknowledged as a novel multiphase flow reactor with great potential in dealing with heavy oil. However, low solids holdup and poor initial gas-solids contacting hinder the promotion and application of downer reactor. Mechanism of increasing solids holdup and approaching high density downer have not been well understood until now. This study starts from the entrance of the downer to in-deep investigate its effects on high density operating condition and to study the contributions of the downer entrance to accomplishing high solids holdup and good gas-solids contacting and in downer reactor. On the other hand, all parts of the entire circulation system have been optimized in order to develop a regulation and control method to obtain and maintain a stable operating of the high density downer. Based on the above work, the mechanism of achieving the high density downer has been demonstrated, a multi-scale computational model has also been established to preciously describe the hydrodynamic characteristics and the scale-up of high density downer has been illustrated all of which can provide the fundamental data and basic theoretical support to develop a technology of downer catalytic cracking of heavy oils and extend the applications of high density downer reactors.

508625 Measurement and Imaging of Gas-Solid Systems Using Electrical Capacitance Volume Tomography

Abstract:

Electrical Capacitance Volume Tomography (ECVT) is a non-invasive imaging technology recently developed for multi-phase flow applications. It is based on distributing flexible capacitance plates on the peripheral of a flow column and collecting real-time measurements of inter-electrode capacitances. Capacitance measurements here are directly related to dielectric constant distribution, a physical property that is also related to material distribution in the imaging domain. Reconstruction algorithms are employed to map volume images of dielectric distribution in the imaging domain, which is in turn related to phase distribution. ECVT is suitable for imaging interacting materials of different dielectric constants, typical in multi-phase flow systems.

ECVT is being used extensively for measuring flow variables in various gas-solid flow systems. Recent application of ECVT include flows in risers and exit regions of circulating fluidized beds, gas-solid bubble columns, trickle beds, moving beds, and slurry bubble columns. ECVT is also used to validate flow models and CFD simulations. The technology is uniquely qualified for imaging phase concentrations in various energy applications as it exhibits favorable features of compact size, low profile sensors, high imaging speed, and flexibility to fit around columns of various shapes and sizes. ECVT is also safer than other commonly used imaging modalities as it operates in the range of low frequencies (~1 MHz) and does not radiate radioactive energy.

High temperature and pressure gas-solid reactors are widely used in industry. Deep understanding of flow behavior in such reactors is limited due to the complex hydrodynamics that exist in the bed. In this effort, ECVT is used to study flow behavior in various gas-solid systems and at different conditions. Case studies that involve bubbling beds, moving beds, and slugging at temperatures up to 900C will be presented. Real-time images and quantitative data acquired in those experiments provide a very unique insight into gas-solid systems at harsh conditions.

508341 Material Distribution of Gas–Solid Fluidized Beds Measured By Electrical Capacitance Tomography – a Big-Data Perspective

Abstract:

Material Distribution of Gas-Solid Fluidized Beds Measured by Electrical Capacitance Tomography - a Big-Data Perspective

Gas-solid fluidized beds are used in many industrial processes, such as coal gasification, power generation, granulation, and polymerization. To measure hydrodynamic characteristics of gas-solid fluidized beds, numerous intrusive and non-intrusive experimental techniques have been developed. The intrusive methods like optical probe, capacitance probe, and pressure measurement are easy to implement but only capable of providing local information concerning the fluid flow. While the non-intrusive techniques such as tomography can be used to visualize the entire flow field without causing any disturbance to the flow. Compared to other industrial process tomography techniques, Electrical Capacitance Tomography (ECT) shows advantages in terms of fast imaging speed, no radiation, robustness, and low cost. Moreover, considering the non-conductive nature of materials in gas-solid fluidized beds, ECT is a suitable measuring technique for hydrodynamic investigation of these reactors.[1]

In ECT, the sensing electronics measure variations in capacitance between pairs of electrodes, which are placed around the periphery of a pipe or vessel under investigation. These measurements are then used to reconstruct the cross-sectional permittivity distribution as a presentation of the material distribution inside the sensing area via a specific image reconstruction algorithm. However, two major difficulties are associated with the reconstruction process. First, the number of measured independent capacitance data is far less than the number of unknown image pixels, and therefore the problem is severely under-determined. Secondly, due to the ill-posed and ill-conditioned property of the reconstruction process, the reconstructed results are sensitive to raw capacitance measurement noise. To address these problems, different reconstruction techniques, including single-step and iterative algorithms, have been proposed and have been reviewed by Yang and Peng[2] and Cui et al.[3]

Although big progress has been made in the image reconstruction with ECT in the last two decades to improve the image quality, i.e., the accuracy of material distribution measurements, the correlation coefficient between the true image and reconstructed image by a certain technique is usually lower than 0.9. This apparently hiders ECT as an excellent technique for gas-solid fluidized beds measurements, especially when a direct comparison between ECT measurements and Computational Fluid Dynamics (CFD) simulations is needed, in which high-accuracy images are necessary.

In the last several years, big-data analysis has gained considerable attention and momentum in both academia and industry, and also see some applications in the field of chemical engineering.[4] The dynamic evaluation method, which we have proposed recently,[5] provides a way for image reconstruction with ECT using a big-data perspective, and is potential to improve the image quality greatly.

To meet different needs, the big-data image reconstruction method is categorized to two schemes: an on-line scheme and an off-line scheme. The procedure of the on-line scheme is as follows. First, CFD simulations of a gas-solid fluidized bed, which is the same as that used in experiments, are performed. Next, a generic sensitivity matrix is obtained from a probability distribution analysis carried out on the material distributions extracted from the CFD simulation results. Finally, material distributions in experiments can be obtained using the modified Tikhonov regularization method equipped with the generic sensitivity matrix. While the off-line scheme is due to fingerprint matching the normalized capacitance between CFD simulations and experiments. As an illustration, Figure 1 shows some typical results obtained by different image reconstruction algorithms and the second-scheme big-data method. As can be seen, the correlation coefficients obtained by the existing image reconstruction algorithms are all lower than 0.9, also some details especially on the boundary between bubble and emulsion phases are blurred. While the best image can be obtained by the big-data method with the correlation coefficient as high as 0.95.

Figure 1. Material distribution reconstructed by different image reconstruction algorithms and the second-scheme big-data method.

Literature Cited

1. Rasel RK, Zuccarelli CE, Marashdeh QM, Fan L-S, Teixeira FL. Towards multiphase flow decomposition based on electrical capacitance tomography sensors. *IEEE Sens J.* 2017;in press.

2. Yang WQ, Peng LH. Image reconstruction algorithms for electrical capacitance tomography. *Meas Sci Technol.* 2003;14:R1-R13.

3. Cui ZQ, Wang Q, Xue Q, Fan WR, Zhang LL, Cao Z, Sun BY, Wang HX, Yang WQ. A review on image reconstruction algorithms for electrical capacitance/resistance tomography. *Sens Rev.* 2016;36:429-445.

4. Qin SJ. Process data analytics in the era of big data. *AIChE J.* 2014;60:3092-3100.

5. Guo Q, Meng SH, Wang DH, Zhao YF, Ye M, Yang WQ, Liu ZM. Investigation of gas-solid bubbling fluidized beds using ECT with a modified Tikhonov regularization technique. *AIChE J.* 2017;in press.

507964 Heat Transfer in a Pressurized Fluidized Bed with Continuous Addition of Fines

Abstract:

Oxygen-fired pressurized fluidized bed combustor technology with downstream carbon capture and sequestration is considered a key approach to clean coal combustion. In such technology, the fluidized bed combustor operates at elevated pressures and houses an in-bed heat exchanger tube bundle. The goal of this work was to investigate the impact of pressure and presence of fine particles (i.e., surrogate for pulverised fuel) on the overall tube-to-bed heat transfer coefficient. Experiments were conducted in a pilot-scale fluidized bed with an inner diameter of 0.15 m under cold flow conditions. A tube bundle consisted of five horizontal staggered rows was completely submerged in the bed (simulating the combustor in-bed heat exchanger tubes). One of the tubes was replaced by a 900 W heating cartridge housed in a hollowed copper rod. Five thermocouples distributed at 45° intervals along the copper rod circumference measured the surface temperature and ensured that local effects were included. Through supplying DC power and monitoring bed and copper rod surface temperature, the overall convective heat transfer coefficient was determined. The bed material was large glass beads of 1.0 mm in diameter while the fines were glass beads of 60 μm in diameter and thus susceptible to entrainment. The fine particles were continuously fed to the fluidized bed over a designed period of time, and then captured downstream by a filter system. Fluidization was conducted at 101, 600 and 1200 kPa with excess gas velocities (U_g - U_{mf}) of 0.21, 0.29 and 0.51 m/s. Fine particle feed rates were 0, 9.5 and 14.4 kg/h. The results showed that average heat transfer coefficients matched the correlation developed by Molerus et al. (1995) within a 5% difference across all conditions when fines were not present. At atmospheric conditions, where slug flow regime occurred, the maximum heat transfer coefficient was at the bottom of the rod, while it moved to the side of the rod at high pressures where the bubbling regime occurred. Overall, the heat transfer coefficient doubled when pressure was increased from 101 to 1200 kPa. The addition of fine particles decreased the average overall heat transfer coefficient by 3-9%, but there was no effect on the angular profile across the tube surface. The fines likely reduced the turbulence adjacent to the tube surface negatively impacting heat transfer via gas convection, which is the primary mechanism.

References

Molerus, O., and Mattmann, W. (1992). Heat transfer mechanisms in gas fluidized beds. Part 3: Heat transfer in circulating fluidzed beds. Chemical Engineering & Technology, 15 (5), 291-294.

508614 Investigation of the Effect of Fluidization on Alkane Catalytic Cracking in Fluidized Bed Reactor By High Temperature ECT Measurement System

Abstract:

Ethylene and propylene are important chemical raw materials. Alkane cracking is one of the major rountes for olefins production. At present, thermal cracking of alkane has been widely used in commercial olefins plant, which normally requires reaction temperature higher than 800 °C, and is considered as one of the most energy consuming processes in chemical industrial sector. Catalytic cracking technology can effectively reduce the alkane reaction temperature and on-purpose produce light olefins. Because of the rapid deactivation of catalyst caused by coke deposition, fluidized bed has been regarded as the well suitable reactor for catalytic cracking. However, the catalyst fluidization in fluidized bed reactor could significantly alter the reaction performance. Own to the lack of efficient high temperature measurement technology for fluidized bed, the influence of fluidization on the alkane conversion and product selectivit under real reaction conditions has not been understood yet.

In our laboratory a high temperature ECT system has been developed for fluidized bed measurment. The high temperature ECT system includes a specially designed high temperature electrode (working temperature above 600°C) and the corresponding algorithm. By use of the high temperature ECT measurement system, we monitored the catalytic cracking reaction of alkanes and investigated the influence of fluidization conditions on the reaction performance.

508024 Turbulent Closure Models for Multiphase Fluids

Abstract:

Turbulent multiphase flows are encountered ubiquitously in the process industry. Examples include flows within hydrocyclones and fluidized bed reactors. The potential for simulating these flows has primarily occured because of advancements in computational hardwarre and software. However, current deficiencies in the turbulences models for mixing seriously limit the practical utility of computational methods as an enabling technology. The proposed presentation will extend a recently developed realizable closure for the Reynolds stress of a single phase fluid to a mixture of immiscible fluids. Most significantly, the theory guarantees realizable behavior of the multiphase Reynolds stress for all turbulent flows in inertial frames and in non-inertial frames.

A CFD-DEM Study of Bubble Properties in Fluidzed Beds Operated with a Single Jet

Siddhartha Shrestha, Jieqing Gan, Aibing Yu and Zongyan Zhou*

Laboratory for Simulation and Modelling of Particulate Systems, Department of Chemical Engineering, Monash University, Victoria 3800, Australia

Abstract

Bubbles properties play an important role in determining the characteristics of gas-solid fluidized beds and thus influence the bed performance significantly. In this work, the bubble properties are studied by the combined approach of CFD and DEM for bubbling gas-solid fluidized beds operated with a continuous central jet. The process of the continuous injection of a central air jet to the bed is successfully reproduced featured with the formation of series of bubbles which rise through the bed and burst at the bed top. The bubbles are categorised as two types: primary/leading bubbles and trailing bubbles. The size of the primary bubble keeps on increasing with bed height while the size of the trailing bubble decreases since trailing bubble is coalescing with the primary bubble. The primary bubbles although formed at different times are of similar shape while the trailing bubbles are generally elongated in shape. The bubble velocity is distinctly different for primary and trailing bubbles.

Keywords: bubble formation, bubble size, bubble shape, single jet, CFD-DEM

1. Introduction

Fluidization technology has been widely used in many industrial applications such as food, pharmaceuticals, energy and environment since they are associated with good heat and mass transfer. The flow behaviour in gas-solid fluidized beds are very complex because of the fluid-particle and particle-particle interactions. Bubbles often exist in fluidized beds and are termed as "motor of fluidization"[1]. They govern hydrodynamics of fluidized beds, induce gas-solid mixing and promote heat and mass transfer [2]. Thus, a good understanding of bubble formation at an orifice and its dynamics is essential to aid in design and development of gas-solid fluidized beds. Various experimental studies have been carried out to portray bubble formation at an orifice [3] and to establish theoretical models [4, 5]. Besides, some experimental studies also used non-invasive techniques to study time evolution of bubble characteristics [3], the pressure wave propagation and attenuation behaviour [6] and bubble induced particle mixing [7]. In recent years numerical techniques have been widely used to study gas-solid fluidized beds and have become a significant approach to characterize bubble behaviour. Numerical techniques for gas-solid systems can be either continuum- or discrete-based with respect to the solid phase. The former is called two-fluid model (TFM), where both gas and solid phases are considered continuous and fully interpenetrating. The latter is referred as discrete element model (DEM) where particle phase is modelled as discrete phase and gas phase as continuous. Several authors have used TFM to study bubble formation at a single orifice and its dynamics [8-12]. However, DEM simulations have advantages over TFM simulations because the solid phase is modelled much more accurately, with respect to friction [13-15]. DEM studies in single jet fluidized beds were carried out to show that the size of a bubble is sensitive to cohesive inter-particle force [15], gas injection velocities, particle properties (size, shape and density) and bed dimensions [14-17], to depict the bubble induced particle mixing [18] and to formulate model for bubble-to emulsion phase mass transfer [19].

* *Corresponding author: zongyan.zhou@monash.edu; Tel: +61 3 9905 0846*

However, only the properties of the single bubble have been studied previously via CFD-DEM. The continuous injection of a central air jet to the bed leads to formation of series of bubbles which rise through the bed and burst at the bed top. In this process different types of bubbles are generated, and their behaviour are significantly different. To date, how the bubble properties vary in the bed when a continuous jet of gas is injected from a single orifice has not been fully investigated and the understanding is still lacking. Therefore, the objective of the present work is to study the bubble properties (size, shape and velocity) and force distribution in gas-solid fluidized beds operated with a continuous central gas jet by using CFD-DEM.

2. Models and simulation conditions

In CFD-DEM, the solid phase is based on the DEM and gas phase is modelled as CFD. The governing equations for particle i with mass m_i and moment of inertia I_i can be written as

$$m_i \frac{dv_i}{dt} = \mathbf{f}_{pf,i} + \sum_{j=1}^{k_i} \left(\mathbf{f}_{c,ij} + \mathbf{f}_{d,ij}\right) + m_i \mathbf{g} \tag{1}$$

$$I_i \frac{d\omega_i}{dt} = \sum_{j=1}^{k_i} (\mathbf{M}_{t,ij} + \mathbf{M}_{r,ij}) \tag{2}$$

where $\mathbf{v}_i$ and $\boldsymbol{\omega}_i$ are the translational and angular velocities of the particle, respectively, and k_i is the number of particles interacting with the particle. The forces involved are: particle fluid interaction force $\mathbf{f}_{pf,i}$, the gravitational force $\mathbf{m}_i\mathbf{g}$, and the inter-particle forces between the particles, which include elastic force $\mathbf{f}_{c,ij}$ and viscous damping force $\mathbf{f}_{d,ij}$. The torques acting on particle i by particle j include: $\mathbf{M}_{t,ij}$ generated by the tangential force and $\mathbf{M}_{r,ij}$ commonly known as the rolling friction torque.

The governing equations for gas phases are given as,

$$\frac{\partial \varepsilon_f}{\partial t} + \nabla \cdot (\varepsilon_f \mathbf{u}) = 0 \tag{3}$$

$$\frac{\partial (\rho_f \varepsilon_f \mathbf{u})}{\partial t} + \nabla \cdot \left(\rho_f \varepsilon_f \mathbf{uu}\right) = -\nabla p - \mathbf{F}_{pf} + \nabla \cdot \left(\varepsilon_f \boldsymbol{\tau}\right) + \rho_f \varepsilon_f \mathbf{g} \tag{4}$$

where $\mathbf{u}$, ρ_f and p are the fluid velocity, density and pressure, respectively. $\mathbf{F}_{pf}$ is the volumetric particle-fluid interaction force in a computational CFD cell of volume, ΔV. $\boldsymbol{\tau}$ and ε_f are the fluid viscous stress tensor and local porosity.

Details regarding the governing equations and the equations to calculate the interaction forces and torques can be referred to [20, 21]. The bed geometry used in this work is a slot model with periodic boundary conditions applied to the front and rear direction. For such a geometry, 2D CFD and 3D DEM are employed as used elsewhere [20]. The bed is produced by generating particles randomly. Particles are generated randomly at different heights and allowed to come to rest to form a bed [21]. The bed is then fluidized by gas from central orifice at the bottom which is of size 4 CFD cells (0.001 m). Table 1 lists the parameters including the bed geometry, physical properties of particles and fluid used in the simulation.

Table 1. Various parameters used in simulation

Variable	Values
Bed geometry	
Bed width (x)	0.02 m
Bed height (z)	0.08 m
Bed thickness (y)	0.0004 m
Orifice width	4 CFD cells

CFD cells (x, z)	320 x 80 cells
CFD cell size (Δx , Δz)	0.00025 m
Particle Properties	
Shape	spherical
U_{mf}, m/s	0.0082
Number of Particles, N_p	200,000
Particle size, d_p	100 μm
Particle density, ρ_p	2450 kg/m^3
Particle-particle sliding friction, μ_s	0.4
Particle-wall sliding friction, μ_r	0.4
Normal and tangential damping co-efficient, c_n	0.3
Young's modulus, E	1 x 10^7 Pa
Poisson ratio, ν	0.3
Time step, Δt	1.6 x 10^{-6}
Gas properties	
Density	1.205 kg/m^3
Viscosity	1.8 x 10^{-5} Pa s

3. Results and discussion

Figure 1 shows the snapshots of the bubble formation process for a central jet velocity of 0.25 m/s. It can be observed that after the introduction of a central air jet to the bed, series of bubbles are formed one after another, the first bubble leaves the distributor at 0.043 s. The bubbles are formed at a certain distance above the distributor and rise through the bed before erupting at the top of the bed. This is consistent with general understanding of this process and the literature reported. For example, according to Tsukada and Horio [22], the jet behavior is transient and bubble formation is periodic in nature and it starts with the gradual formation of the bubble-like void at the top of the jet.

The previous studies have shown that after the formation of the first bubble, the air through the central jet is not able to form the next bubble immediately. Instead, an elongated void or jet is formed at the distributor [15] which merges with the bubble above, similar jet behavior is observed as shown in Figure 1 (t = 0.071 s). In the present work, since series of bubbles are formed, several such instances are recorded (Figure 1, t = 0.0126 s, t = 0.212 s, 0.343 s, etc). Therefore, the bubbles are categorised as two types: primary/ leading bubbles and trailing bubbles as shown in Figure 2. The primary bubbles are defined as the bubbles whose motion are not affected by the bubbles ahead them but they affect the succeeding bubbles. While the trailing bubbles are those bubbles whose motion is affected by the primary bubbles. For example, Bubble 1 (B1) is a primary bubble and B2 is a trailing bubble, similarly, B3 is a primary bubble and B4 and B5 are trailing bubbles (Figure 1 and 2) etc. The trailing bubbles catch up with the leading bubbles at some height in the bed thereby coalescing. The coalescence observed for the bubbles formed mainly consists of a leading-trailing type of coalescence [23]. Leading-trailing coalescence describes the scenario where two bubbles form consecutively rather than simultaneously. As they rise, the first bubble slows due to its interaction with the trailing bubble. The trailing bubble will then 'catch' and coalesce with the leading bubble. The coalesced bubble will move up the remaining height of the bed as a single bubble or may coalesce with other bubbles. Figure 2 plots bubble nose position against time. Several process of bubble formation together with bubble coalescence is depicted. There were three positions where coalescence occurred as shown in Figure 2. The first position is near the distributor just after bubble detachment at a bed height of 0.005-0.008 where after the detachment of bubble, and the bubble is sequentially followed by a jet which merges to form a larger bubble. The second coalescence occurs at a bed height of 0.01-0.015 m where a leading bubble was caught up by a trailing bubble as shown. The final coalescence occurs near the top of the bed at bed height of > 0.022 m where the merged bubbles erupt concurrently. The bubble breakup is not so significant since at lower height, the bubble recoalesces after bubble splits, and at the top of the bed, the bubble breaks up to erupt.

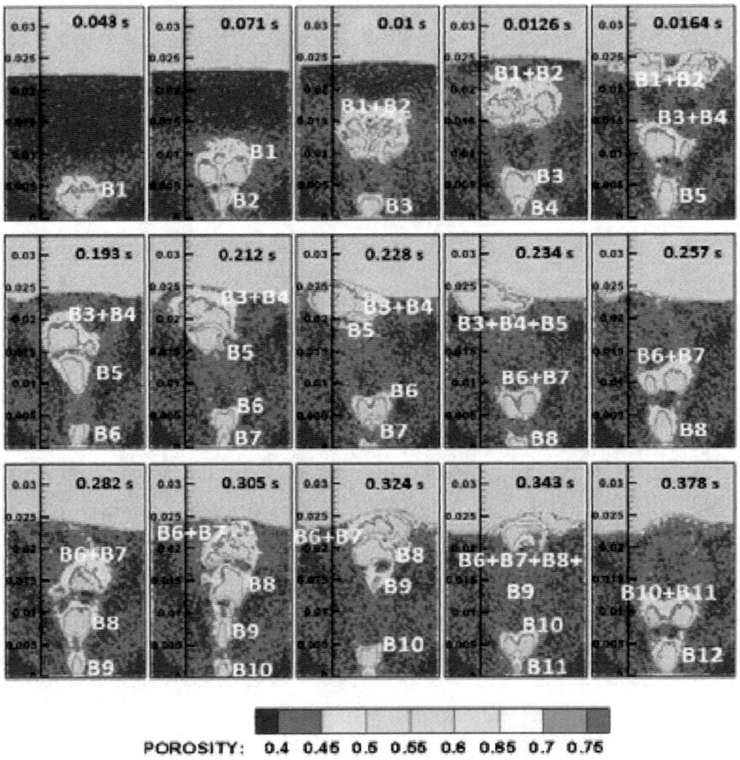

Figure 1: Bubble formation and trajectory snapshots for jet velocity of 0.25 m/s.

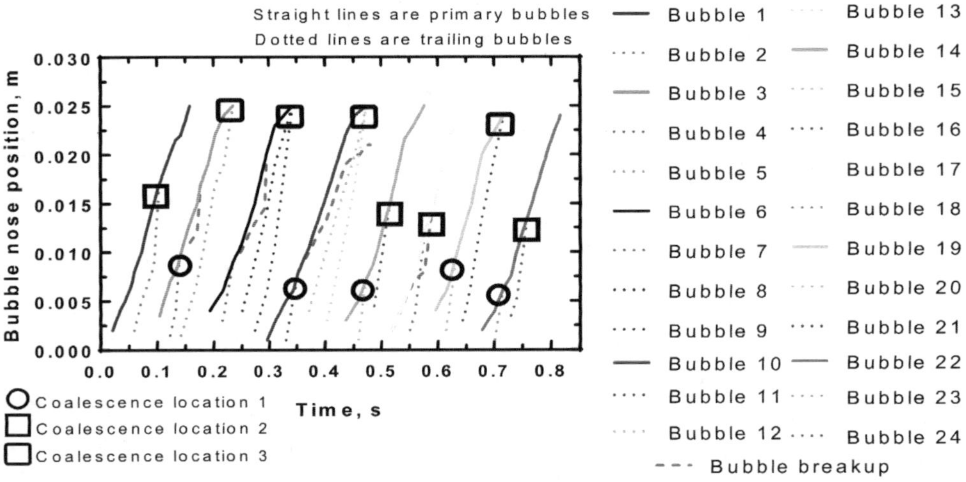

Figure 2: Motion of chain of bubbles for jet velocity of 0.25 m/s.

Figure 3 shows the bubble diameter at different positions. Figure 3a, b and c shows the change in bubble diameter with bubble location for both primary and trailing bubbles. It can be seen that the increase in the bubble diameter is because of the bubble coalescence while the decrease in bubble diameter is due to bubble splitting. In this process the size of the primary bubble keeps on increasing while the size of the trailing bubble decreases since trailing bubble is coalescing with the primary bubble. Figure 3d shows the change in bubble diameters of 3 primary bubbles at different bubble position. It can be seen that the bubble diameter and its change for the primary bubbles at different location are almost alike indicating the growth of primary bubbles are similar.

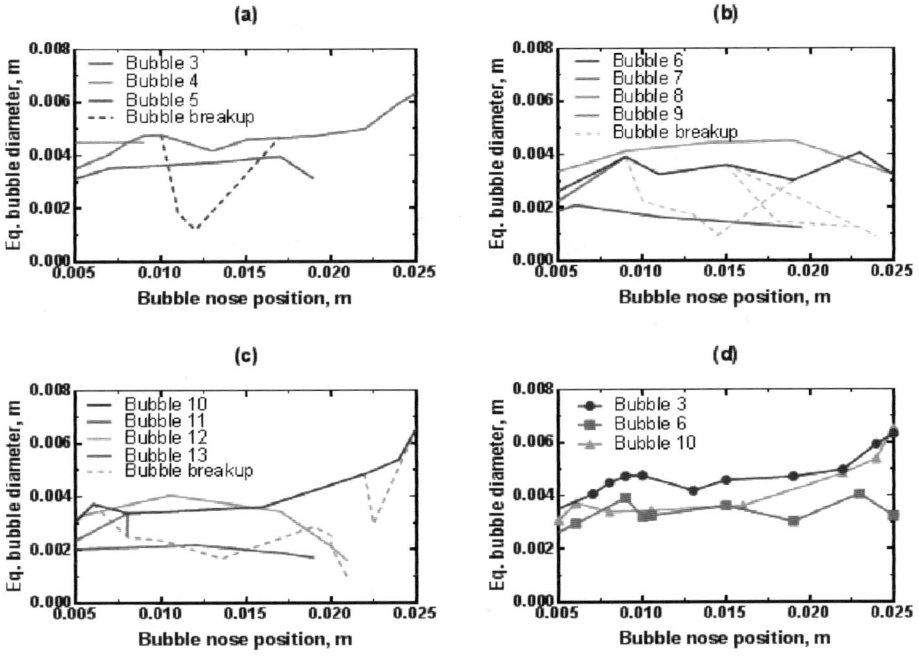

Figure 3: Bubble diameter at different positions.

Figure 4 shows the bubble shape characterized by bubble aspect ratio for different position of the bubble. The figure clearly shows the differences in the bubble shape for the primary and trailing bubbles. The primary bubbles although formed at different times are of similar shape. As the bubble rises through the bed it expands in the vertical direction thereby becomes more and more elongated i.e. prolate shaped (aspect ratio >1) at lower bed height. After the bubble is detached, the bubble stretches in the horizontal direction due to development of wake resulting in the decrease of aspect ratio. At the mid-level of the bed the bubble is almost circular (aspect ratio = 1). As the bubble rises towards the top of the bed it stretches more in horizontal direction and becomes oblate shaped (aspect ratio < 1) before eruption. However, the shape of the trailing bubbles follow different trend. The trailing bubbles are generally elongated in shape (aspect ratio >>1), this is because their motion is under the influence of the primary bubbles. When the trailing bubbles move into the vortex of the primary bubbles they accelerates and catches up to finally merge with the primary bubble which results in an elongated bubbles (For example, Figure 1 0.193 s (B5), 0.305 s (B8)).

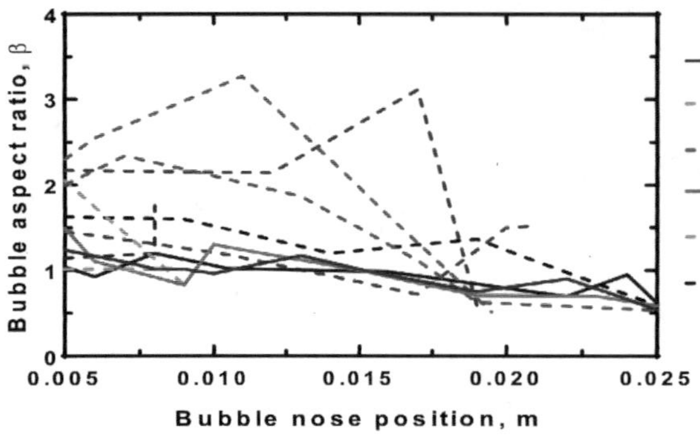

Figure 4: Bubble shape at different positions.

Figure 5 shows the bubble velocity for different position of the bubble. It is evident from the figure that the bubble velocity is distinctly different for primary and trailing bubbles. The bubble velocity for the primary bubble initially increases as it rises through the bed then remains constant (at mid-level of the bed) until it falls rapidly at the top of the bed during eruption. On the contrary, the trailing bubbles have very high bubble velocity around 1.5-2 times the velocity of primary bubbles. This is because the trailing bubbles are in acceleration to catch up and merge with the primary bubble which is at higher level.

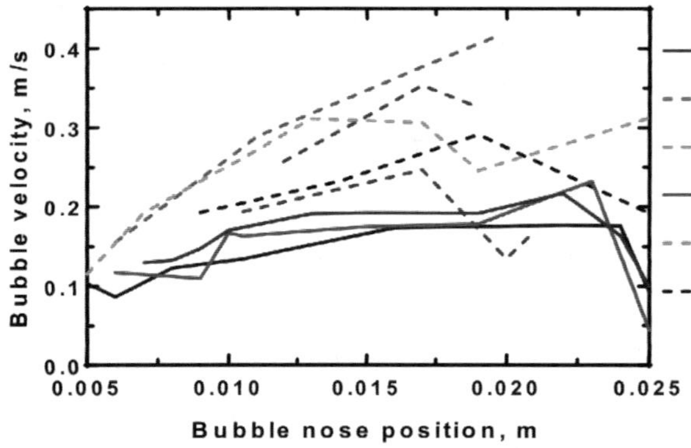

Figure 5: Bubble velocity at different positions.

One of the important forces in the process of fluidization is the interaction between particles and interstitial fluid which consists of the fluid drag force and pressure gradient force. This forces vary spatio-temporally. As the example, the spatial distribution of various forces (top - drag force; middle - pressure force; and bottom - particle-fluid interaction force) at different times is shown Figure 6. It should be noted that the force ratio is used by normalizing forces with particle gravity *mg*. It can be observed that these forces around gas bubbles are significantly different from the other areas. The drag force, pressure gradient force and their sum the particle fluid interaction force are larger in the dense areas around the gas bubbles than in the loose areas in the bubbles or far from the bubbles due to the differences in porosity. These forces are maximum in between the bubbles and minimum in the bubbles.

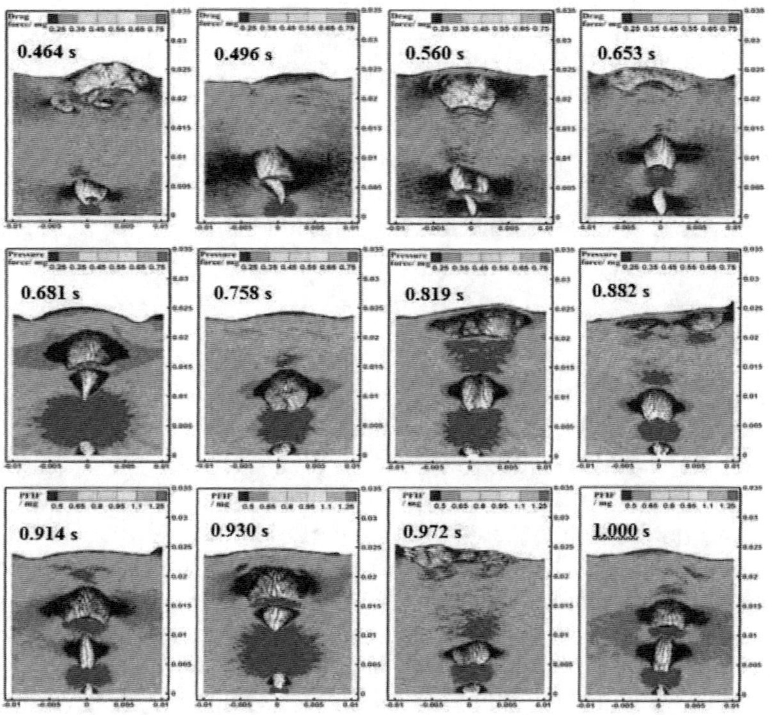

Figure 6: Spatial distribution of different forces (top: drag force, middle: pressure force and bottom: particle fluid interaction force) at jet velocity of 0.25 m/s.

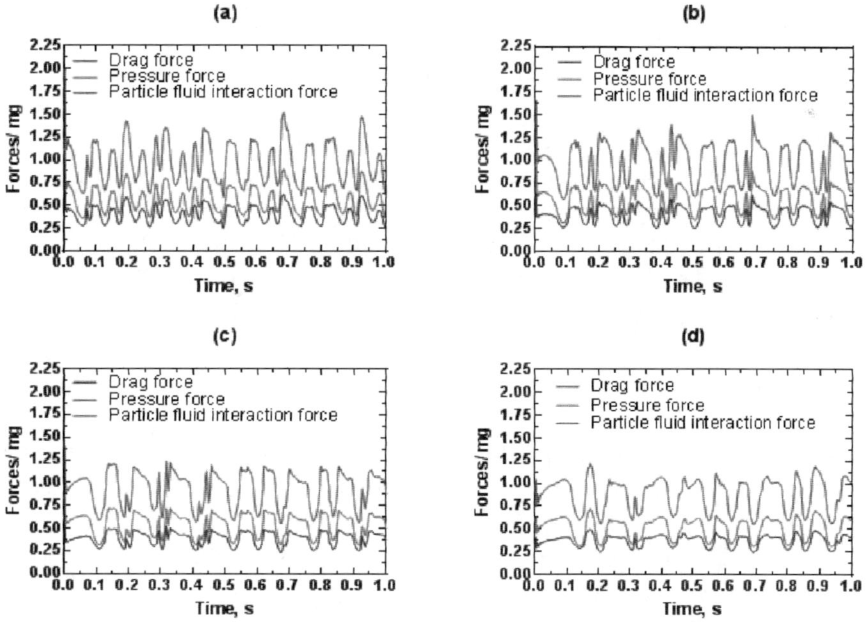

Figure 7: Different forces acting on the bed at different times for different bed height (axial direction): a) H = 0.005 m, b) H =0.01 m, c) H = 0.015 m and d) H = 0.02 m.

Figure 7 shows the temporal variation of drag force, pressure gradient force and particle fluid interaction force at different bed heights. It can be seen that the pressure gradient force is greater than the drag force at any level of the bed. Moreover, the magnitude of the forces decreases as the bed height increases which is due to the loss in the energy retained by the bubbles as the bed height increases. In addition, the peak depicted in the figures are due to the bubble passage. The maximum represents when the bubble has passed the point while the minimum represents when the bubble centroid is at that point. It can be observed that the peak gets plateaued as the bed height increases. This is because of the bubble coalescence. At lower level of bed there is no significant bubble coalescence, however as the bed level increases there is significant bubble coalescence present. Therefore, the gap in between the bubble increases leading to increased presence of solid material with less porosity thereby increasing the magnitude of the relative forces resulting in the formation of plateau as the bed height increases.

Conclusion

Formation of a series of bubbles in a fluidized bed operated with a single continuous jet was investigated using CFD-DEM to study the bubble characteristics such as bubble size, shape and velocity. The bubbles are categorised as two types: primary/leading bubbles and trailing bubbles. The primary bubbles are defined as the bubbles whose motion are not affected by the bubbles ahead them but they affect the succeeding bubbles. The trailing bubbles are those bubbles whose motion is affected by the primary bubbles. The trailing bubbles catch up with the leading bubbles at some height in the bed thereby coalescing. The size of the primary bubble increases at the cost of trailing bubbles. Both shape and velocity of the trailing bubbles are significantly different compared to primary bubbles. Trailing bubbles are elongated in shape and have higher bubble velocity. The force analysis showed that the drag force, pressure gradient force and their sum the particle fluid interaction force are higher in the dense areas in between the gas bubbles than in the bubbles or far from the bubbles.

References

[1] M. Rüdisüli, Characterization of rising gas bubbles in fluidized beds by means of statistical tools, PhD. Thesis ETH Zurich (2012).

[2] A. Busciglio, G. Vella, G. Micale, L. Rizzuti, Analysis of the bubbling behaviour of 2D gas solid fluidized beds: Part I. Digital image analysis technique, Chem. Eng. J., 140 (2008) 398-413.

[3] A.D. Lotfabad, S. Movahedirad, M. Sadeghi, Bubble Formation on a Single Orifice in a Gas Solid Fluidized Bed Using Digital Image Analysis, Iran. J. Chem. Eng., 13 (2016) 61.

[4] D. Harrison, L. Leung, Bubble formation at an orifice in a fluidized bed, Chem. Eng. Res. Des., 39 (1961) 409-414.

[5] H.S. Caram, K.-K. Hsu, Bubble formation and gas leakage in fluidized beds, Chem. Eng. Sci., 41 (1986) 1445-1453.

[6] K. Nosrati, S. Movahedirad, M.A. Sobati, A.A. Sarbanha, Experimental study on the pressure wave attenuation across gas-solid fluidized bed by single bubble injection, Powder Technol., 305 (2017) 620-624.

[7] Z. Amiri, S. Movahedirad, M. Shirvani, Particles mixing induced by bubbles in a gas ‐ solid fluidized bed, AIChE Journal, 62 (2016) 1430-1438.

[8] J. Kuipers, W. Prins, W. Van Swaaij, Theoretical and experimental bubble formation at a single orifice in a two-dimensional gas-fluidized bed, Chem. Eng. Sci., 46 (1991) 2881-2894.

[9] J. Nieuwland, M. Veenendaal, J. Kuipers, W. Van Swaaij, Bubble formation at a single orifice in gas-fluidised beds, Chem. Eng. Sci., 51 (1996) 4087-4102.

[10] R.P. Utikar, V.V. Ranade, Single jet fluidized beds: experiments and CFD simulations with glass and polypropylene particles, Chem. Eng. Sci., 62 (2007) 167-183.

[11] M. Lungu, J. Wang, Y. Yang, Numerical simulations of flow structure and heat transfer in a central jet bubbling fluidized bed, Powder Technol., 269 (2015) 139-152.

[12] Y. Zhang, R. Xiao, M. Ye, Z. Liu, A numerical study of the bubble induced pressure fluctuation in gas-fluidized beds, Powder Technol., 314 (2017) 387-399.
[13] D. Gera, M. Gautam, Y. Tsuji, T. Kawaguchi, T. Tanaka, Computer simulation of bubbles in large-particle fluidized beds, Powder Technol., 98 (1998) 38-47.
[14] O.O. Olaofe, M.A. van der Hoef, J.A.M. Kuipers, Bubble formation at a single orifice in a 2D gas-fluidized bed, Chem. Eng. Sci., 66 (2011) 2764-2773.
[15] J.K. Pandit, X. Wang, M. Rhodes, A DEM study of bubble formation in Group B fluidized beds with and without cohesive inter-particle forces, Chem. Eng. Sci., 62 (2007) 159-166.
[16] L.-w. Rong, J.-m. Zhan, C.-l. Wu, Effect of various parameters on bubble formation due to a single jet pulse in two-dimensional coarse-particle fluidized beds, Adv. Powder Technol., 23 (2012) 398-405.
[17] S. Shrestha, Z. Zhou, A CFD-DEM study of single bubble formation in gas fluidization of spherical and non-spherical particles, EPJ Web of Conferences, EDP Sciences, 2017, pp. 15026.
[18] Z. Amiri, S. Movahedirad, M. Shirvani, O. Vahidi, The role of bubble injection characteristics at incipient fluidization condition on the mixing of particles in a gas-solid fluidized bed at high operating pressures: A CFD-DPM approach, Powder Technol., 305 (2017) 739-747.
[19] L. Tan, I. Roghair, M. van Sint Annaland, Discrete particle simulations of bubble-to-emulsion phase mass transfer in single-bubble fluidized beds, Particuology, 33 (2017) 80-90.
[20] Z. Zhou, D. Pinson, R. Zou, A. Yu, Discrete particle simulation of gas fluidization of ellipsoidal particles, Chem. Eng. Sci., 66 (2011) 6128-6145.
[21] J. Gan, Z. Zhou, A. Yu, CFD–DEM modeling of gas fluidization of fine ellipsoidal particles, AIChE Journal, 62 (2016) 62-77.
[22] M. Tsukada, M. Horio, Gas motion and bubble formation at the distributor of a fluidized bed, Powder Technol., 63 (1990) 69-74.
[23] K. Lim, J. Zhu, J. Grace, Hydrodynamics of gas-solid fluidization, International journal of multiphase flow, 21 (1995) 141-193.

504801 Statistical Analysis on Large-Scale Direct Numerical Simulation of Gas-Solid Flow

Abstract:

Particle-resolved direct numerical simulation (DNS), where the computational grid for the fluid phase is reduced to below the particle size, the flow around each particle is fully resolved, and the fluid-solid interaction force is directly obtained by integrating the viscous force and pressure on the particle's surface. It has been regarded as the most accurate numerical method for simulation of gas-solid flow. However, the main disadvantage is its huge computational cost resulting from small grid size and time step limited by Kolmogorov length and time scales. Only hundreds of particles scale is reported for DNS of gas-solid flow in the latest literature, which is very different from the number of particles in the real gas-solid flows. Therefore, small-scale simulation results are hardly to be established a true and reliable constitutive law for gas-solid flow. In order to solve the problem of computational speed and scale, we use the lattice Boltzmann method (LBM) with good parallelism to replace the traditional N-S equation to describe the gas phase, an immersed boundary method in framework of LBM has been adopt to realized the fluid-solid coupling to avoid a stair-step representation of the solid particles' surfaces (Wang et al., 2010; Zhou et al., 2011), and the multi graphics processor units (GPUs) parallel computing is also implemented. Taking advantage of the inherent parallelism of LBM and the attractive Flops/Price ratio of GPU, we have implemented 576 GPUs parallel computing on a Mole-8.5 system and conducted the largest scale DNS of gas-solid suspensions so far, with 1,166,400 solid particles in an area of 11.5cm x 46cm for a two-dimensional system and 129,024 solid particles in a domain of 0.384cm x 1.512cm x 0.384cm for a three-dimensional system (Xiong et al., 2012). The scale of DNS data has been reached the size in traditional computational grid, which implies the really meaningful statistical results from large-scale DNS of gas-solid flows were obtained for the first time. The effects of mesoscale structure on the interaction force between gas and solid phases (Zhou et al., 2014) and the statistical properties of particles (Liu et al., 2017) were explored, which may provide the corresponding constitutive relation and detailed microscopic information for discrete particle simulation and two-fluid model.

References

1. Limin Wang, Guofeng Zhou, Xiaowei Wang, Qingang Xiong, Wei Ge. 2010. Direct numerical simulation of particle-fluid systems by combining time-driven hard-sphere model and lattice Boltzmann method. Particuology, 8: 379-382.
2. Guofeng Zhou, Limin Wang, Xiaowei Wang, Wei Ge. 2011. Galilean-invariant algorithm coupling immersed moving boundary conditions and Lees-Edwards boundary conditions. Physical Review E 84, 066701.
3. Qingang Xiong, Bo Li, Guofeng Zhou, Xiaojian Fang, Ji Xu, Junwu Wang, Xianfeng He, Xiaowei Wang, Limin Wang, Wei Ge, Jinghai Li. 2012. Large-scale DNS of gas–solid flows on Mole-8.5. Chemical Engineering Science 71: 422–430.
4. Guofeng Zhou, Qingang Xiong, Limin Wang, Xiaowei Wang, Xinxin Ren, Wei Ge. 2014. Structure-dependent drag in gas-solid flows studied with direct numerical simulation. Chemical Engineering Science, 116: 9-22.
5. Xiaowen Liu, Limin Wang, Wei Ge. 2017. Meso–scale statistical properties of gas–solid flow—a direct numerical simulation (DNS) study. AIChE Journal 63:3-14.

509635 Capillary Interaction in DEM Simulations of Wet Particulate Materials

Abstract:

Processing stages such as granulation, encapsulation, coating, drying, particle melting are present in a variety of industries, including the manufacture of food, pharmaceuticals, catalyst, mechanical part products. The presence of some liquid between the particles determines forces and complex particle formation or breakage whose understanding and characterization is far from complete. Discrete Element Method (DEM) simulations possess the ability to carry out analysis of dense particulate material under actual industrial processing conditions, being able to include interaction contributions at the individual particle level.

In the present work, microscopic models of particle-particle capillary interaction for inclusion in advanced DEM simulations are investigated. Liquid bridge formation, interaction force, stability and rupture are investigated as a function of the liquid volume, liquid wetting properties (e.g. contact angle), particle separation distance, particle size ratio. The extensive set of results are obtained adopting a rigorous computation of the equilibrium bridge shape and characteristics. Comparison with existing approaches, such as the toroidal approximation, will be finally presented.

507290 Pressure Distribution in a Three Phase Moving Bed

Abstract:

Heterogeneous reactions widely exist in chemical industries. Dispersed principles of different heterogeneous reactions result in the developments of various multi-flow reactors, e.g., the fluidized bed for polymerization, the trickle bed for hydrocracking of cat-cracked gasoil and vacuum gasoil. Therefore, confronting so many kinds of heterogeneous reactions, and constantly demands of increasing the product qualities and decreasing the process energy consumption, developing new multi-flow reactors is always necessary and significant. This work aims at developing a totally new gas-liquid-solids three phase moving bed reactor on the basis of gas-solid moving bed and trickle bed. The advantages of the three phase moving bed reactor may include: 1) gas, liquid and solid phase are close to the plug flow; 2) the residence time of catalyst particles is easy to control, and the regeneration operation of catalyst is convenient; 3) the cold gas quench can be designed between different stages of the reactor to easily control the reactor temperature increase, and the separation of liquid products can also be applied between different stages to easily control the reaction rate.

A cold-model experimental apparatus of the three-phase moving bed was established, which mainly consists of four parts: the inlet system, the reaction unit, the discharging and separation system, and the inner structures for even flow distribution. The pressure distribution in the moving bed reactor was investigated. The results showed that the pressure drop linearly increased with the length of the reactor in the situation without liquid, and the increasing rates were higher at larger gas velocities. Moving of solid particles further increased the pressure drop. As liquid was injected into the reactor, it made the further increase of pressure drop. The uneven distribution of liquid distorted the linearly distribution of pressure drop. The building up of cold-model experiments and the experimental results will provide guidance for the scale-up of three-phase moving bed reactors.

506783 Numerical Study of Gas-Liquid Flow inÂ Trickle Bed Using MPS Method

Abstract:

Numerical study of gas-liquid flow in trickle beds using MPS method

Kei Sakakura[1*] ,Ryou Tanabe[1] and Yasunobu Kaneko[1]

Technology & Engineering Center, Idemitsu Kosan Co., Ltd Chiba, Japan.

**E-mail: kei.sakakura@idemitsu.com*

Abstract

1.Introduction

Trickle-bed reactors are widely used in chemical and energy fields due to simple geometry and easy reactors operation. However, the overall performance of reactors are deeply affected by catalyst bed (packing configuration, porosity, particle shape, size, coating), flow maldistribution, mass and heat transfer rate and reaction kinetics. Some of the important issues in microscopic modeling of gas-liquid flow through packed beds are closure models for interphase drag, wetting, and capillary forces. It is important to understand how liquid-gas interacts with solid surfaces with complex shapes. To clarify that, Liquid spreading in gas-liquid concurrent trickle beds is investigated using MPS approach considering the wall contact angle. Liquid phase is simulated by MPS as lagurangian particles, while gas phase is solved as eulerian method using mesh. To evaluate the MPS model, complex geometry of packed beds with sphere, quadrillobe and quadrihollow types of catalyst are obtained by DEM simulation. The MPS model is shown to achieve good agreement with experimental data and to predict accurately, tendencies of liquid spreading and tracer ink residence time through the complex geometry. .

2.Experimental setup

The experimental were conducted using 0.15m width,0.5m tall,0.015m thickness rectangle cold-model packed actual three demetallization catalysts whose shapes are sphere, quadrilobe, quadrihollow type, and water and air connected flow meter were supplied concurrently from top of packed bed. Each flow rate was adjusted as same superficial velocity as that of actual trickle bed reactor. After the liquid and gas flow was fully developed in the bed, black ink was injected in liquid as tracer for visualization of the liquid flow. Liquid dispersion rate and black ink passage time thorough the bed were measured by video camera for CFD validation.

3.Modeling description and simulation condition

Packing geometries are set up from catalyst packing simulation using DEM (Discrete Element Method). Each catalyst is composed by small DEM particles which are fixed in a catalyst. When catalysts contact each other in packed bed, small particles are used to judge the collision and predict the collision force and angular moment to solve the motion of the catalysts. After that, packing geometries are converted to CAD data(stl file) for MPS simulation.

MPS method is one of the particle methods which can be applied to fluid flow. Since the mesh is not required, large deformation of liquid free surfaces can be calculated without mesh distortion even in complex geometry. The governing equations are represented by lagrangian description and it is not necessary to discretize the convection terms. In this study, liquid movement is simulated by MPS with consideration of surface tension force and gas phase is solved by conventional fundamental equation using mesh. Therefore, lagragian and eularian method are coupled.

4.CFD Results

In packed bed composed of sphere type catalyst, the liquid flow though the bed does not dispersed well as compared with quadrilobe and quadrihollow type. On the other hand, quadrilobe type shows the good dispersibility to the wall direction due to high packing density and high fluidity caused by stick- like shape of catalyst. And CFD results show reasonably good agreement with the experiment results regarding the liquid dispersion and liquid passage time through the beds.In all the packed beds the gas velocity thorough the bed is almost uniformed and shows higher velocity at the wall due to higher void rate.

Keywords

DEM MPS trickle flow

507933 Theoretical and Experimental Study of Anaerobic Fluidized Bed Reactor: Case of Pharmaceutical Wastewater Treatment

Abstract:

The successful design and operation of an anaerobic fluidized bioreactor depend on the ability to predict the fundamental properties of the system, including hydrodynamics, individual phase mixing, and heat and mass transfer properties. In this context, this study focuses on an experimental and theoretical study of such a reactor for treating pharmaceutical wastewaters.

The experimental setup consists of a mesophilic lab-scale reactor, with three kinds of support (sand, glass beads and PVC beads) for biofilm development, knowing that biological fluidized bed reactor has two types of bed expansion, the first one is due to an increase of the superficial velocity and the second to the microbial growth in the bed.

The predicted and measured responses of hydrodynamic and biological variables are investigated and experimental data were used to apply known models used for bed expansion and particle mixing characteristics, pressure drop, predicting the biofilm development, and biomass hold-up empirical values of biofilm detachment coefficient for the case of a synthetic water biotreatment containing the methyl 4-hydroxybenzoate, which shows promising results and that will be also discussed.

508058 Innovative Design for Improved Contact of Hydrocarbon Feedstock with Catalyst in a Downer Reactor

Abstract:

Down flow reactor (downer) in a circulating fluidized bed configuration such as Fluid Catalytic Cracking (FCC) unit is achieving importance in the industry due to its inherent capability of reducing the back mixing of catalyst with reacting hydrocarbon stream. The flow of gas and solids in a downer reactor approaches to ideal plug flow against the flow in conventional upflow in a circulating fluidized bed (riser) with significant back mixing. This makes the residence time distribution in a downer reactor narrower compared with a riser and hence, higher selectivity towards olefinic products and lower yield of undesired products.

However, the downer reactor lacks in the configuration for better contact of hydrocarbon stream and catalyst resulting in lower conversion of feedtstock. An innovative design has been developed to improve the initial contact of hydrocarbon and catalyst in a downer reactor. In this design, the reactor is the annular section of two concentric pipes and the catalyst allowed to fall freely into the annular downer reactor as annular curtains. A cold flow pilot plant has been installed based on in-house design for generating the hydrodynamic data. The current paper describes about the distribution of catalyst in the annular downer reactor confirming its superiority in contacting the hydrocarbon stream with catalyst.

Particle Measurement in High-Temperature Gas based on Light Scattering

Li-feng LU[1,3], Xiao-lin WU [1,3], Zhong-li JI [1,2], Ming-xing WANG[1,3]

[1] Department of Chemical Engineering, China University of Petroleum, Beijing, P. R. China

[2] Department of Mechanical and Transportation Engineering, China University of Petroleum, Beijing, P. R. China

[3] Beijing Key Laboratory of Process Fluid Filtration and Separation, China University of Petroleum, Beijing, P. R. China

Abstract: In order to achieve the particle measurement at 650 °C, a single particle measurement method and device suitable for high-temperature gas were developed. The influence of gas temperature on optical sensor of optical particle counter (OPC) was investigated in detail based on geometric optics theory and heat transfer theory. The thermal expansion of aerosol tube along the airflow direction was caused by the heat conduction between the tube and the gas, resulting in the position of the light apertures to shift to obscure the partial light beam. The invar material can be adopted to eliminate the above mentioned disadvantages. According to the principle of gas state equation, the optical parameters (refractive index) of the gas at different temperatures change with the change of its state parameters. Based on the geometrical optics theory and the geometric mathematical relationships, the theoretical and experimental dynamic models of the optical measuring volume with the change of the gas temperature were established. The established models were then used to analyze the influence of gas temperature on the optical measuring volume. The results showed that the influence of the temperature on the optical system is negligible. Meanwhile, particle detection experiments were conducted at different temperature to achieve a good performance of OPC.

Keywords: High-temperature gas; Optical measuring volume; Dynamic model

1 Introduction

With the continuous improvement of environmental protection and the safety requirements for industrial production, the technology of high-temperature purification is drawing more and more attention, especially in the fields of clean coal power generation, coal gasification, petrochemical industry and metallurgy [1]. For example, nuclear power plants often produce a large number of low radioactive solid wastes, and the combustible wastes usually account for a considerable proportion. The incineration and pyrolysis are effective treatment techniques for the combustible waste, during which the temperature of flue gas is 600~800 °C. With the aim to purify the flue gas, the German Karlsruhe Nuclear Research Institute, the French nuclear power station and the Japanese nuclear power station have installed the silicon carbide ceramic filter tube, aluminum silicate fiber and alumina ceramic filter tube at gas temperatures above 650 °C, respectively, and investigated their properties properties [2-4]. These filters have a direct influence on the purification quality of gas and the particle concentration in the downstream. In an industrial cracker system, the particle concentration is required to reach 0.014 mg/m^3 after the filters; the Integrated Gasification Combined Cycle (IGCC) requires the particle size to be smaller than 1 um with a particle concentration of less than 2.2 mg/m^3 and in the range of 1~10 um with less than 5.8 mg/m^3, and to remove the dust size of larger than 10 um [5]. In a Shell Coal Gasification Process (SCGP), the particle concentration is required to be less than 20 mg/m^3 [6]. In order to detect the concentration and size of the particles in pipes in real time for calculating the separation efficiency of evaluating the purification performance of the filtration system, the particle detection technology is needed to be improved [7].

The widely-used particle detection devices in industry are mainly based on the principle of light scattering [8-10], and the particle size and concentration are obtained through an inversion of the scattering light intensity and particle size, which are suitable for the conditions of below 200 °C. The method of first cooling and then detecting can be adopted for the particle detection under higher-temperature conditions, but droplet condensation and sublimation will inevitably occur and result in an incorrect particle samples.

With the aim to achieve the particle detection at 650 °C, an optimization was conducted on the existing particle detector (OPC) in this paper. And based on geometric optics theory [11] and heat transfer theory, a detailed research was performed on the influence of gas temperature on OPC optical sensor, including the dynamic change of the aerosol tube and the optical measuring volume (OMV). Finally, the experiments were conducted on the on-line particles measurement of different temperatures.

2 Experimental setup and method

Figure 1 gives the experimental setup for detecting particles in high-temperature gas, mainly including the heating unit, optical detection unit and gas purification and processing unit.

The heating unit consists of the vibration particle feeder and the air heater. The gas brings the particle out of the feeder, and is heated up to 1000 °C by the heater, which is monitored quickly by a PID temperature control instrument with easy operation.

The optical detection unit is used to achieve the experimental measurement of the characteristics of the OPC optical system and the on-line detection of particles, including the optical sensor and the OPC control host. The optical sensor is shown in Figure 2, which is composed of an aerosol tube and optical system. Aiming to avoid the influence of high-temperature gas on optical system, the aerosol tube (shown in Figure 3) has been developed, and is in a distance of 15 mm to the optical lens. In addition, the insulation coating with lower thermal conductivity (-80-1700 °C) is adopted to treat the outer wall of the aerosol tube. A K-type micro thermocouple was bent and installed into the light aperture to monitor the gas temperature at the center of the aerosol tube in real time without affecting the airflow and optical paths. Figure 4 shows the optical system of the sensor, which is composed of an incident light path and an observed scattering light path. Both the paths are arranged in 90° with the same optical lens and elements [12-14], and the intersected area is the center of the aerosol tube to form an OMV [15]. The intersected area is enough small and allows particles to pass through it one by one [16, 17]. The OPC host consists of the light source system (wavelength: 300-700 nm) [18], the photoelectric detection system and the signal analysis and processing system. In the particle detection, particle is illuminated by the incident light when passing through the OMV, and the scattering light is collected and transmitted to the photoelectric detector to achieve the conversion of a pulse voltage signal from an optical signal. Finally, the tested particle size is calculated based on the calibration curve of particle size and voltage amplitude [14, 19].

The gas purification and processing unit is composed of a vacuum pump, a flow controller, a dust collector and a heat exchanger. The high-temperature gas form the aerosol tube is completely cooled during the heat exchanger, the flow rate of which is adjusted by the flow controller. In order to avoid the damage of particles on the pump, the dust is captured and collected by a filter in the dust collector.

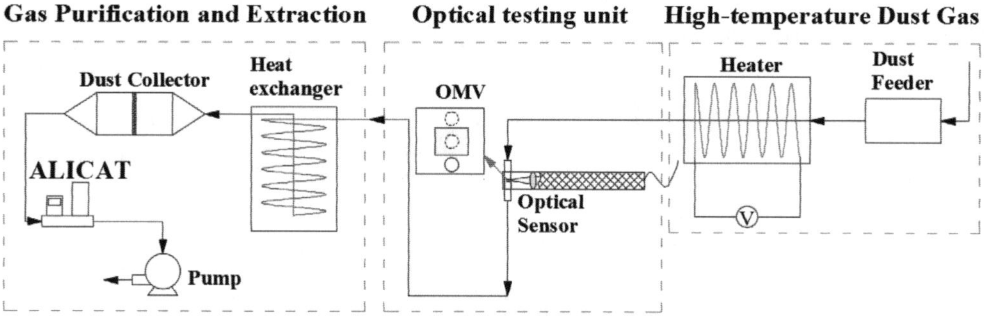

Figure 1. Experimental setup for detecting particles in high-temperature gas

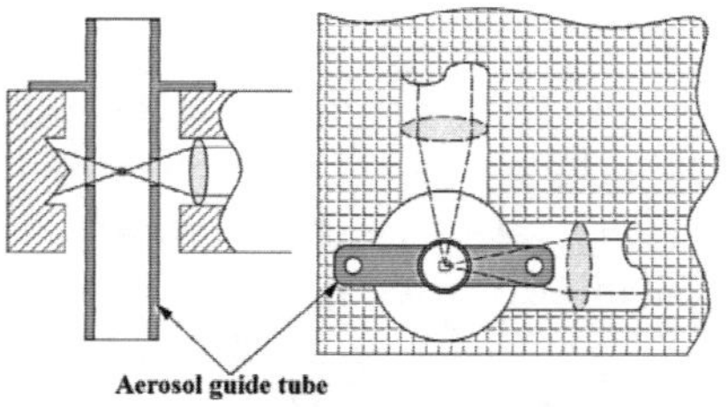

Figure 2. Optical sensor

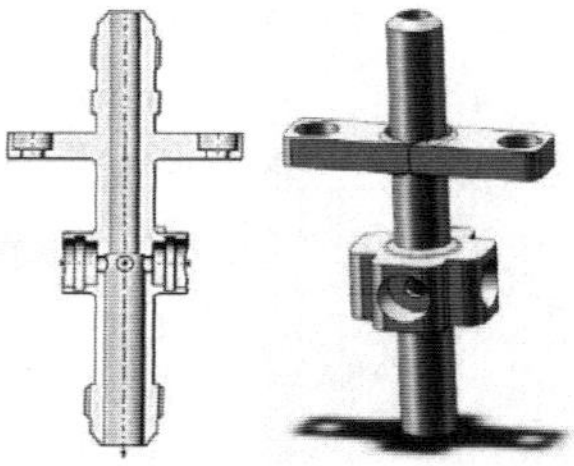

Figure 3. Aerosol tube for high-temperature gas

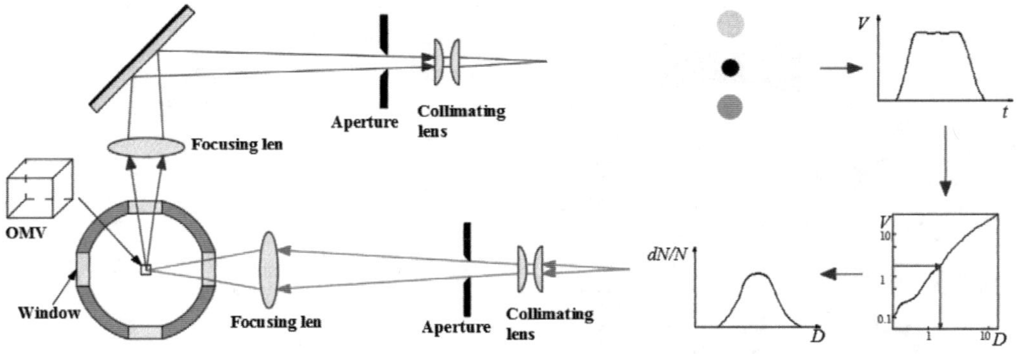

Figure 4. Optical system and principle of OPC

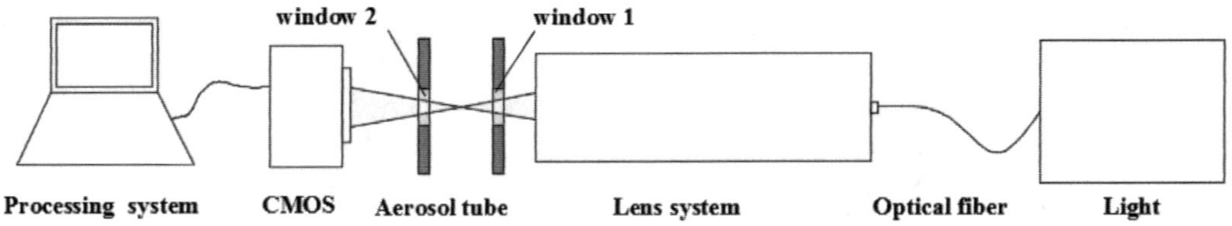

Figure 5. Evaluation and analysis device for light beam

3 Analysis and discussion

3.1 Influence of temperature on aerosol tube

The aerosol tube shown in Figure 3 has four apertures along the center axis, in which the glasses with a thickness of 2 mm were installed, as shown in Figure 4. When the gas flows in the tube, it can reduce the damage of high-temperature gas to the optical system elements. But the heat resistance of the tube has a direct influence on the imaging of the optical system at the center of the aerosol tube.

Figure 5 gives an evaluation and analysis device for light beam, including the light source system, the light path system, the aerosol tube and the light beam analyzer (CINOGY CMOS-1201, Germany), which can be equipped at the optical testing unit as shown Figure 1. For the gases at different temperatures, the light passes through the glasses in the aerosol tube and images on the CCD of CMOS-1201 to detect the spot of the light beam in real time.

After eliminating the background noise, the spot image at different temperatures can be acquired. When the aerosol tube is made of 304 stainless steel, the measurement results are shown in Figure 6. It can be found that the shape of the light spots is distorted and shows the imperfection of the imaging with an increase in gas temperature. That's due to thermal expansion along the airflow direction caused by the change of gas temperature, which results in the shift of light apertures position to obscure the partial light beam. A bad result must be acquired for smaller particle size and lower particle count efficiency because of the decreased incident light. Meanwhile, the shape of the OMV was changed to greatly reduce the performance of OPC detection by a failure of eliminating the error of edge detection in optical system.

With the aim to eliminate the above mentioned disadvantage, the invar material with a linear expansion coefficient of 1.6×10^{-6} m/°C was adopted for high-temperature experiment. The results are shown in Figure 7. At different temperatures, no deformation was found in the shape of light spots, indicating that the material can meet the testing requirement for high temperature up to 650 °C with a good heat resistance.

Table 1. Comparison of the application conditions of the two materials

Material	Light divergence angle /°	Gas temperature /°C
304 (Stainless steel)	<10	<200
4J36 (Invar)	<30	<700

The application conditions of the two materials are

summarized in Table 1. Compared with the stainless steel, the invar material is of great significance for the development of high-temperature detection device.

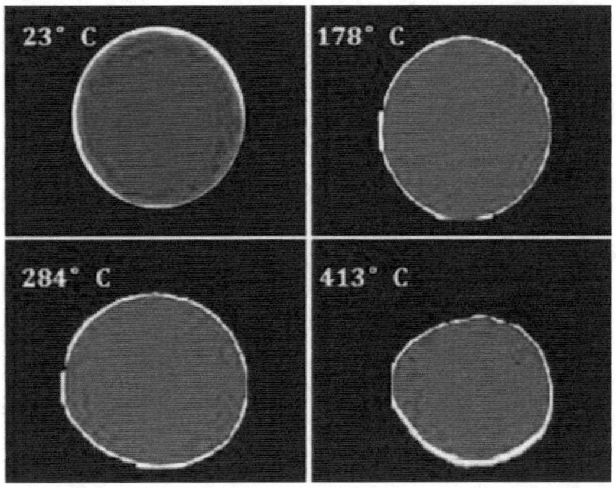

Figure 6. The spot images measured at four different temperatures of gas for 304 stainless steel.

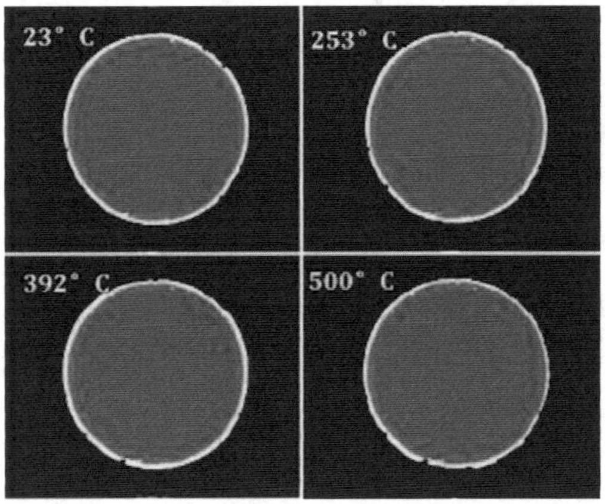

Figure 7. The spot images measured at four different temperature of gas for 4J36 (Invar).

3.2 Influence of temperature on optical system

In general the optical sensing system in the particle detection device is only applied at atmospheric temperature, but can also satisfy the temperature up to 150 °C. Under high-temperature condition, the gas volume is going to expand with the reduction in the gas density. The transmission characteristic of light beam could be changed, resulting from the decrease in the optical performance of the sensor. In order to investigate the influence of temperature on optical system, the specific causes are analyzed by theoretical method, and the extent of its influence is determined through experimental verification.

3.2.1 Theoretical method

By calculating the refractive index of the gas at different temperatures, the change of the OMV with the temperature can be obtained.

a) Calculation of gas parameters

The gas medium is air, so its refractive index can be calculated by the formula proposed by Rueger at the working temperature, given by:

$$n_{ph} = \frac{273.15}{1013.25} \cdot P \cdot N_{sph} \cdot \left(a_0 + \frac{a_1}{T} + \frac{a_2}{T^2} + \frac{a_3}{T^3} + \frac{a_4}{T^4}\right) \times 10^{-6} + 1 \quad (1)$$

Where P is the gas pressure; a_0, a_1, a_2, a_3 and a_4 are the correlation coefficient and constant; N_{sph} is the gas refractive index at atmospheric temperature, given by:

$$N_{sph} = (n_{sph} - 1) \times 10^6 = 287.6155 + \frac{1.62887}{\lambda^2} + \frac{0.01360}{\lambda^4} \quad (2)$$

The wavelength range is 400-700 nm. The gas refractive indexes at different temperatures in the range of 25-650 °C are calculated by Eq. 1 and 2. The results are shown in Figure 8. It can be found that the refractive index of air decreases with the increase in the gas temperature. And it can be inferred that the light divergence angle at high temperature is going to increase compared with that at atmospheric temperature. Consequently, the position of the beam focus is offset, leading to larger optical measuring volume, as shown in Figure 9.

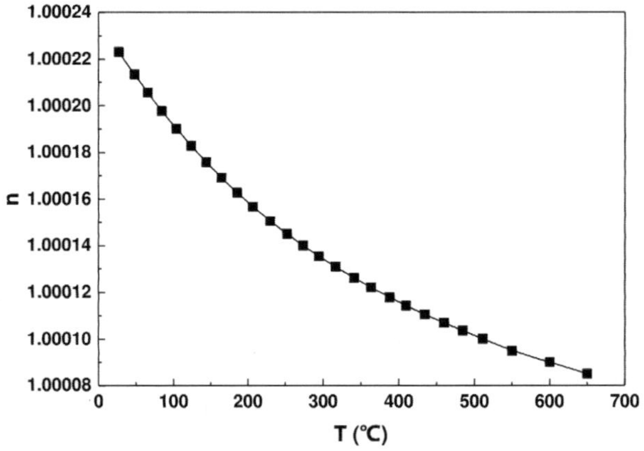

Figure 8. Gas refractive index at different temperature

In order to analyze the above unfavorable influence, a model of the OMV and gas temperature is needed to be built, mainly including the functions of the determination of the optical parameters and the calculation of the dynamic variable.

b) Calculation of optical parameters

On the basis of the Lorentz-Lorenz formula and actual gas state equation, the relationship between gas pressure and refractive index is given by:

$$\frac{n^2-1}{n_0^2-1}=\frac{\rho}{\rho_0}=\frac{PT_0Z_0}{P_0TZ} \quad (3)$$

Where T_0, P_0, ρ_0, Z_0 and n_0 are the gas temperature, pressure, density, compression factor and refractive index under atmospheric condition, respectively; T, P, ρ, Z and n are the gas temperature, pressure, density, compression factor and refractive index under high-temperature condition, respectively.

The light divergence angle is calculated by the refraction law shown below:

$$n_0' \sin\alpha = n \sin\beta \quad (4)$$

Where n_0' and α are the air refractive index and light divergence angle under atmospheric condition; β is the light divergence angle under high-temperature condition.

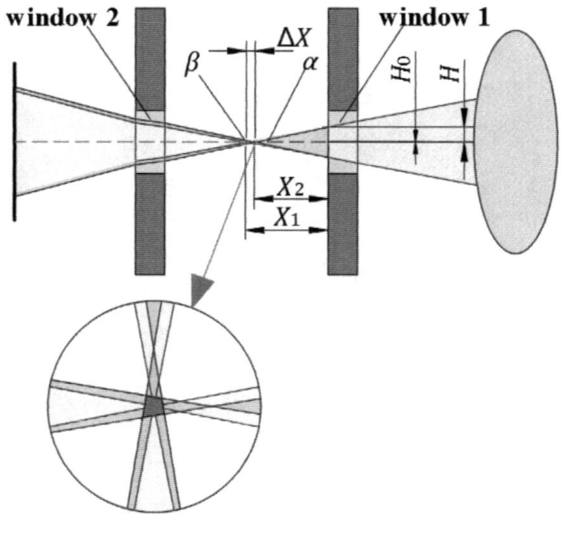

Figure 9. Schematic of the enlargement of the OMV

c) Calculation of dynamic variable

Based on the geometric optical theory and the mathematic relationship, the distance between the light focus position and the glass window 1 is triangular with the equivalent spot radius of the inner side of the glass window 1. According to X_1 and X_2 (the distance between the light focus position and the inner side of the glass window 1 under atmospheric condition and high-temperature condition), the theoretical model of the OMV with gas temperature was built. X_1 and X_2 are calculated by:

$$X1=\frac{H-H_0}{\tan\alpha} \quad (5)$$

$$X2=\frac{H-H_0}{\tan\beta} \quad (6)$$

The change of the light focus position (ΔX=$X1$-$X2$) can be calculated by Eq. 5 and Eq. 6.

The change of the equivalent spot diameter of the OMV can be calculated based on the triangle relation, as shown below:

$$\Delta D = 2\Delta X \tan\beta \quad (7)$$

Eq. 6 and Eq. 7 are the formulas of the theoretical dynamic model of the OMV with the change of the gas pressure.

Using the model, the optical parameters at different gas temperatures in the range of 25 ~ 700°C were calculated with three light divergence angles (4.02°, 9° and 13°, tested under atmospheric condition, respectively). The results are presented in Figure 10. It can be found in Figure 10 that ΔX increases with the gas temperature with approximate two patterns. In addition, ΔD changes by a smaller margin (less than 1 um). However, ΔD and ΔX satisfy a trigonometric relation. It can be also noted that the enlargement rate of the OMV volume (δ) has a small increase with the increase in gas temperature. Under the constant temperature, δ is going to increase for the larger light divergence angle, but it is still relatively smaller (<1%) compared with the OMV at atmospheric condition. Consequently, the influence of the gas temperature on the OMV can be ignored based on the above theoretical analysis.

3.2.2 Experimental method

In order to verify the reliability of theoretical analysis, an experimental dynamic model of the OMV with gas temperature variation was established based on the image analysis and the mathematical geometry relations, including the measurement of the change of light spot, the calculation of the geometric relation and the calculation of the relative variable.

By the refraction law, the changes of light spot on the inner and outer side of the glass window 2 under high-temperature condition (ΔD and $\Delta D'$) meet ΔD=$\Delta D'$. Taking the light focus position as the central axis, a triangle relation was formed by the distances between the light -

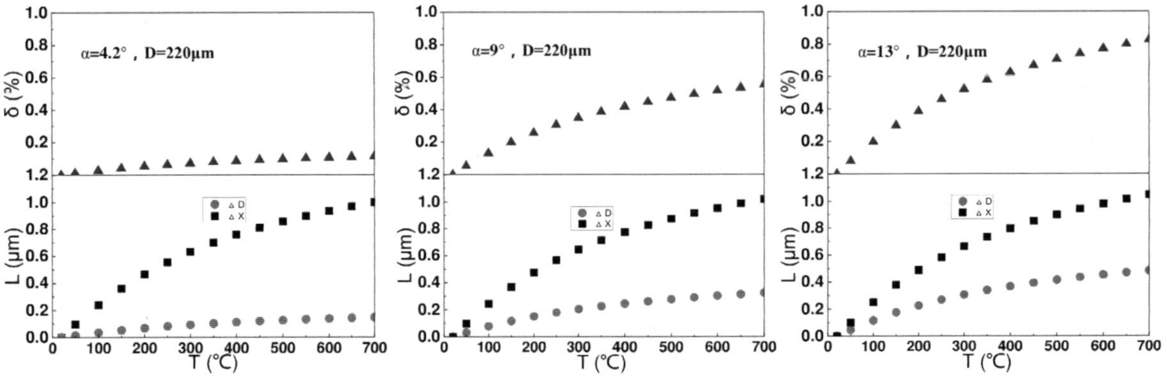

Figure 10. Results calculated based on theoretical dynamic mode with three light divergence angles (4.02°, 9° and 13°)

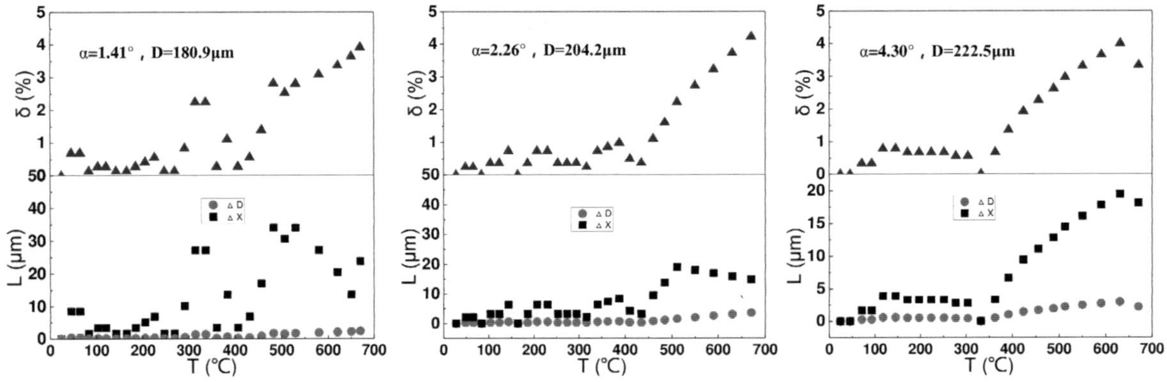

Figure 11. Results calculated based on experimental dynamic model

focus position and the inner side of the glass windows 1 and 2 (x and y), the difference of the light spot equivalent radius on the inner side of the glass windows 1 and 2 (H and h, $h = H - \frac{\Delta D}{2}$) and the equivalent radius of the OMV (H_0). Therefore, we have the following equations.

$$\frac{H - H_0}{h - H_0} = \frac{y}{x} \tag{8}$$

$$x + y = k \tag{9}$$

$$\beta = \arctan\left(\frac{h - H_0}{x}\right) = \arctan\left(\frac{H - H_0}{y}\right) \tag{10}$$

where β is the light divergence angle under high-temperature condition; k is the distance of the inner sides of the glass windows 1 and 2.

The changes of ΔX and ΔD under high-temperature condition to those under atmospheric condition are shown as follows:

$$\Delta X = \frac{y - x}{2}, \text{ or } \Delta X = y - \frac{k}{2}, \text{ or } \Delta X = \frac{k}{2} - x \tag{11}$$

$$\Delta D = 2\Delta X \tan\beta = \Delta D' \tag{12}$$

Eq. 11 to Eq. 12 are the formulas of the experimental dynamic model.

Using the above established experimental dynamic model, the ΔX, ΔD and δ of four cases in Table 2 were calculated, and the results are shown in Figure 11. It can be found that these three parameters increase obviously with the gas temperature increasing, but cannot meet the two function growth mode as the results of theoretical model. That may be caused by the reasons listed below: 1) the fluctuation of gas flow in the pipeline causes the unstable gas temperature field; 2) the beam quality analyzer is slightly influenced by other vibration sources; 3) the aerosol tube is affected by the gas flow fluctuation. But according to the experimental results, the maximum values of ΔX and ΔD are very small (less than 50 μm and 5 μm at the temperature below 700 °C, respectively) compared with the equivalent diameter of the OMV. Consequently, the ratio of the volume of the OMV is less than 5%. To sum up, the gas temperature has greatly weakened influence on the OMV, which can be ignored in actual technical development.

Table 2. Optical parameters of experimental model

number	light divergence angle /°	spot equivalent diameter /μm
1	1.41	180.9
3	1.94	285.1
4	2.26	204.2
5	4.30	222.5

3.2.3 Experiments

The performance of the optical sensor is investigated at different temperatures up to 650°C. The experimental results showed that the temperature of the optical sensor is around 50°C with the gas temperature in the range of 200-300 °C, and is going to increase about 10 °C at the gas temperature of 400-500 °C. The temperature is in the controllable range with a maximum value of 70 °C under the gas temperature of 650 °C. Therefore, the developed optical sensor can be used for the detection of particles in high-temperature conditions for a long time.

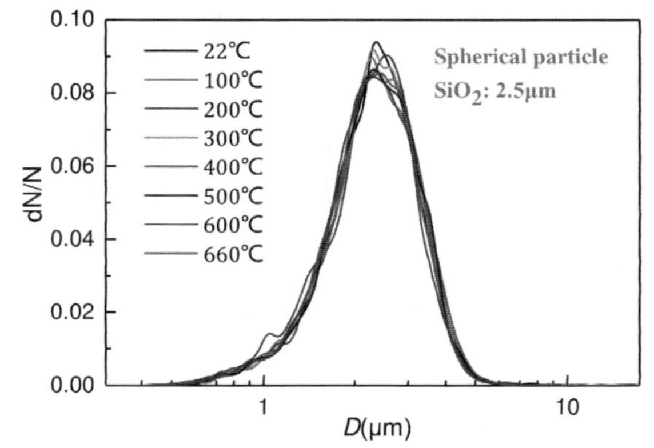

Figure 12. The results obtained under high-temperature condition

In the experiment of particle detection, the standard silica spherical particles with a strong temperature resistance ($D_{particle}$=2.5 μm) were tested at different gas temperatures. The testing time is 60 s. Figure 12 shows the test results. It can be found that all of the particle size distributions are in good agreement and have a good normal distribution, which meets the test results of single particle size, and the peak particle size is about 2.5 μm. Consequently, the optical sensor worked well for the particle detection under higher-temperature condition.

4 Conclusion

The emission of high-temperature flue gas causes serious air pollution and partial waste of available heat. In order to meet the relevant emission standards, additional purification units (high-temperature ceramic filters, etc.) are installed in the industrial process to collect the dusts in the flue gas. As the purification units are important for controlling the dust concentration, it is necessary to calculate the separation efficiency through a real-time measurement of the dust concentration in the flue gas in the upstream and downstream of the purification equipment.

In present work, a single particle measurement method and device suitable for high-temperature condition were developed. In order to complete the particles measurement at 650 °C, the beam images in optical sensing system under different temperature conditions were measured in real time by CMOS sensor. It was found that the beam images were deformed when the light divergence angle increased to more than 10°, and the deformation became more serious as the temperature increased. Depending on the principle of heat transfer, thermal expansion along the airflow direction caused by heat conduction between the duct and the gas is the main cause of this phenomenon. By changing the aerosol tube material t o one with an ultra-small linear expansion coefficient (Invar: 4J36), the optical system can be applied to the condition with temperature up to 700 °C, and in particular the light divergence angle extends to 30°.

According to the principle of gas state equation, the optical parameters (refractive index) of the gas at different temperatures change with the change in its state parameters. According to the results calculated by the formula of Ruegur's atmospheric refractive index, it was found that the refractive index of the gas decreases with the increase in temperature, which leads to the increase in the divergence angle of the beam in the aerosol tube, and shift of the position of the optical beam. Based on the geometrical optics theory and the geometric mathematical relationships, the theoretical dynamic model of the optical measuring volume with the change of the gas temperature was established. By calculating the amount of the beam focus offset and the volume of the optical measuring volume at different temperatures, it was found that the optical measuring volume increases with the temperature increasing. However, when the gas temperature was 650 °C, the increase rate of optical measuring volume was extremely small, so the influence of the temperature on the optical system can be negligible.

The performance of the whole detection system was tested by using 2.5 μm standard spherical silica particles. The satisfying measuring results were obtained and they could be applied to the actual on-site measurement at the temperature of 650 °C.

Acknowledgement

The authors thank the supports of "The National Key Research and Development Program of China (NO.2016YFB0601100)"; and would also like to thank the laboratory and staff of the Beijing Key Laboratory of Process Fluid Filtration and Separation, China University of Petroleum.

Reference

[1] J. Meyer, H. Umhauer, K. Reuter, A. Schiel, G. Kasper, M. Förster, Concentration and size measurements of fly ash particles from the clean gas side of a pressurised pulverised coal combustion test facility, Powder Technology, 180 (2008) 57-63.

[2] H. Leibold, F. Dirks, V. Rüdinger, Particulate emissions from a LLW incinerator and off-gas cleaning with a new type of ceramic candle filter, Waste Management, 9 (1989) 87-94.

[3] B.L. Perkins, Incineration facilities for treatment of radioactive wastes: a review, Incinerators, (1976).

[4] D. Fino, N. Russo, A. G. Saracco, V. Specchia, Multifunctional Filter for Treatment of the Flue Gases from Municipal Waste Incinerators, Industrial & Engineering Chemistry Research, 44 (2005) págs. 9542-9548.

[5] Scheibner, Wolters, Schumacher hot gas filter long-term operating experience in the nuon power buggenum IGCC power plant, (2002).

[6] J.M. Wang, Shell coal gasification technology and its application in China, Coal Quality Technology, (2006).

[7] Q. Xu, Z. Ji, L. Yang, Performance assessment of high temperature flue gas on-line particle size analyzer, Ciesc Journal, 63 (2012) 3506-3512.

[8] B.Y.H. LIU, V.A. MARPLE, K.T. WHITBY, N.J. BARSIC, Size distribution measurement of airborne coal dust by optical particle counters, Aihaj, 35 (1974) 443-451.

[9] R. Maus, F. Löffler, Size analysis of airborne bacteria by an optical particle counter and an impactor, Journal of Aerosol Science, 25 (1994) 141-142.

[10] B. Sachweh, H. Büttner, F. Ebert, Improvement of the resolution and the counting accuracy of an optical particle counter by fast digital signal recording, Journal of Aerosol Science, 20 (1989) 1541-1544.

[11] E.W. Max Born, Principles of Optics, Cambridge University Press, Cambridge, 1999.

[12] S. Hochrainer, Further developments of the particle counter sizer PCS-2000, Journal of Aerosol Science, 31 (2000) 771-772.

[13] B.S. Li, Z.L. Ji, L. Mei, An approach for correcting particle size distributions measured by the optical particle counter welas 2000, Aip Conference, 2010, pp. 686-689.

[14] J. Stier, M. Quinten, Simple refractive index correction for the optical particle counter PCS 2000 by Palas, Journal of Aerosol Science, 29 (1998) 223-225.

[15] H. Umhauer, Particle Size Distribution Analysis by Scattered Light Measurement Using an Optically Defined Measuring Volume, Journal of Aerosol Science, 14 (1983) 765-770.

[16] B. Sachweh, H. Umhauer, F. Ebert, H. Buttner, R. Friehmelt, In situ optical particle counter with improved coincidence error correction for number concentrations up to 107 particles cm-3 quantification and elimination of the effect of particle shape and structure, Journal of Aerosol Science, (1998) 1075-1086.

[17] D.I.J. Raasch, D.I.H. Umhauer, Errors in the Determination of Particle Size Distributions Caused by coincidences in optical particle counters, Particle & Particle Systems Characterization, 1 (1984) 53-58.

[18] H. Buettner, Measurement of the Size of Fine Nonspherical Particles with a Light-Scattering Particle Counter, Aerosol Science & Technology, 12 (1990) 413-421.

[19] D. Holve, S.A. Self, Optical particle sizing for in situ measurements Part 2, Applied Optics, 18 (1979) 1632-1645.

508245 An Experimental Investigation of Single Droplet Drying Above Boiling Point

Abstract:

An experimental method has been developed to enable dried particle structure, properties and drying rate to be predicted based on droplet drying history, particularly focusing on effects driven by boiling including bubble nucleation, growth and collapse. The method involves a single droplet drying rig which will be used to investigate drying kinetics above boiling point and across a broad range of drying conditions, drop size and composition space. The drying rig was improved as compared to previous apparatuses reported in literature, in that it is designed to measure droplet drying rates and morphological changes at high temperatures and under conditions that more closely resemble those found in industrial applications. The aim is to conduct drying experiment for droplets as small as 300μm diameter, in an air stream which temperature may be as high as 350°C.

Drying kinetics and morphological evolution will be determined, particularly focusing on the second drying stage as well as inflation/deflation periods. Initially, the constructed drying rig is checked and evaluated by conducting experiments on droplets of pure water. Afterwards, experiments will be conducted on solutions of different materials such as Hydroxypropyl Methylcellulose (HPMC), sodium sulphate and sucrose. An assessment of morphologies, transformations and drying behaviour under isothermal drying would be completed at a wide space of process condition, initial water content, initial drop size and composition space. The investigation will look to link material properties, particle size, initial moisture content and drying history to morphology. Results will be compared with single droplet drying models reported in literature.

An initial investigation has focused on mapping the morphological drying behaviour of HPMC, sodium sulphate and sucrose. It has shown that the three materials show different behaviour; sucrose showed deflation and inflation cycles, sodium sulphate droplets supersaturated and crystallized while HPMC showed inflation and rupture. Differences in drying behaviour is due to difference in material and transport properties. Therefore, the next stage of this work is focused on conducting some physical properties measurement at different water concentrations and temperature. In addition, different materials will be dried on the rig and the drying behaviour will be determined. The experimental data from this work will be used to develop, a universal morphology regime map which will link material properties, particle size, initial moisture content and drying history to morphology.

508482 Characterising Powder Flow in Dynamic Processes

Abstract:

Flow inconsistency of cohesive powders is a major issue for process industries. It is often difficult to predict and control flow of cohesive powders to enable efficient and reliable process operations particularly for dynamic processes such as mixing and filling. Many quantitative and qualitative flow measurement techniques exist that are used for measuring powder flowability under quasi-static conditions. However, there are few available techniques capable of assessing powder flowability under dynamic conditions. The FT4 Powder Rheometer (Freeman Technology), which drives a rotating blade along the vertical axis through a column of powder contained within a cylindrical vessel, is one of the few available devices that is capable of measuring powder flow in a dynamic state [1]. The flow resistance of the rotating blade as it penetrates the powder bed is characterised by flow energy, however, there is lack of understanding in how to relate this measurement to process design.

This work focuses on determining the relationship between the torque and shear stress in the FT4. The effect of particle size on the external flow profile around the blade has been investigated experimentally for glass beads using a high speed camera. The material investigated shows a decrease in downward specific flow energy with increasing particle size; indicating that the increased shear resistance in the bulk for smaller particle sizes exists under dynamic conditions. Particle Image Velocimetry (PIV) analysis indicates an increase in the volume of particles displaced by the blade with increasing particle size for both upward and downward blade movement. An increase in impeller tip speed leads to a reduction in flow energy for this material, and a reduction in the volume displaced by the blade. The prevailing stresses and strains as well as the volume of particles displaced by the blade are estimated by Discrete Element Method (DEM). DEM allows for the determination of a relationship between the flow energy and/or shear stress with particle properties and blade velocity and direction. Furthermore, Positron Emission Particle Tracking (PEPT) analysis of the FT4 has been carried out to determine the radial and axial displacement of particles with respect to the bed height, and to validate the DEM results.

508671 Impact ofÂ Non-Spherical Projectiles on Granular Media

Abstract:

Recently, several studies involving numerical and experimental methods have focused on the study of impact dynamics in both dry and wet granular media. An example of such instances can be an animalâ€™s foot impacting sand during its locomotion. Understanding the physics behind the interaction between granular media and animal feet can be helpful in optimizing locomotion parameters for bio-inspired robots. However, most of the studies involving impact dynamics considered spherical projectiles under different conditions, while characteristic models should involve more complex shapes. Impacting different geometries with conserved density, volume, and velocity on a granular bed may experience contrasting drag forces upon penetration. This is due to the difference in the surface areas coming in contact with the granular media. In this work, non-spherical geometries of following shapes were considered: cylinder, triangular prism, hexagonal prism and cuboid. These geometries were impacted on a loosely packed non-cohesive dry granular bed with the same impact velocities. The effect of surface area in contact with the granular media was analyzed for impact, with the help of penetration depth measurements. We also tested whether the power law for spherical projectiles is applicable for non-spherical projectile impacts.

508664 Computational and Experimental Shear Cell Study with Rigid Cylindrical Particles

Abstract:

Flow properties of particles are commonly assessed through shear cells. Both the Schulze ring shear tester (SRST) and Jenike shear tester (JST) measure the interparticle stresses which are important to the design of many granular flow processes. This work uses precision-cut rigid cylinders of various aspect ratios and sizes to compare the measured stresses between both testers. It is found that the shearing results for both testers are similar, which is important for future work involving flexible biomass materials.

Previous work has sought to use shear cell results to validate various aspects of discrete element method (DEM) simulations. The SRST is recreated in a DEM simulation with rigid glued-sphere cylindrical particles to assess the contact detection method by comparing the experimental and computation results. Also, a parameterization study is performed to examine the effects of coefficient of restitution, coefficient of friction, and other particle properties on the measured steady-state stress. Overall, this serves as an efficient validation of the contact detection method which will be used to simulate complex particle shapes.

508865 Defluidization Behaviour of Industrial Reactive Powders at High Temperature

Abstract:

Fluidized bed reactors are used in a wide range of industrial chemical processes, and they are particularly useful in high temperature systems thanks to their ability to provide high heat transfer rates and rapid solids mixing which lead to isothermal and more controllable conditions. Their applications spread the environmental, chemical, energy and process industries. For example, they are a key technology in the petroleum industry, catalytic chemicals synthesis processes, combustion and gasification of solid fuels (coal, wastes and biomass), metals productions and many more.

Use of fluidized beds in combustion-type operations may lead to particles sintering and bed agglomeration, which again may result in defluidization, unscheduled process downtime and additional costs. However, the details of the physical mechanisms behind such phenomena are still not entirely understood. Despite the countless researches that have been done so far, such processes still disclose unrevealed features and challenge the researchers worldwide.

The objective of this paper is to assess the effect of high temperature on the flow properties and fluidization behaviour of mixtures of carbon coke and two different titanium ores, namely synthetic and natural rutile. To this end, a simple approach for the early detection of agglomeration in the regimes of incipient fluidization -where the carbon combustion causes defluidization- will be followed. Such a monitoring approach is based on the simultaneous measurements of local temperatures and the pressure drop across the particles bed.

507739 Manufacturing and Characterization of Spherical Blend PBT-PC Particles for Additive Manufacturing

Abstract:

Polymer powder-based additive manufacturing technologies often yield excellent mechanical and functional part qualities. One of these technologies is selective laser sintering (SLS), a process employing micron-sized polymer particles, which are selectively fused by a laser [1]. While hardly any boundaries regarding part design exist, there are quite some restrictions concerning the variety of commercially available SLS materials. For the time being, the most widely used polymeric material for SLS is polyamide 12 (PA12) with a market share of roughly 95% [2]. Hardly any blend materials are available; mostly , the processing of multi-material systems via SLS is achieved by simply mixing different polymer powder materials [3].

In this contribution, a new approach to blend particles is introduced. Via co-grinding and subsequent rounding, manufacturing of spherical particles consisting of semi-crystalline polybutylene terephthalate (PBT) and amorphous polycarbonate (PC) thermoplastic is possible. The polymers are jointly comminuted and mixed in a planetary ball mill and the product particles are rounded in a heated downer reactor [4,5]. The thereby achieved intermixed, but not transesterified, PBT-PC composite particles are characterized regarding size and morphology via electron microscopy. Intermixing of the polymers is investigated via staining techniques and Raman spectroscopy. Thermal properties of the blend particles are analyzed via DSC, also in regard to transesterification in SLS process-like conditions. Furthermore, flowability and size distribution of the blend particles are examined. This way, we show the high potential of our process route [4] for the production of novel polymer blend particle systems for SLS.

References:

[1] I. Gibson, D. Rosen, B. Stucker, Additive manufacturing technologies, 2. ed., Springer, New York, NY, 2015.

[2] T.T. Wohlers, Wohlers Report 2016, Wohlers Associates, Fort Collins, Col., 2016.

[3] D. Drummer, K. Wudy, F. Kühnlein, M. Drexler, Polymer Blends for Selective Laser Sintering: Material and Process Requirements, Phys. Procedia. 39 (2012) 509–517. doi:10.1016/j.phpro.2012.10.067.

[4] J. Schmidt, M. Sachs, C. Blümel, B. Winzer, F. Toni, K.-E. Wirth, W. Peukert, A novel process route for the production of spherical LBM polymer powders with small size and good flowability, Powder Technol. 261 (2014) 78–86. doi:10.1016/j.powtec.2014.04.003.

[5] J. Schmidt, M. Sachs, S. Fanselow, M. Zhao, S. Romeis, D. Drummer, K.E. Wirth, W. Peukert, Optimized polybutylene terephthalate powders for selective laser beam melting, Chem. Eng. Sci. 156 (2016) 1–10. doi:10.1016/j.ces.2016.09.009.

506658 Creating Tuneable Agglomerates via 3D Printing

Abstract:

It is essential to offer a universally accepted way to systematically evaluate agglomerate breakage propensity and mechanisms. Discrete Element Method (DEM) simulation is a commonly used tool but is limited by the lack of identical, controlled agglomerates to test and validate simple models, let alone replicate the complex structure of real industrial agglomerates.

Our research presents a systematic approach to produce 3D printed agglomerates with tuneable properties. Agglomerates with different mechanical properties and structures were designed and printed by an Objet500 Connex 3D printer. The tensile and shear strength of inter-particle bonds was characterized using Instron tester. Breakage results of agglomerates with simple cubic structures and the design of agglomerates with irregular structures were presented. This 3D printing method will allow more rigorous testing of agglomerate breakage models. Discrete Element Method (DEM) has been used to simulate the agglomerate crushing using the Timoshenko Beam Bond Model (TBBM) with bond properties matching the 3D printed agglomerates. For both two agglomerate structures used in this research, the simulation and experiment showed similar breakage patterns. However, the simulation results significantly underestimated the compressive loads of spherical random structured agglomerates. In order to improve the simulations of compressive granule breakage, the influences of bond geometry definitions and non-linear behaviour of the bond contacts need to be explored.

508699 Effect of Particle Size Distribution and Wettability on Penetration Behavior, Granule Formation and Granule Properties in Single Drop Granulation

Abstract:

In wet granulation, single drop granulation (where one drop forms one granule) can simplify the complexity and reveal the fundamentals of the interaction between the liquid and the powder bed. Combining drop penetration time, granule formation mechanisms and the characterization of granule properties is a novel study to achieve a comprehensive map of the single drop granulation process, which is relevant in particulate applications like pharmaceuticals, detergents, agricultural chemicals and catalysts.

In this work, single drop impact of liquid on a static powder bed was studied to investigate the granule formation mechanism, droplet penetration time, and the characterization of granule properties (morphology, internal structure and content uniformity). Water was used as the liquid binder and two pharmaceutical powders, microcrystalline cellulose (MCC) (mean particle size ~54.8 Î¼m, contact angle with water ~36Â°) and acetaminophen (APAP) (mean particle size ~9.74 Î¼m, contact angle with water ~89°), were mixed to make heterogeneous powder beds with varying compositions. The complete drop impact and penetration was recorded with a high-speed camera. The granule morphology, its internal structure and content uniformity were characterized by the prism method with image analysis, micro-CT and UV-vis spectrometry, respectively. Besides the current MCC and APAP system, MCC and APAP with other particle sizes and relative contact angles will also be studied. It is believed that the mean particle size of the powder bed is predominant in influencing the granule formation mechanism, drop penetration time, and granule properties.

508017 Atomically Deposited Sintering Aids: Assessing the Effects of Alumina Particle ALD on the Sintering and Performance of SOFC Electrolytes and Dental Ceramics

Abstract:

The crystal structure of zirconium oxide (ZrO_2) can be tailored by the incorporation of aliovalent dopants, most commonly, yttrium oxide (Y_2O_3). The concentration of dopant dictates the thermodynamically stable crystal phase where 3 mol%
Y
conducting cubic phase. To achieve suitable mechanical properties and ionic conductivity, Y-TZP and YSZ ceramics must be near theoretical density, requiring high sintering temperatures. Previous researchers have successfully lowered the required Y-TZP and YSZ sintering temperatures through incorporation of aluminum oxide (Al_2O_3) particles. While this technique has proven to reduce the sintering temperature in both systems, the presence of particulate Al_2O_3 inclusions in densified parts reduces the homogeneity of the material which may have deleterious effects on densification and microstructure. In this work, we report on the use of atomic layer deposition (ALD) to conformally coat individual Y-TZP and YSZ particles with the desired amount of Al_2O_3. Constant-rate-of-heating (CRH) experiments were conducted using a horizontal push-rod dilatometer to extract the activation energy of sintering and also gain mechanistic insights into the active diffusion mechanisms as a function of ALD film thickness. Transmission electron microscopy (TEM) and scanning electron microscopy (SEM) were used to characterize the as-deposited thin Al_2O_3 films and the effects of the Al_2O_3 films on grain growth and microstructure during densification. Ionic conductivity and flexural strength tests were conducted to determine the effect of Al_2O_3 thin films on relevant material properties. Al_2O_3 incorporation by ALD is shown to have unique phenomenological effects on the densification of Y-TZP and YSZ as compared with conventional Al_2O_3 incorporation using particle-based approaches.

508644 Effect of Particle Size Distribution and Operating Parameters on Conduction and Convection Heat Transfer Mechanisms in Rotary Drums

Abstract:

Granular materials undergo process steps that include transportation, drying, heating and chemical or physical conversion. In several instances, these materials need to be heated or cooled during processing, and rotary drums (kilns) are the most commonly used process equipment for this purpose. It is not always possible to conduct experiments using different materials and consider all process conditions to study the thermal mechanism due to the inherent complexity of these processes. Therefore, computational modeling and simulations have been extensively used to understand the heat transfer mechanism in rotary drums. To maintain simplicity and to overcome the challenges of simulating polydispersed particles, a common practice is to assume that the polydispersed particles with a narrow size distribution behave like monodispersed particles thermally. However, no effort has been made to validate this assumption. This work focuses on understating the effect of polydispersed particles on the conduction and convection heat transfer mechanisms in indirectly heated rotary drums. It also aims to investigate the effect of rotation speed, fill level, particle size and their interaction effects on the conduction and convection heat transfer mechanisms in rotary drums. For this, the discrete element method (DEM) (using MFIX-DEM, an open-source computational fluid dynamics solver suite) simulation technique is used to analyze the thermal behavior of spherical silica beads, which are the most common catalyst support material.

507974 Selective Hydrogenation of Citral over Supported Pt Catalysts on Various Substrates

Abstract:

Highly dispersed platinum (Pt) nanoparticles (NPs) were deposited on various substrates by atomic layer deposition (ALD) for the reaction of citral hydrogenation. The substrates included multi-walled carbon nanotubes (MWCNTs), silica gel (SiO_2), commercial Î³-Al_2O_3, and ALD-prepared porous Al_2O_3 particles (ALD-Al_2O_3). Based on TEM and CO-chemisorption analysis, the Pt NPs on all substrates were around 1.3 nm, and Pt/MWCNTs, Pt/ALD-Al_2O_3 and Pt/Î³-Al_2O_3 showed similar Pt dispersion (75%-85%). But the Pt dispersion on SiO_2 was low (63%). In the reaction of citral hydrogenation, Pt/SiO_2 showed the highest activity due to the strong acidity of SiO_2 and the conversion of citral reached 82% after 12 hr with a selectivity of 58% of unsaturated alcohols (UA), including geraniol and nerol. Among all catalysts, Pt/MWCNTs showed the highest selectivity of UA (65%) due to its unique structure and electronic effect. In the cycling test, Pt/MWCNTs and Pt/ALD-Al_2O_3 catalysts were more stable than Pt/SiO_2 which could be attributed to the different interactions between the catalysts and the solvent.

508345 Numerical Simulation of Particle Classification in a Classifier Based on Coanda Effect

Abstract:

Efficient and precise classification of particles has been an important issue in many industrial applications as well as in academia. A relatively new particle classifier based on 'Coanda effect' was developed and reported to have many favorable features (Liang 2010). They include simultaneous multiple classification, simple design with no moving parts, and easy scale up to higher production rates. However, this classifier has attracted little attention so far, and our understanding of underlying mechanism associated with classification performance needs to be improved.

The purpose of this study is to investigate the detailed fluid flow and particle transport characteristics in a Coanda classifier using CFD (Computational Fluid Dynamics). The main focus is placed on the effects of working conditions (inlet flow velocity, particle diameter) and numerical method for turbulence dispersion of particles on the classification performance. As a numerical method, three different approaches are considered: k-ε model, k-ε model combined with DRW (Discrete Random Walk) model, LES (Large Eddy Simulation). In the first approach, standard k-ε model is adopted to obtain the mean flow and then particle trajectory is calculated based on the mean flow without consideration of turbulence dispersion. However, in the second approach using DRW model, the instantaneous flow velocity is constructed under the assumptions of isotropic turbulence in order to include turbulence effect. Finally, in LES, the instantaneous flow field is dynamically and accurately obtained at each time step. Numerical simulations are carried out for two-dimensional mean flow in case of k-ε model, but for three-dimensional unsteady flow in LES. Particles are tracked using a Lagrangian approach based on one-way coupling. We consider talc particles with density of 2,860 kg/m^3 and diameter of 1 to 20 μm.

The Coanda classifier is mainly composed of three inlet channels, three outlet channels, and Coanda block. Inlet channels are divided into one main inlet and two control inlets. The main air jet containing particles is supplied with the velocity of 35 m/s through the main inlet. Two control air jets with velocities of 32 m/s (lower) and 40 m/s (upper) are also supplied through the control inlets for enhanced control.

The trajectories of moving particles obtained from the mean flow (k-ε model) are analyzed for different diameters. The main jet containing particles deflects toward the curved Coanda block due to Coanda effect as expected. Fine particles ($d \leq 2$μm) follow the flow streamlines and enter the lower outlet channel. However, coarse particles ($d \geq 8$μm) are detached from the streamlines and go into the upper outlet channel. This observation agrees well with the previous result (Liang, 2010).

Even though the mean flow and corresponding particle trajectories provide useful information, realistic particle passage do not take a form of single curve but they change in time due to flow turbulence. Therefore, particle trajectories are simulated again, turbulence dispersion being considered with DRW model. Due to turbulence dispersion, particles of the same diameter are scattered around the particle pathline obtained from the mean flow. As a result, particles of the same diameter can not be collected into a single outlet. For example, some particles (about 5%) of 1 μm enter the middle outlet even though most of them (about 95%) enter the lower outlet designed for collecting fine particles. This means that flow turbulence deteriorates the accuracy of classification.

Finally, LES simulations are performed to understand dynamic behaviors of fluid flow and particle transport. Instantaneous flow fields such as vorticity contours clearly show unsteady fluid motions inside the classifier. It is also shown that the generation of unsteady vortices leads to the scattering of particle trajectory.

A more detailed analysis including the classification efficiency will be given in the final presentation.

Acknowledgements

This research was supported by the National Research Foundation of Korea (NRF) under grant number 2017M2A8A4018482 and 2015R1D1A1A0105 9675.

References

Liang, S. (2010). *Chem. Eng. Comm.*, **197**, 1016-1032.

503491 Numerical Simulation of the Secondary Air Distribution Layer in the Gas Solid Fluidized Bed Based on Flunt Software

Abstract:

Due to the continuity beneficiation in the practical production, the gangue would accumulate on the distributor forming the gangue layer, which was called Secondary Air Distribution Layer and has vital effect on the gas distribution. In this paper, the ventilation characteristics of the secondary air distribution layer was analyzed. In addition, FLUENT software was used to study the effect of the modeled layer for the fluidity characteristics (e.g., pressure drop, bed height, volatility, etc.) in the gas-solid fluidized bed with the complex layer. The results showed that the complex layer could promote the secondary distribution of the airflow in the bed, which effectively inhibited the production of large bubbles and reduced the bed pressure drop fluctuation. Thus, itâ€™s of significance to provide synergy effect on the fluidization stability in the fluidized bed. Based on the above analysis, the introduced complex structure would be convenient to achieve the coal separation effectively.

507949 CFD-DEM Simulation of the Fluidization of Non-Spherical Particles in Fluidized Bed

Abstract:

In chemical and energy industries, fluidization concerning non-spherical particles (such as rod-like particle, disk-like particles, ellipsoidal particles, etc.) is often encountered, but research in this area is minimal. Yet understanding fluidization of non-spherical particles is crucial in many relevant industrial process. To investigate the fluidization of non-spherical particles, a CFD-DEM coupling algorithm is employed in this research to simulate a bubbling fluidized bed. The non-spherical particles are modeled by super-ellipsoids, multi-sphere models, and other particle models. The simulations results show that, the fluidization of non-spherical particles has significant differences from that of the spherical particles. The local void fraction and particle's projected face area perpendicular to fluid flow direction are directly related to the orientations of the non-spherical particles, which has a significant influence on the particle drag force and fluidization. Due to the significance of particle orientation, we further investigate the factors that influence the distribution of particle orientations. The results show that particle orientation depends on the fluidization time, fluidization velocity, and structure of the fluidized bed.

508320 CFD and DEM Simulation of the Cold Spray Process for Surface Coating with Fine Particles

Abstract:

The coating and functionalization of technical surfaces has been become increasingly important in the last two decades for various applications, since of its manifold possibilities for process optimization, as friction reduction, corrosion resistance or improved wetting behavior. Cold spray is a thermal spraying method for surface coating. It was developed in the late 1980s in Novosibirsk [1] and has gone through continuous optimization on certain applications since then. For example showed our previous study a reduction of the friction properties on steel pins, microstructured with fine Titania particles by an adapted spraying method [2]. In the cold spray process particles are dispersed in a gas stream, which is heated up to several hundred degrees and accelerated to supersonic velocities in a Laval nozzle. The particles impact on a substrate with about 800 m/s and form a strong bonding due to the high kinetic energy which is transferred to deformation and heat in the contact zone. In contrast to other thermal spraying methods, the temperatures occurring during the process are far below the melting temperatures of the used materials, which makes the process suitable also for thermal sensitive powders. The quality of the produced coating layers depends highly on the properties of the powder and substrate as well as on the nozzle design and the process parameters as initial temperature and particle velocities. Due to the high gradients, the high occurring velocities and the small size of the particles, typically in the range of 1 to 50 µm, the contact zone of impacting particles can only be measured indirectly and with destructive methods. Therefore, still not all phenomena which lead to a strong bonding in the contact area and a high quality coating are completely understood. For the study of these mechanisms two different simulation approaches have been used:

a) Simulation of particle and gas flow with Computational Fluid Dynamics (CFD)

For a better understanding of the spray process, the nozzle, the adjacent flow areas and the substrate were computed with a CFD method. The trajectories of single particles were calculated by an Euler-Lagrange approach. For the flow the k-Ï‰ SST model was used. Local temperatures in the substrate were calculated by the law of heat conduction. The impact conditions dependent on the nozzle geometry and the particle properties. With the local temperature distribution in the substrate and in the particle, precise impact conditions could be estimated. A study has been conducted to vary the parameters of the spray process, temperature and pressure as well as the particle properties as shape factor, density and size. Moreover the nozzle geometry has been varied in the simulation to optimize the spray process. The simulation results are compared to images of coated surfaces obtained by Scanning Electron Microscopy (SEM) for validation.

b) Simulation of particle-surface impact behavior with Discrete Element Method (DEM)

The DEM is widely used for investigations of particular processes. The advantage of this method is the detailed calculation of particle-particle or particle-wall contacts to obtain the important parameters as forces and deformations. The contact force in the particle wall contact is also dependent on the surface morphology. Therefore the surface topography of two real technical surfaces, a polished and an unpolished Titanium substrate were measured by Scanning Probe Microscopy (SPM) imaging with a nanoindenter (TI Premier, Hysitron). The surface topography was discretized and implemented in a DEM model. A parameter study has been conducted to determine the distribution of forces and contact angles occurring at the particle impact. Different particle sizes and impact velocities were studied.

The cold pray process dependents on various parameters which affect each other. Since of the high gradients and the small scales occurring during the process experimental measurements are often difficult. With this numerical study different parameters could be considered uncoupled of the other parameters. The results can be used for a better process understanding and an optimization of commercial cold spray systems.

[1] Alkhimov, A. P., Papyrin, A. N., Kosarev, V. F., Nesterovich, N. I., Shuspanov, M. M.: Gas dynamic spraying method for applying a coating; US Patent, 5302414; 1994

[2] Buhl, S., Schmidt K., Sappok D., Merz, R., Godard, C., Sauer, B., Kerscher, E., Kopnarski, M., Antonyuk, S., Ripperger, S.: Surface structuring of case hardened chain pins by cold-sprayed micro particles to modify friction and wear properties; Particuology 21, 2015

508568 Two-Fluid Validation of Constituive Models for the Simulation of Cylindrical Particles

Abstract:

Two-fluid simulations are a popular tool in the modeling of large processes. The constitutive models that close the governing equations are based in granular kinetic theory and typically assume that the particles are spherical. In reality, very few processes utilize spherical particles, hence there is a need for constitutive models that describe more complex shapes. Previous work used discrete element method (DEM) simulations to create a solid stress[1,2] and collisional dissipation rate[3,4] model for cylindrical particles that can replace the current spherical models. These new models have been added to MFiX and this work focuses on their validation. Cylindrical particles of various aspect ratios and solid densities were used in cratering and hopper discharge experiments. Simulations are designed to recreate the experiments in order to compare the results for crater depth and hopper discharge rate. The model effectiveness is assessed with regards to assumptions made concerning particle rotation and preferential alignment.

1. Guo Y, Wassgren C, Ketterhagen W, Hancock B, James B, Curtis J. A numerical study of granular shear flows of rod-like particles using the discrete element method. *J Fluid Mech*. 2012:1-26.

2. Berzi D, Thai-Quang N, Guo Y, Curtis J. Stresses and orientational order in shearing flows of granular liquid crystals. *Phys Rev E*. 2016;93(4):40901.

3. Buettner KE, Guo Y, Curtis JS. Using the Discrete Element Method to develop Collisional Dissipation Rate Models that Incorporate Particle Shape. *AIChE J*.

4. Berzi D, Thai-Quang N, Guo Y, Curtis J. Collisional dissipation rate in shearing flows of granular liquid crystals. *Phys Rev E*. 2017;95(5):50901.

509002 The Comparsion of Coarse Grained CFD-DEM for Simulating the Dense Bubbling Fluidized Bed" Requires Immediate Action

Abstract:

CFD-DEM is an effective tool to study the hydrodynamics in dense bubbling fluidized bed. However, the expansive cost of calculating the particles motion is not easy for its application in industrial application or dense particles system. Thus several coarse grained models (CGM) were put forward to simplifying the calculation of particles and decreasing the particles motion, such as PSM, SPA, MSPA,CGEE, EMMS-DPM where a group of particles are set as a parcel or coarse particle to reduce the computation. In this study the experiment and simulation test conducted to test five coarse grained methods. The result indicated that the models increasing the gravity force, pressure gradient force and drag force linearly with the coarse grained ratio (CGR) can be effective to ensure similar macro behavior between original system and coarse graining system. The simulation data of the solid fraction distribution and bed expansion ratio reflected good agreement as experiment. But the bubbles structure became less clear as increasing the CGR. Furthermore, the power spectrum density (PSD) and Wavelet analysis of the pressure signal were used and showed that SPA and EMMS-DPM gave better agreement result in term of dominant frequency and energy ratio at mesoscale feature. Finally the computational efficiency for different models was compared and the models of EMMS-DPM, PSM and CGEE have higher computation efficiency than SPA and MSPA relatively if the collision model of spring-damping were adopted.

509153 Stability Analysis of Uniform Gas Solids Flow

Abstract:

Uniform distribution of gas solids flow should be ensured to gain sufficient performance of fluidized beds. The highly chaotic entropy dissipation across solids bed will trigger non-uniform fluidization, although distributor plays an important role in obtaining homogenous gas stream. Hence, total solids pressure loading (Î¦$_T$), the pressure drop ratio between solids bed and distributor, is introduced here to present their interaction. The boundary between uniform and non-uniform fluidization can be detected by stability analysis based on Prigogineâ€™s minimum of entropy production principle, and this criterion for N parallel paths can be simplified into calculating the second derivative of pressure drop through one path with respect of Î¦$_T$. Furthermore, a phase diagram, illustrating the effects from operational parameter (Î¦$_T$), geometrical characteristics of distributor (C_d) and property of solids (Ga) on instability of uniformity, is attained to draw some useful conclusions of robust uniform fluidization.

509309 Numerical Simulation of Particle Sedimentation Related to Nuclear Safety By CFD-DEM Algorithm

Abstract:

Numerical Simulation of Particle Sedimentation related to Nuclear Safety by CFD-DEM Algorithm

Byoungcheol Hwang[1], Kiyofumi Moriyama[1], and Hyun Sun Park[1*]

[1]Division of Advanced Nuclear Engineering (DANE), POSTECH, Pohang, Korea

**corresponding author: hejsunny@postech.ac.kr*

Abstract

The Discrete Element Method (DEM) is a technique to treat particle-particle interactions, and is used in many industries such as civil engineering, powder processing, pharmacy, and so on. Coupling of this method with Computational Fluid Dynamics (CFD) is needed in problems where the particles interacts with fluid significantly, such as bubble column reactor, mineral processing, and so on.

The simulation of important phenomena in nuclear power plant accidents that involves significant core damage and melting, i.e. so called severe accidents, is also one of the topic where the CFD-DEM is useful. After the Fukushima accident in 2011, the severe accident analysis and preparation of counter measures to mitigate the accident consequences became regulatory requirements in many countries. One of the measures for the cooling and stabilization of the molten core is the flooding of the cavity under the reactor vessel where the dropping core melt is caught in a water pool. It is expected that the molten core breaks up into particles of several millimeter sizes and accumulates on the floor, then makes a particle bed, which is hopefully coolable. In this situation, the assessment for the coolability of the particle bed is very important, and a number of experimental and analytical approached suggest that the internal/external structure of the particle bed affects strongly the coolability. The CFD-DEM is a promising candidate for an analysis method that can clarify such details of the particle bed formation.

We tested the applicability of the multiphase CFD-DEM code, an open-source developed by CFDEM®coupling in 2011, for the simulation of particle sedimentation and particle bed formation under water. Our focus in the present work is on the prediction of the particle falling characteristics and particle bed shape. Although the real condition during a severe accident includes very high temperature particles (>1000K) and boiling water (two-phase flow) in a large scale, we started with a simulation of much simple experiment cases without heat transfer or gas phase, with spherical particles only as the first step for the validation of the method. The experimental data with solid particles of various diameters (4 mm, 6 mm) and densities (3600, 6000, 7800 kg/m^3), particle release nozzle diameters (20, 30, 40 mm) were referred to. The results of numerical simulation were compared with the experimental data in terms of particle falling time, maximum particle dispersion angle, particle average falling velocity, particle bed distribution, and so on. The particle falling time was well predicted by the simulation with the maximum deviation 5%. The maximum particle dispersion angle by the simulation significantly deviated from the experimental data probably due to the influences of initial disturbances. The particle average falling velocity was overestimated. The final particle bed profile by the simulation showed agreed with the experimental observation accurately in the peripheral region, but the height of the center part was underestimated. The fluid dynamic drag model and the particle-particle contact model would need improvement for better accuracy. The present work confirmed the validity of CFD-DEM algorithm on the application for the particle sedimentation in a simple liquid pool. The particle-pool interaction including the gas phase will be the next step.

Keywords: CFD-DEM; particle-fluid interaction; particle sedimentation; nuclear safety; severe accident

509440 Application of Extedned Discrete Element Method to the Melting Process of a Packed of Particles

Abstract:

Application of eXtedned Discrete Element Method to the melting process of a packed of particles

Mehdi Baniasadi, Maryam Baniasadi, Bernhard Peters, Xavier Besseron

Université du Luxembourg, Faculté des Sciences, de la Technologie et de la Communication, Luxembourg

e-mail: mehdi.baniasadi@uni.lu, maryam.baniasadi@uni.lu, bernhard.peters@uni.lu, xavier.besseron@uni.lu, web page: http:/www.xdem.de

ABSTRACT

The phenomena of melting takes place in many industrial processes, such as metal processing, environmental engineering and thermal energy storage systems. The problem of modeling of solid-liquid-gas flow including physical and chemical conversions of particles such as melting is the subject of intense research in industrial processes. These systems are difficult to model with numerical analysis due to the complex interaction between the different fluids and granular phases condering heat, mass and momentum excahnge. Basically models used to predict multiphase flows with particles are Eulerian-Eulerian and Lagrangian-Eulerian. The difference between the two models is based on different treatment of the movement of the particles. The particles are tracked individually in Lagrangian-Eulerian approach while they are considered as a continuum medium in Eulerian-Eulerian. Lagrangian-Eulerian is closer to the real physical processes rather than the Eulerian-Eulerian model for granular flows with particleâ€"particle interaction[1].

In this paper, our main focus is on the modelling of the melting process for a packed bed of particles subject to a fluid flow using a numerical technique based on Lagrangian-Eulerian so-called eXtended Discrete Elemnet Method (XDEM)[2]. In this approach, the particles are resolved as discrete phase coupled via heat, mass and momentum transfer to the surrounding fluid phase. The Discrete Element Method (DEM) coupled by heat transfer is used in order to evaluate temperature distribution, melting rate as well as velocity, oriantion and position for each particle individually. In addition the approach allows to be linked with continues numerical approaches such as Computational Fluid Dynamics (CFD) in order to solve fluid fields. The model has been validated by comparing predicted results with existing experimental data for melting of a single ice particle as well as a packed bed of ice particles in the horizontally flowing water. Fluid-particle and particle-particle intractions has also been considered. In addition, the model has the capability to be extended to the packed bed of particles with different size and properties to produce different liquid phases.

REFERENCES

[1] Dierich, F., Nikrityuk, P. A. & Ananiev, S. 2D modeling of moving particles with phase-change effect. *Chem. Eng. Sci.* **66,** 5459â€"5473 (2011).

[2] B. Peters, Thermal Conversion of Solid Fuels (WIT Press, Southampton, 2003).

514248 Real-Time Interpretation of Chaotic Characteristics and Instability of Dense Gas Solid Fluidization

Abstract:

The establishment of instability monitoring system targets at analyzing the chaotic characteristics and evaluating the instability of the fluidization process for dense gas solid fluidization. Experimental pressure fluctuation and pressure drop measurements were interpreted from dense gas solid fluidization system based on chaotic time-series analysis. The pressure fluctuation signals were filtered in terms of attenuating the noise signals. The filtered pressure signals were analyzed with respect to attractor reconstruction. Based on the reconstructed attractor, the maximum Lyapunov exponent and entropy were investigated and estimated under different bed heights and air velocities. Results show that the maximum Lyapunov exponents udner different bed heights and air velocities are positive, which indicates that dense gas solid fluidization exhibits chaotic characteristics. With the increase of bed height, estimated entropy of dense gas solid fluidization increases with the increase of bed height under identical air velocity, while it varies slightly with the increase of air velocity when the air velocity alters from minimum fluidization velocity to 1.36 times of the minimum fluidization velocity.

508510 Glucan Particles As Potential Carriers of Natural Flavonoids for Treatment of Idiopatic Inflammation Diseases

Abstract:

One of the novel methods of pharmaceutical formulation is the loading of the drug into biocompatible microparticles, preferably extracted from natural sources. This approach offers several potential advantages, *e.g.* increased bioavailability (the ratio of the administered drug, which actually enters the bloodstream) compared to conventional drug delivery methods, enabling the option of oral administration even for parenteral drugs. Multiple principles of the loading process are possible – drug encapsulation, adsorption onto the particle surface or precipitation inside the pores a hollow particle. The presented work focuses on precipitation of a drug from an organic solvent inside glucan particles. Glucan microparticles are ellipsoid porous shells, which are obtained from cell walls of *Saccharomyces cerevisiae* (baker´s yeast). The particle size is 2-5 micrometers. Cell walls of glucan particles are mainly composed of β-1,3-glucans. In order to prevent inflammatory reaction in organism, the yeast organelles are removed by a series of alkaline and acidic extraction steps. Subsequently, present lipopolysacharides are extracted using different organic solvents.
Lymphatic cells are mainly present in inflammation area in organism where they are responsible for progress of inflammation. Glucan particles (GPs) interact with lymphatic cell´s receptors on their surface and after that β-1,3-glucans are phagocytosed by the lymphatic cells. Therefore, GPs represent an effective oral delivery system for transporting of biological active substances. Curcumin is an anti-inflammatory flavonoid, which occurs in plants of family *Zingiberacaea.* In our project, curcumin was delivered to an inflammation in organism using GPs. Loading of curcumin into GPs was enabled via precipitation of curcumin from organic solvents. Curcumin dissolved in organic solvents was added to the GPs and through pores penetrated into cavity of glucan shell. Next, the solvent was removed and the curcumin was precipitated in glucan microparticles. GPs with the flavonoid inside were observed using confocal microscopy and SEM (scanning electron microscopy). The achieved content of curcumin was successfully analyzed by other methods, which can not be mentioned directly due to article in preparation. Furthermore, the curcumin-loaded GPs were tested *in vitro* on macrophage like cells and *in vivo* on laboratory mice in order to evaluate their effectivity. The obtained results show that these composites are promising materials for biomedical applications such as efficient treatment of inflammatory diseases (e.g. idiopatic inflammatory bowel disease, rheumatoic arthritis).

508601 Use of Advanced Imaging Techniques in Tablet Disintegration Study

Abstract:

Certified testing of tablet disintegration and dissolution provides us useful information. However, there are many more that are hidden due to techniques' restrictions. Such information is e.g. mechanism of disintegration, change of surface morphology or uniformity of the substances' distribution in the tablet. Therefore, there are many techniques used to obtain the remaining information such as Magnetic Resonance Imaging (MRI), Fourier Transform Infra-Red (FTIR) or Raman spectroscopy and microscopy, Terahertz spectroscopy, Scanning Electron Microscopy (SEM) and other.

Presented work uses tablet made by direct compression of binary mixture of ibuprofen and lactose (based on previous results there was 0.2 - 0.5 w/w of ibuprofen content used) pressed at different compaction pressures (low, medium and high). The tablets were subjected to several solutions and were studied by techniques that have insight into the tablet (MRI) and ones that can see the surface (FTIR and SEM).

MRI experiments provided us with several interesting and several expectable results. As the content of ibuprofen was increasing so was increasing the dissolution time and the wetted layer was decreasing. Similar results were observed when the compaction pressure was increased. When the tablets were submerged to dissolution limiting solutions (saturated lactose solution and solution saturated with both lactose and ibuprofen) the disintegration was severely limited and some tablets began to increase their volume. However, in case of 0.2 w/w ibuprofen tablets it didn't matter whether there was dissolution discriminating solution or not, it disintegrated anyway.

FTIR microscopy showed, that not only the amount, but also the shape and size of the particles from which the tablets were compacted can be determined. These findings were further supplemented by the dry state information obtained by Scanning Electron Microscopy after different dissolution times. Further the FTIR-ATR imaging technique helped to understand the basic mechanisms of the tablet disintegration and to study the dissolution and water penetration into the model ibuprofen-lactose tablets.

508632 Correlation Study between Liquid Penetration and Mechanical Properties of Pharmaceutical Tablets

Abstract:

Main phenomenon influencing treatment of patient in pharmacy is amount of active pharmaceutical ingredient dissolved in medium such as water or blood. Even though this isa key step to achieve desirable medical effect it is not the first one occurring.The first step is governed by surface properties such as wettability and roughness and the first contact point is followed by liquid penetration which is highly dependent not onlyon type of used liquid, ionstrength, liquid density but also on mechanical and structural properties of tablets such as porosity, structure strength or particle bonds.Through many analytical methods we can describe each of thesesteps individually. Aim of this study is to find proper correlation amongseveralmethods such as sessile drop method, texture analysis (TA) and magnetic resonance imaging (MRI) of tablet during dissolution.

For experiments binary mixture of tabletwas used. Tablets consist of lactose (hydrophilic) and ibuprofen (hydrophobic)with variable content (from 0.2 to 0.5 w/w ratio) and were also variable in compaction pressure used on them (from low â€“ 5 kN through 10kN as middle to 18 kN as high).Three different solutions were used according to their saturation with primary substances to prevent the tablet from dissolving. Surface properties were evaluated on contact angle and soaking rate of liquid, both measured by sessile drop method. Texture analysis was performed by recording the material resistance during dry and wet state of tablet. Then the data was processed and trends were obtained.Tablets during the MRI experiments were submerged into dissolution medium and the proceeding wetting front was measured. Later the images were processed via image analysis and trends were obtained.

Results obtained from surface experiments show that higher the ibuprofen content is the more hydrophobic surface appears to be and therefore the contact angle is increasing. Contact angle is independent on manufacturing pressure but soaking rate on the other hand depends strongly.MRI images show that this technique is not suitable for rapidly disintegration tablets, where the calculated error is within experiment conditions. On the other hand, tablet with prolonged dissolution are well measured. Results from MRI experiments also state that local saturation with substances supress dissolution of tablet. Texture analysis of tablets show that they undergo immediately structural weakening after being poured with solution independently on type of solution. Rate of this structural change mainly depends on solution's acidity. Thenthe tablet decreases its resistance in time which can be interpreted as relaxation rate. With increment of ibuprofen or saturation with primary substances this rate decrease, which means that dissolution is suppressed. Correlation between MRI and relaxation rate was done. Results shows that the correlation is mainly dependent on texture analysis measurement. On the other hand, correlation between MRI and soaking rate appears to be more tentative.

Experiment characterising surface, structural and liquid penetration properties were investigated by three different techniques. Obtained results were evaluated and put to mutual correlation. From liquid penetration point of interest there is correlation between MRI and contact angle method. Also, a correlation between MRI and texture analyser was found.

508642 In Situ Drug Amorphisation By Microwave Irradiation Stabilized By Mesoporous Silica

Abstract:

One of the common issues in the pharmaceutical industry is the poor solubility of the newly developed active pharmaceutical ingredients (APIs). The solubility might be significantly enhanced by changing the form of the API from crystalline to amorphous form. The amorphous state is characterized by lower thermodynamic stability and thus better dissolution properties leading to better bioavailability. In addition, the amorphous API in a dosage form (e.g. tablets) must also remain stable and not recrystallize for at least the shelf life of the drug product (usually 2 - 3 years). Various methods are being utilized to prepare and stabilize amorphous formulations in the industry, however they usually involve difficult formulation steps or suffer from poor efficiency, low stability or can be patent obstructed.

In this work a stable amorphous formulation is prepared simply using the standard wet and dry granulation process or direct tableting. Crystalline form of API is used during the initial process. The amorphous form is then prepared *in situ* and â€œon demandâ€ either during manufacturing, or by the end user by applying microwave radiation (MW).

To stabilize the API in an amorphous form after the MW, multi-structured mesoporous silica particles were added to the formulation. The particles were prepared using an emulsion synthesis of octylamine in tetraethyl orthosilicate. Using ethanol as an inhibitor instead of the commonly used nitric acid as a catalyst reduces the silica primary nano-particles nucleation rate. The formed nuclei grow larger and then self-assemble around the octylamine droplets and form hollow micro-particles. The prepared microparticles were characterized using the Scanning electron microscopy (SEM) and the Transition electron microscopy (TEM). To characterize the porous structure, nitrogen sorption (BET) analysis was conducted. The results clearly show the particle assemblies contain macropores (cavities), mesopores and micropores and possess very high specific surface area and also very large pore volume. Together with the thermal and chemical stability of silica these properties make the particles well suitable for preserving the melted API in an amorphous form after the MW.

As a proof of concept, initially powder mixtures and later simple tablets composed of Ibuprofen (model API), silica micro-particles and a filler excipient were experimentally investigated. The *in situ* amorphization of ibuprofen was conducted using a commercial microwave oven. The heating properties of the oven, i.e. the spatial distribution of microwave heating was characterized. Various microwave power input and varying composition of the mixtures (excipient/Ibuprofen ratio) were tested with respect to the melting point of the Ibuprofen. The change from the crystalline to amorphous structure due to MW heating was analyzed by x-ray diffraction (XRD), differential scanning calorimetry (DSC) and FTIR-ATR techniques. The dissolution profiles of ibuprofen from both MW untreated and treated tablets were measured using the High Performance Liquid Chromatography (HPLC).

Based on these initial results, fully formulated tablets were prepared with various API dosage strength using a tablet press. Same compression force and same dwell time was used for all of the tablets. The final tablet composition includes excipient capable of absorption of the MW, which ensures a uniform heating in the whole volume of the tablet. Next it contains the silica microparticles, which by capillary forces absorb the API once melted and stabilize it after cooling down.

Lastly, the tablets contain standard filler, lubricant and disintegrant excipients. After the MW application the tablets change their structure due to the melted API intake into the silica micro-particles, forming visible pores in the tablet structure. This has a notable impact on the dissolution and hardness of the resulting tablets. For the final tablet characterization SEM, Texture Analysis, FTIR and Magnetic Resonance Imaging were used.

508646 Manufacture of Personalised Medicines By API Printing on Porous Tablets

Abstract:

Personalised medicine offers an opportunity to produce tailor-made medicaments for patients. In order to realize this vision, it is necessary to find an alternative approach to the conventional tablet design and production. One that will replace the existing mass production of drugs offering only a few dose options of the active pharmaceutical ingredient (API), while at the same time it must meet the strict reproducibility and a defined relationship between composition and release rate of active substances requirements.

One of such approaches is to print the API on demand on a solid substrate directly when it becomes needed. Appropriate dosing of API and the possibility to even combine several APIs into a single medicament improves the variability of the dosages available for prescription and reduces the overall number of pills the patient has to take. The goal of this work is to find a simple, automated and cost-effective method of preparing personalised medicaments, which could take place directly in the pharmacy shops or at home.

The method in this work consists of a precise dosing of a dissolved API into an already prepared substrate â€“ porous placebo tablets with high sorption capacity. The tablets are prepared using direct compression in a tablet press from a simple physical mixture of a binder, filler, disintegrant and lubricant. As a filler, mesoporous silica particles were used. The high pore volume provides the free space for API sorption into the tablets and the small size of the mesopores ensure the API is deposited in an amorphous form.

The prepared particles used in the formulation were characterized using Transmission electron microscopy (TEM), Dynamic light scattering (DLS) and nitrogen sorption analysis (BET). The prepared tablets hardness and uniformity was characterized both prior and after loading. The tablets after loading were characterized using the X-ray powder diffraction and the Fourier transform infrared spectroscopy to determine the API form and deposition uniformity, followed by dissolution tests to confirm the applied dosage strength and to measure the release profile. So far, several BCS Class II APIs were successfully infused into the substrates using precise on-demand dosing. Impregnation of various doses of API were accurate. Later characterization revealed no crystalline API were present and the dissolution profiles were very satisfactory.

509441 Understanding Phase Transition of Acetaminophen in the Bulk and Surface of Acetaminophen

Abstract:

Pharmaceutical solids processing often lead to the generation of new phases in the bulk and on the surface of the processed drugs. Phase transformation of the API can potentially change the physiochemical properties and influence the drug performances including solubility, dissolution and stability. To detect these changes, analytical tools such as XRD are usually employed. However, these techniques remain suitable for bulk analysis. Whereas, phase transition on the surface remains hard to detect due to the lack of literature, researches in this area and the absence of surface analytical techniques. The aim of this work is to investigate different phases (pure crystalline, partially crystalline partially amorphous and amorphous) of Acetaminophen drug by combining bulk and surface analytical techniques. For bulk analysis, FTIR and XRD have been employed. The former shows no significant differences between Acetaminophen various phases. Whereas the latter confirmed the presence of different phases indicated by a remarkable variation in the absorbance intensity the peak width. Where, high and sharp peaks obtained for the pure crystalline Acetaminophen, fewer peaks observed for semi crystalline Acetaminophen and a single broad peak observed for the amorphous Acetaminophen. For the surface analysis, Nitrogen K near-edge X-ray absorption fine structure (NEXAFS) spectra of a series of NEXAFS indicates that the precise determination of IPs and Ïf* shape resonance energies is sensitive enough to distinguish between the different phases of Acetaminophen. DFT calculations provides information on the molecular orbitals involved during electronic excitation occurred as a result of NEXAFS experiment.

509599 Treatment of Cystinosis through Vitamin E Modified Silicone Hydrogel

Abstract:

Cystinosis is an orphan disease caused by a genetic mutation that leads to deposition of cystine crystals in many organs including cornea. Ophthalmic manifestation of the disease can be treated with hourly instillation of cysteamine eye drops. The hourly eye drop instillation is tedious to the patients leading to poor compliance and additionally, significant degradation of the drug occurs within one week of opening the bottle, which further complicates this delivery approach. Our work focuses on designing a contact lens to treat the disease with improved efficacy compared to eye drops. Our goal is to design a lens that is safe and that can deliver a daily therapeutic dose of cysteamine to the cornea while retaining drug stability. We show that cysteamine diffuses out rapidly from all lenses due to its small size. The rapid release of cysteamine and other ophthalmic drugs can be prolonged significantly by integration of vitamin E aggregates inside of the polymer matrix of silicone hydrogels. The increase in release duration increases as square of the volume fraction of the loaded vitamin E fraction which leads to high efficacy and suggests that the mechanism of attenuation in drug release is formation of nanosized barriers. The increase in duration depends on the type of the contact lens. The barriers are not effective in single-phase hydrogel lenses while the effect in the bi-phasic silicone hydrogels depends on the type of the contact lens. Vitamin E incorporation increases the release duration of both ACUVUE®OASYS® and ACUVUE® TruEye™ but the effect is more pronounced in TruEye™ likely due to the low solubility of vitamin E in the lens matrix and higher aspect ratio of the barriers. This result along with the high aspect ratio seen through SEM imaging in silicone hydrogels suggests that barriers could be forming at the interface of the silicone and hydrogel phases. The presence of vitamin E has an additional beneficial effect of reduction in the oxidation rates, likely due to a transport barrier between the oxygen diffusing through the silicone channels and drug located in the hydrogel phase. Based on this study, both Acuvue®OASYS® and ACUVUE® TruEye™ can be loaded with vitamin E to design a cysteamine eluting contact lenses for effective therapy of cystinosis. The lenses must be worn for about 4-5 hr. each day, which is less than the typical duration of daily-wear. Vitamin E loaded contact lenses could be very useful for treating many other ophthalmic diseases as well.

517683 Systems Integration for Dry Granulation Based Continuous Pharmaceutical Tableting

Abstract:

Over the past decade, the pharmaceutical industry has seen significant progress in the mechanistic understanding of relevant unit operations, availability of multiple sensor technologies for inline monitoring of critical quality attributes and advances in data analytics to support manufacturing decision-making. These have enabled a systematic realization of continuous manufacturing for pharmaceutical tablets using the direct compaction route. However, the limitations and difficulty in handling of some particulate materials provide the impetus for investigating granulation routes to the continuous manufacturing of tablets.

In this work, the integration of dry granulation into a continuous tableting line is investigated. Specifically, the Process Analytical Technology (PAT) tools required for inline and at-line monitoring of the tableting line, and the semi-mechanistic models for the roller compactor and the tablet press are discussed. While the Kawakita model is used for the tablet press, the roller compaction process model is consisting of an integration of material balance and actuator dynamics from Hsu et. al (2010) with the modified Johansonâ€™s model presented by Liu & Wassgren (2016). The PAT tools, mechanistic equations, and automation system are integrated for measurement validation and data accuracy using data reconciliation in a structured ISA-95 network using DeltaV. The experimental data is presented from the operation of the continuous dry granulation line-based pilot plant at Purdue University.

509248 Modeling Granular Material Mixing and Segregation Using a Multi-Scale Model

Abstract:

A multi-scale model is presented for predicting the magnitude and rate of powder mixing and segregation in blenders. The model combines diffusion and percolation correlations from small-scale discrete element method (DEM) simulations with advective flow field information from large-scale finite element method simulations. The multi-scale model predictions closely match results from both large-scale DEM-only simulations and experiments, but they are much more computationally efficient than DEM-only simulations. Therefore, in contrast to DEM, the multi-scale model is well suited for predicting blending in industrial-scale systems.

Modelling Granular Media using Dynamical Density Functional Theory

T. Hurst, B. Goddard & R. Ocone

Motivation

Granular media is ubiquitous in industrial and physical problems. One of the characteristic qualities of granular media at the microscopic level is inelastic hard body collisions, which drive macroscopic behaviour. Modelling granular media is a complex and interesting problem. Solid-like phenomena make it difficult to approximate a granular system with a continuum model, while chaotic behaviour and inelastic collapse in hard particle systems introduce modelling challenges on the microscopic scale. We investigate uses of dynamical density functional theory (DDFT), augmented with data gained from microscopic models, for granular media simulation.

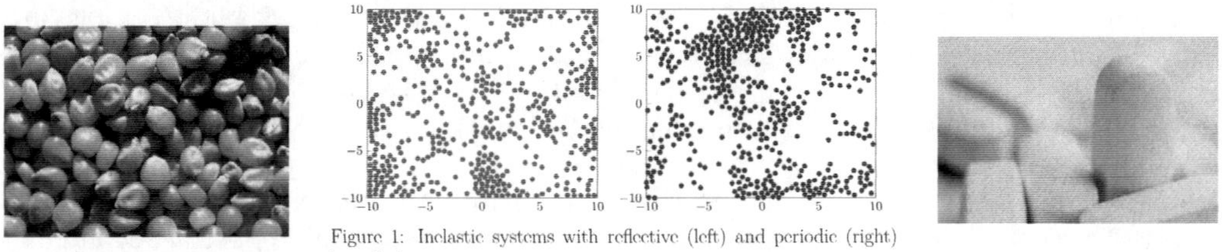

Figure 1: Inelastic systems with reflective (left) and periodic (right) boundaries

Dynamical Density Functional Theory

To derive a DDFT, we start with a set of Langevin equations for N hard particles of diameter σ in d dimensions with positions $\boldsymbol{r}^N$ and momenta $\boldsymbol{p}^N$:

$$\frac{d\boldsymbol{r}^N}{dt} = \frac{\boldsymbol{p}^N}{m}, \quad \frac{d\boldsymbol{p}^N}{dt} = -\nabla_{\boldsymbol{r}^N} V(\boldsymbol{r}^N, t) - \gamma \boldsymbol{p}^N + (m k_b T \gamma)^{1/2} \boldsymbol{G}(t), \tag{1}$$

on the hard sphere domain, $\mathcal{D}_N = \{(\boldsymbol{r}^N, \boldsymbol{p}^N) \in \mathbb{R}^{2dN} : \forall i, j, \|\boldsymbol{r}_i - \boldsymbol{r}_j\| \geq \sigma\}$, where $V(\boldsymbol{r}^N, t)$ includes external and interparticle potentials, and $\boldsymbol{G}(t)$ is an optional stochastic force term. The Kramer's equation is then a partial differential equation for its distribution function $f^{(N)}(\boldsymbol{r}^N, \boldsymbol{p}^N, t)$:

$$\frac{\partial f^{(N)}}{\partial t} + \frac{1}{m}\boldsymbol{p}^N \cdot \nabla_{\boldsymbol{r}^N} f^{(N)} + \nabla_{\boldsymbol{r}^N} V \cdot \nabla_{\boldsymbol{p}^N} f^{(N)} = \gamma \nabla_{\boldsymbol{p}^N} \cdot \boldsymbol{p}^N f^{(N)} + \gamma k_b T \nabla^2_{\boldsymbol{p}^N} f^{(N)}. \tag{2}$$

We integrate out all but one position and momenta in the Kramers equation, to find:

$$\frac{\partial}{\partial t} f^{(1)} = \mathcal{L}_1 f^{(1)} + \mathcal{L}^{2,\dots,N}_{\text{part}}(f^{(k)}) + \mathcal{L}^{(1)}_{\text{diss}} f^{(1)} + \mathcal{L}_{\text{coll}}(f^{(2)}), \tag{3}$$

where $\mathcal{L}_1$ introduces effects from external potentials, $\mathcal{L}^{(1)}_{\text{diss}}$ effects due to dissipation and friction, $\mathcal{L}^{2,\dots,N}_{\text{part}}$ effects from interparticle potentials, and

$$\mathcal{L}_{\text{coll}}(f^{(2)}) = -\sigma^{d-1} \int_{\mathbb{R}^d} d\boldsymbol{p}_1 \int_{\mathbb{R}^d} d\boldsymbol{p}_2 \int_{S_1^{d-1}} (\boldsymbol{p}_1 - \boldsymbol{p}_2) \cdot \omega f^{(2)}(\boldsymbol{r}_1, \boldsymbol{r}_1 + \sigma\omega, \boldsymbol{p}_1, \boldsymbol{p}_2, t) d\omega, \tag{4}$$

is a general collision term that arises due to the hard-sphere domain, when assuming only binary collisions. By taking the first two $\boldsymbol{p}$-moments and upon additional assumptions we arrive at a closed set of equations to model the dynamics of the system, a DDFT [1, 2]:

$$\frac{\partial}{\partial t}\rho = -\nabla \cdot (\rho \boldsymbol{v}), \tag{5}$$

$$\left(\frac{\partial \boldsymbol{v}}{\partial t} + \boldsymbol{v} \cdot \nabla \boldsymbol{v}\right) + \gamma \boldsymbol{v} = -\frac{1}{m}\nabla \frac{\delta \mathcal{F}[\rho]}{\delta \rho} - \sigma^{d-1} g_\sigma \int_{S_1^{d-1}} \rho(\boldsymbol{r} + \sigma\boldsymbol{\omega}/2)\rho(\boldsymbol{r} - \sigma\boldsymbol{\omega}/2)$$
$$\times \boldsymbol{\omega} \cdot \left\{\frac{k_B T}{m}\mathbf{1} + \boldsymbol{v}(\boldsymbol{r} + \sigma\boldsymbol{\omega}/2) \otimes \boldsymbol{v}(\boldsymbol{r} + \sigma\boldsymbol{\omega}) - \boldsymbol{v}(\boldsymbol{r} + \sigma\boldsymbol{\omega}/2) \otimes \boldsymbol{v}(\boldsymbol{r} - \sigma\boldsymbol{\omega}/2)\right\} d\omega, \tag{6}$$

where $\mathcal{F}$ is the well-studied Helmholtz free energy operator, which can include particle-particle interactions and external potentials, ρ is the density, $\boldsymbol{v}$ is the local velocity and g_σ is the value of the correlation function at σ, which must be determined. We choose to model g using statistical data from microscopic simulation.

DDFT Simulations

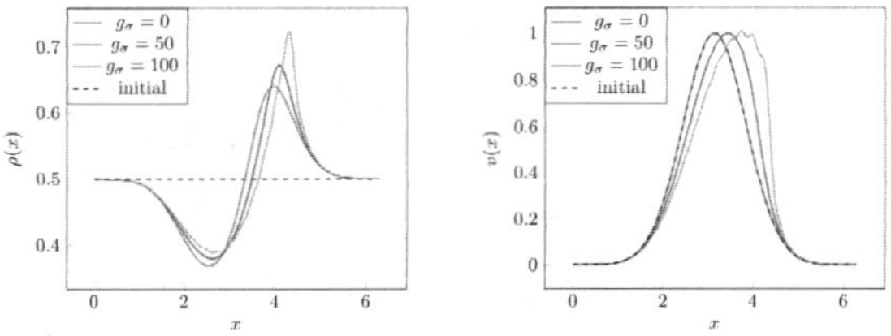

Figure 2: A simple 1D system neglecting all but the collision operator in (5-6), for different values of g. Left: density $\rho(x)$, right: local average velocity $v(x)$.

Data-Driven Modelling of g

A system of N hard spheres of diameter σ in d dimensions with positions $\boldsymbol{x}^N$ and velocities $\boldsymbol{v}^N$ obey

$$\frac{d\boldsymbol{r}^N}{dt} = \boldsymbol{p}^N, \quad \frac{d\boldsymbol{p}^N}{dt} = V(\boldsymbol{r}^N, t), \tag{7}$$

with a pairwise collision condition: if particles i and j are in contact, i.e. $\|\boldsymbol{r}_i - \boldsymbol{r}_j\| = \sigma$ and $(\boldsymbol{r}_i - \boldsymbol{r}_j) \cdot (\boldsymbol{p}_i - \boldsymbol{p}_j) < 0$, we instantaneously change the velocities via the rule:

$$\boldsymbol{p}_i^{out} = \boldsymbol{p}_i^{in} - \frac{1+\alpha}{2}\boldsymbol{\nu}_{ij} \cdot (\boldsymbol{p}_i^{in} - \boldsymbol{p}_j^{in})\boldsymbol{\nu}_{ij}, \quad \boldsymbol{p}_j^{out} = \boldsymbol{p}_j^{in} + \frac{1+\alpha}{2}\boldsymbol{\eta}_{ij} \cdot (\boldsymbol{p}_i^{in} - \boldsymbol{p}_j^{in})\boldsymbol{\eta}_{ij}, \tag{8}$$

where $\boldsymbol{\eta}_{ij} = \frac{\boldsymbol{r}_i - \boldsymbol{r}_j}{\|\boldsymbol{r}_i - \boldsymbol{r}_j\|}$ and $\alpha \in [0, 1]$ is the coefficient of restitution. For simple V, the Newton equations can be modelled using Event Driven Particle Dynamics (EDPD). The cell method and bounded increasing priority queues [3] reduce the computational effort to $\mathcal{O}(1)$ per collision. Using statistical methods for synthetic data (e.g. emulation [4]) we gain an understanding of the correlation function g near equilibrium, which we can include in (6-7).

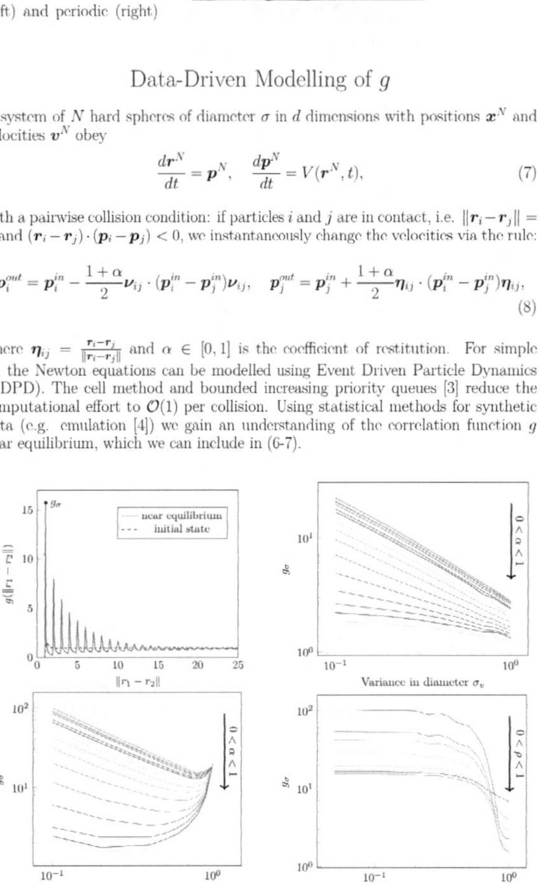

Figure 3: Top left: $g(\|r_1 - r_2\|)$ near equilibrium for a 1D system with $\rho = 0.5$ and $\alpha = 0.5$, using 10000 samples of EDPD simulations. The other three graphs show how g_σ changes for different parameters.

Future Work

We wish to include additional moments in the DDFT derivation, simulate DDFT in higher dimensions including interparticle potentials, and use EDPD in higher dimensions to calculate g.

[1] Archer, A., *J.Chem.Phys.*, **130**, 014509, (2009).
[2] Goddard, B. et. al., *Phys.Rev.Lett*, **109**, 1-5, (2009).
[3] Paul, G., *J.Comp.Phys.*, **221**, 615-625, (2011).
[4] O'Hagan, A., *Reliab.Eng.Syst.Saf.*, **91**, 1290-1300, (2006).

The support from EPSRC (grants EP/N034066/1, EP/L025159/1 and EP/L016508/01) is kindly acknowledged.

509555 A Study on Partially Wetted Particle Collisions with a Wet Wall

Abstract:

For the production of poly-ethylene, a gas-phase polymerization reaction is executed in a fluidized bed reactor. The polymerization reaction produces an excess of heat and the reactor is unable to efficiently remove the excess heat. The introduction of an Inert Cooling Agent (ICA) solves the temperature problem but also has a downside. Introducing the ICA in the reactor influences the particle collisions, leading to substantially different flow behavior. Besides the reduction of the restitution coefficient, the introduction of the liquid can result in adhesion instead of collision. This adhesion results in the formation agglomerates in the reactor and reduces the efficiency of the process. Due to the complex physics of this problem, Computational Fluid Dynamics (CFD) is a suitable method to gain insight in this phenomenon.

In order to study this complex system of partially wetted particle collisions a Direct Numerical Simulation (DNS) is used for incompressible flow. To model the coupling between the solid phases and the fluid phase, a second order accurate Immersed Boundary Method (IBM) is used (Deen et al.,2012). The interface between the fluids is tracked by a Volume Of Fluid (VOF) method (Baltussen et al., 2017). The contact-line dynamics are solved by imposing a contact angle as a boundary condition (Patel et al., 2017).

The simulated results will be compared with dry particle collisions on a wet wall (Tang et al., 2017) and with wet particle collisions on a wet wall.

References:

1. Deen, N.G., Kriebitzsch, S.H., van der Hoef, M.A. and Kuipers, J.A.M., Direct numerical simulation of flow and heat transfer in dense fluidâ€"particle systems. Chemical Engineering Science, 81 (2012): 329-344.

2. Baltussen, M. W., Segers, Q. I. E., Kuipers, J. A. M., and Deen, N. G., Cutting bubbles with a single wire. Chemical Engineering Science 157 (2017): 138-146.

3. Patel, H.V., Das, S., Kuipers, J.A.M., Padding, J.T. and Peters, E.A.J.F., A coupled Volume of Fluid and Immersed Boundary Method for simulating 3D multiphase flows with contact line dynamics in complex geometries. Chemical Engineering Science, 166 (2017): 28-41.

4. Tang, Y., Buck, B., Heinrich, S., Deen, N.G., Kuipers, J.A.M., Interface-resolved simulations of normal collisions of spheres on a wet surface. American Institute of Chemical Engineers, (2017).

508789 An Idea of DEM-CFD Coupling Model Decoupling Spatial Averaging Scale from Computational Cell Size

Abstract:

DEM-CFD coupling model is widely used for the numerical simulation of dense gas-solid flows. Because it is a mesoscopic approach, fluid motion is obtained by solving spatially-averaged continuity and momentum equations. It is usual that computational cell size used for CFD calculation is same with the spatial-averaging scale of governing equations and the results obtained by the model can change depending on the computational cell size. Fine computational cell is desired for accurate CFD calculation, on the contrary, sufficiently large averaging scale is required to fulfil the requirement of spatial averaging. These are contradictory and it is important to adopt a proper cell size carefully to obtain reasonable results. It could be a big problem in the case particles with different size exist and particle size changes over time and position.

In the present study, an idea of DEM-CFD coupling model that decouples spatial averaging scale from calculation cell size used for CFD calculation is discussed. We present the concept of proposed model first and develop discussions on its feasibility and validity.

508739 A Fundamental Framework for Simulating Large-Scale Particulate Systems Using the Direct Simulation Monte Carlo Method

Abstract:

Abstract: A hybrid direct simulation Monte Carlo (DSMC) and multiphase particle in cell (MPPIC) scheme is developed to accurately simulate large-scale particulate systems. The MPPIC method tracks parcels that represent a large number of real particles, and can be used to simulate large-scale particulate processes with complex particle-scale physics. Despite the efficiency of the MPPIC method, its usefulness for dynamic systems is limited due to evaluation of the solids stresses and solids heat flux. The DSMC is a fundamental modeling approach that is used to calculate collisions between parcels. In doing so, the DSMC method effectively evaluates the collision integral in the Enskog equation. The DSMC method therefore does not require simplifying assumptions that are often used to derive continuum closures, such as closeness to equilibrium and small Knudsen numbers. The validity of the Enskog equation breaks down as the solids phase reaches the packed limit. Consequently, pure DSMC simulations show errors such as over-packing in dense regions. To eliminate errors associated with over-packing, the MPPIC solids stress model is used when the solids concentration is close to the packed limit and the DSMC method is used for dilute and moderately dense regions. Simulations using the hybridized MPPIC-DSMC method are completed for a homogeneous cooling system, fluidized bed, and rider flow. The new hybrid simulation technique is compared to more traditional methods, such as two fluid modeling, MPPIC, and the discrete element method.

506236 Comparison of CFD Models in Predicting the Fluidization Behavior of Geldart B Particles

Abstract:

A particular flow regime occasionally observed when fluidizing Geldart B particles in small scale units is slugging. This flow regime, characterized by high pressure fluctuations and big bubbles that can cover the entire section of the fluidizing column (square-nosed slugs), may result in difficulty of extrapolation of the hydrodynamic results obtained for small units into larger scales.

CFD models represent a potential tool for the design of gas-fluidized beds. Despite the efforts of the scientific community, the ability of these models in predicting the hydrodynamic behavior of these multiphase systems is still limited. Robust and reliable CFD tools that can predict the hydrodynamics of group B particles slugging in lab or pilot scale units is extremely useful for beforehand definition of the operating conditions and geometry that can assure cold flow model experimentation reproducing the same hydrodynamics of large scale reactors.

In this work, two different commercial CFD tools have been compared in their ability of predicting hydrodynamic features of slugging gas-solid fluidized beds. The study is of special interest to chemical looping combustion process that employs Geldart B particles but also to fluidized bed combustion or gasification processes. Different parameters such as bed expansion, pressure fluctuations, voidage and solid velocity obtained experimentally have been compared with CFD results and the effects of simulation parameters on the predicted flow structures have been discussed.

507540 Evaluation of Drag Models for Gas-Solid Fluidization of Geldart a Particles in All Flow Regimes

Abstract:

Coarse grid two-fluid simulation of gas-solid fluidized bed reactors based on the kinetic theory of granular flow exhibits a significant dependence on drag models, especially for Geldart A particles. Inappropriate selection of drag models leads to inaccurate fluidization behavior predictions. This study focused on the evaluation of drag models derived from different methods for 3D two-fluid model simulations of gas-solid fluidization of Geldart A particles in bubbling fluidization, turbulent fluidization, fast fluidization and dilute phase transport regimes. Eight candidate drag models were selected, which included five heterogeneous drag models and three homogeneous drag models that belong to six types. Comparison with available experimental data clearly demonstrates the need for modification of homogeneous drag models to account for the effect of mesoscale structures (bubble and cluster). The Sarkar drag model and EMMS drag models are found to achieve superior predictions in all fluidization regimes, while the other drag models are found to be capable of predicting fewer fluidization regimes. The results of this work provide a guideline for choosing appropriate drag models for simulating Geldart A particles and provide suggestions on developing more reliable and general drag models applicable in all flow regimes.

508171 Two Sphere Aggregation in a Shear Thinning Fluid

Abstract:

Particulate flows occur throughout nature and are encountered in all industries. It is important to understand how particles migrate in liquids since this has a consequence on the transport and processing of particle suspensions. A simple system of two spheres settling at low particle Reynolds in a Newtonian fluid is well described. The method requires the calculation of forces acting on each sphere, including the contribution from gravity, and the force associated with the second sphere, which can be determined by the Stokes stream function.

Unlike Newtonian liquids, non-Newtonian liquids provide shear dependent fluid properties which create opportunity to transport some of the more challenging particulate suspensions, i.e. dense phase transport of nuclear sludge particulates. The shear responsive viscosity of a non-Newtonian fluid potentially allows for particle resuspension under high shear, and particle stability under low shear. However, it has been observed that the low Reynolds settling of spheres in a shear thinning fluid leads to the formation of particle chains, with particle aggregation shown for a minimum of two spheres. This is in contrast to the low Reynolds settling of two spheres in a Newtonian fluid, where the particles settle at an equivalent velocity and maintain a constant particle-particle separation.

Previous research has attributed the particle aggregation to a “corridor of reduced viscosity” which forms behind the leading sphere, causing the trailing sphere to experience a reduced viscosity allowing faster settling to occur, eventually leading to particle aggregation. Despite the importance of particle settling in non-Newtonian fluids, limited work has been completed. Attempts to predict the sphere velocities have been made by empirical fits. The current study aims to replicate the methodology applied to determine the settling velocities of spheres in Newtonian fluids and extend the approach to non-Newtonian fluids.

The fluid velocity around a 1.5 mm stainless steel ball bearing settling in a 50 mm × 50 mm × 300 mm square glass column (Vitrocom) was visualised using Particle Imaging Velocimetry (PIV, Dantec Dynamics). A DualPower 65-15 laser was used to illuminate 1-20 μm PPMA-Rhodamine B tracer particles (25 mg in 100 μL water per 1 L solution). A FlowSense EO 2M Camera captured the spheres settling and the images were processed using the DynamicStudio software to produce vector maps. These vector maps were further processed to elucidate the velocity profile along the line of particle motion. The settling velocities of the two spheres at different separations were also calculated directly from the location of the sphere during settling. Using a Point Grey Chameleon3 camera at 30 fps and a custom MATLAB® script, the spheres were identified using peak analysis and the sphere settling velocity determined using locations over 9 sequential images.

The Newtonian method was validated using a 95 wt% glycerol + 5 wt% water solution. The experimental fluid velocity decayed more rapidly than that predicted by the Stokes stream function velocity. The fluid velocity given by Stokes was half the sphere velocity ($V = 0.5V_S$) at a distance of 2.88 sphere radii, whereas the PIV data decayed to the same sphere velocity in 2.36 sphere radii, although this was attributed to the finite column size and non-zero Reynolds number. A DigiDEM™ simulation was executed with parameters equivalent to the experiment. The resulting velocity profile was in good agreement with the experimental velocity profile. Using the PIV fluid velocity profiles, sphere velocities at different separations were calculated and compared to experimentally determined velocities. There was no significant difference with the calculated velocities within the scatter of the experimental data.

A TA Instruments DHR-2 rheometer was used to characterise the flow properties of a 0.25 wt% xanthan + 60 wt% glycerol + 40 wt% water solution. The prepared solution was found to be highly shear thinning, viscoelastic and exhibited a long recovery time, but low thixotropy. A steady state stress sweep showed a flow curve which could be described by the Carreau model. A stress hold test was used to measure the recovery of the fluid after shearing the fluid at 10 Pa for 2 s which mimicked the settling of a sphere past a single point. Fitting the data using an exponential decay, the time constant of the fluid was found to be 173 s.

Since the Newtonian fluid method does not account for the time dependent viscosity, a viscosity ratio was included in the gravitational force term. Using the fluid velocity around a single sphere in a non-Newtonian fluid, the sphere velocities were calculated using the new method (non-Newtonian fluid) for different particle separations. These calculated velocities were in good agreement with the experimentally determined sphere velocities.

The new method for non-Newtonian fluids uses i) the single sphere velocity, ii) doublet velocity, iii) fluid velocity profile around a single sphere and iv) the time dependent viscosity to successfully predict the settling velocity of two spheres settling along their line of centres in a shear thinning fluid. These four parameters are much easier to obtain than completing a number of settling experiments at different particle separations. Furthermore, it has been shown that the fluid velocity profile scales to the sphere radius and density, so this method may be extended to predict the velocities of different sized spheres with little additional work. This is particularly promising as most particulate suspensions in industry are non-uniform, and there is very little research published which considers the settling of different sized spheres.

Binary Solid Mixture Separation in a Cold Flow Spouted Bed for Chemical Looping Applications

Michael Bobek[1,2], Steve Rowan[1,2], Jingsi Yang[1,2], Justin Weber[1], Ronald W. Breault[1]

[1]*Nattional Energy Technology Laboratory, Morgantown, WV, USA 26507*

[2]*Oak Ridge Institute for Science and Education, Morgantown, WV, USA, 26507*

ABSTRACT

Fluidized beds are used in many industries for their favorable characteristics of good solids mixing, high rates of heat and mass transfer, and large throughputs. One process of interest is chemical looping combustion (CLC), which is a promising technology for electrical generation with inherent carbon capture. One difficulty in the development of this technology is the removal of ash and unburnt carbon (or char) from the oxygen carrier material prior to re-entering the air reactor. In an effort to address this issue, several cold flow experimental efforts are being undertaken at the U.S. Department of Energy's National Energy Technology Laboratory (NETL) towards the development of effective particle separation techniques. This work focuses on the study of the hydrodynamics and elutriation characteristics of a mixture consisting of fine (<50μm) particles (serving as a char analog) and larger (~600μm to ~800 μm) particles (serving as the carrier analog) in a modified spouted bed. In this modified system, the spouted bed has an annular gap between the wall and spout cone. A sweep gas within this gap is used to aid in the separation, with the smaller particles being removed via elutriation.

INTRODUCTION

As our society continues to grow in population and develop across the globe, the energy demands are expected to increase significantly. By 2040 the energy consumption of the world is expected to exceed 740 quad BTU, a 40% increase to 2010 levels [1]. Along with that increased energy consumption comes increased pollution and concerns of excess carbon emissions to the atmosphere. The energy demands and environmental concerns are the main motivators for research into cleaner and more efficient energy sources. While alternative energy sources are vital to the societies energy portfolio in the future, hydrocarbon fuels still provide most of the energy consumed. This leads to the need for efficient methods of utilizing these sources as well as reducing pollution and environmental impact from their use. Focusing on limiting environmental impact and reducing carbon emissions, clean energy technologies and carbon capture are essential for the future.

There are three standard methods for carbon capture: treating the fuel before the combustion to decarbonize it, pure oxygen combustion, or post combustion separation of CO_2 from the flue gasses. These methods all have large energy penalties and costs associated with them [2] [3]. Chemical looping combustion (CLC) is an alternative combustion process that provides inherent CO_2 separation with minimal energy losses [4]. CLC relies on the use of a metal oxide as an oxygen carrier that moves between two fluidized bed reactors, a generalized schematic of the system is shown in Figure 1. In the fuel reactor the metal oxide is reduced and the fuel is combusted, the carrier then moves to the air reactor where the oxygen in air is used to oxidize (regenerate) the carrier.

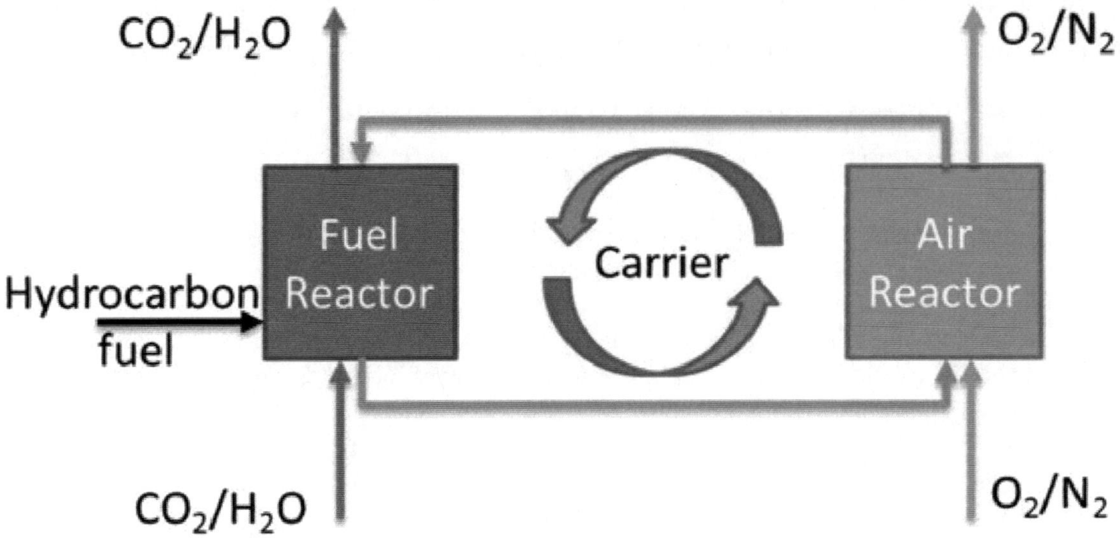

Figure 1: Chemical looping schematic

Based on system studies performed at NETL [5], one aspect contributing to the economic viability of an industrial scale CLC reactor is the need to separate the char and ash from the oxygen carrier. In the Stevens et. al. study the concentration of char and ash in the exit stream of the fuel reactor in question is between 0.5-1%, of that, roughly 80% would need to be separated from the carrier. NETL is currently investing several particle separation methods, such as aerodynamic separation in a fluid bed separation, in order to determine a viable option to meet this design requirement.

SETUP

The simulants of the oxygen carrier and char/ash mixture used for this study are 800 μm glass beads for the oxygen carrier and 40 μm DayGlo ZQ-17 plastic particles for the char/ash. The glass beads have a specific gravity of 2.4, providing a reasonable middle ground of density and size comparison to the metal oxides typically used as carrier materials [3]. The DayGlo particles have a specific gravity of 1.2, which also falls into the range of density and particle size expected for typical char/ash particles resulting from coal combustion [6].

The riser design examined in this work is a spouted bed configuration, shown in Figure 1. The binary solids mixture is fed with a spout gas to the middle of the riser, once in the main riser the fines are elutriated upward and the coarse particles fall through an annular gap between the riser wall and the bed. A secondary sweep gas is also fed such that the gas velocity through the annular gap exceeds the terminal velocity of the fine particles, but is less than that of the coarse particles. The particles are fed into the spout gas stream via an eductor. The feed pipe has a diameter of 2.54 cm, the annular gap has a total area of 0.0025 m^2, and the main riser has a diameter of 15.24 cm and a total height of 144 cm, 92 cm of which are above the spouted bed cone. The large particles fall into a sealed collection cannister that sits on a scale, allowing for real-time weight measurements of collected solids. The small particles are elutriated out the top and collected in a filter bag connected to the exit.

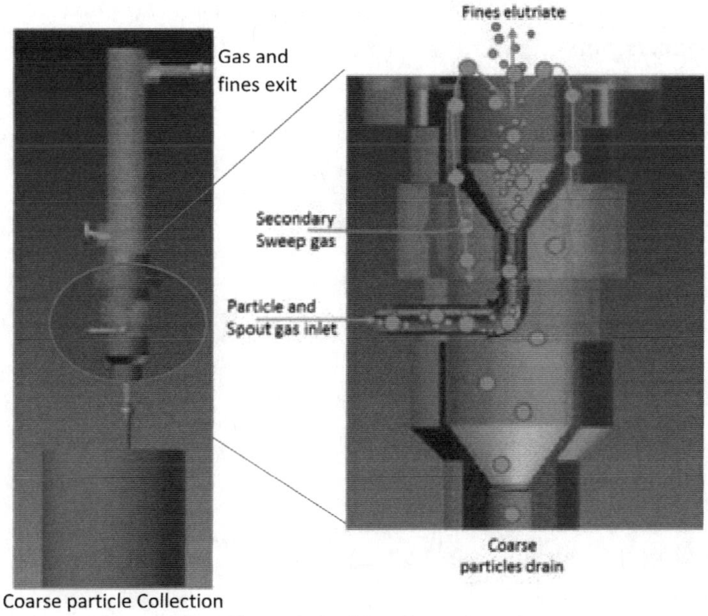

Figure 2: Bed Configuration

For all the experiments, the particle mixture was set at 99% glass beads and 1% DayGlo by mass. Each mixture was loaded into a hopper, and fed via drop tube to the eductor. The spout gas, sweep gas, and mass feed rates were varied between experiments, the full list of test conditions can be seen in Table 1, with experiments grouped by similar flow conditions. These flow rates were initially chosen based on the terminal velocities, U_t for the simulant particles. The glass beads have a $U_{t,gb} = 4.8$ m/s and the DayGlo particles have a $U_{t,dg} = 0.14$ m/s, which are determined from correlations suggested by Haider and Levenspiel [7]. The spout velocity (U_{sp}) ranged from 8 – 10 m/s, ensuring it is greater than the glass bead's terminal velocity ($U_{sp} > U_{t,gb}$). The sweep gas velocity (U_{sw}) ranged from 0.7 – 1 m/s, and the averaged velocity in the main riser (U_{riser}) ranged from 0.34 – 0.41, putting both the sweep velocity and riser velocity between the two particles respective terminal velocities ($U_{t,gb} > U > U_{t,dg}$). The spout gas and sweep gas flow rates were set with Alicat mass flow controllers, and the mass feed rate was determined as an average based on the feed time and mass fed for each experiment. The separation efficiency was determined by sieving the particles collected from the bottom exit of the riser and comparing the concentration of fines found in that mixture to the initial concentration of 1%, shown in Equation 1 and Equation 2.

$$\text{sieve results} = \frac{total\ mass\ -\ glass\ beads}{total\ mass}$$

Equation 1: Final concentration

$$\text{Separation efficiency} = \frac{initial\ - final}{initial} = \frac{0.01\ -\ sieve\ results}{0.01}$$

Equation 2: Separation efficiency

Table 1: Experimental conditions and separation efficiency.

Day-Glo mass%	sweep gas (slpm)	spout gas (slpm)	mass feed (g/s)	separation efficiency
1	150	300	20	0.9394
1	150	300	9	0.9185
1	150	300	7	0.9001
1	150	250	14	0.8988
1	150	250	12	0.9001
1	150	250	6	0.8788
1	150	250	16	0.9195
1	100	300	10	0.9394
1	100	300	17	0.8386
1	100	300	8	0.9394
1	100	300	20	0.9393
1	100	275	25	0.9262
1	100	275	11	0.8793
1	100	275	14	0.9010

RESULTS

The results from the tests were very promising with all the conditions exceeding the goal of 80% separation. In Figure 3, the separation efficiency is compared to the spout gas flow rate in standard liters per minute, the red line at 0.8 is representative of the target separation efficiency. Averaging and taking the standard deviation of the values for each flow rate, shown in Table 2, two things can be determined; 1) the average separation efficiency shows a very minimal increase as the spout gas increases, and 2) the standard deviation increases as the spout gas increases. From this it would appear the most consistent results will come from the lower flow rates for the spout gas, overall though the effect of the spout gas on separation efficiency is minimal.

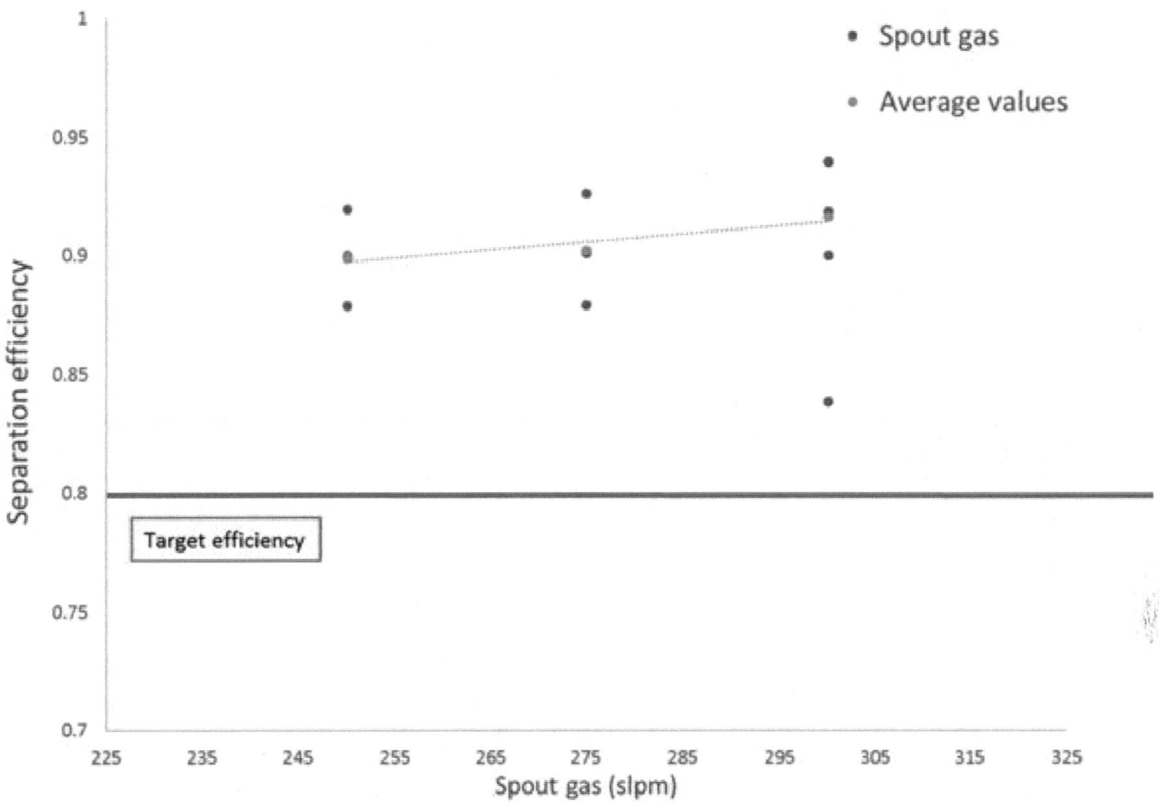

Figure 3: Spout gas vs separation efficiency

Table 2: Averaged Spout gas data

Flow rate	Separation efficiency	Standard deviation
250	0.899	0.017
275	0.902	0.024
300	0.916	0.038

Two different sweep gas conditions have been examined so far, the results are shown in Figure 4 The separation efficiency is compared to the sweep gas flow rate in standard liters per minute, the red line at 0.8 is representative of the target separation efficiency. Table 3 shows the averaged values for the two sweep gas values, along with the standard deviation. From the table it can be see that the sweep gas has very little effect on the separation efficiency in this range. The effect that can be seen is that decreasing the sweep gas increases the standard deviation, implying higher sweep gas flow rates will have more consistent separation efficiencies.

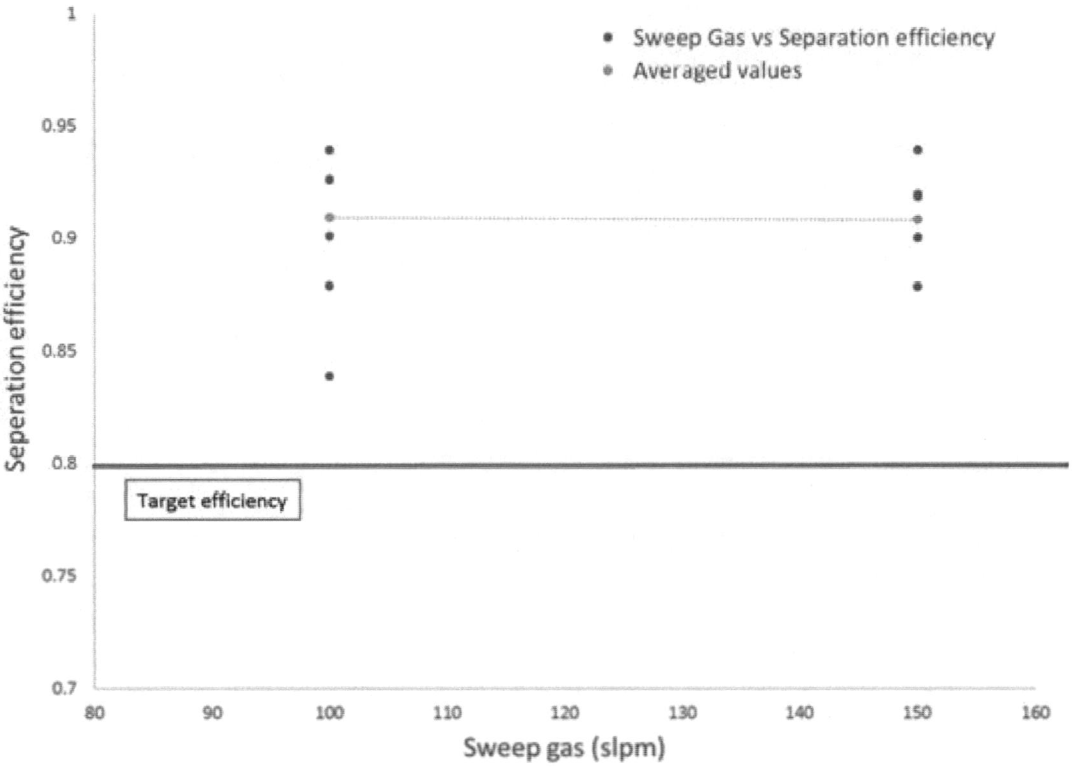

Figure 4:Sweep gas vs Separation Efficiency

Table 3:Averaged Sweep gas data

Flow rate	Separation efficiency	Standard deviation
100	0.909	0.039
150	0.908	0.020

The final parameter tested was the mass feed rate. Figure 5 Shows the mass feed rate in grams per second vs separation efficiency. The minimum mass feed rate was 6 g/s and the maximum feed rate achieved was 25 g/s. Since the feed rate was controlled manually via a gate valve, exact and consistent feed rates were difficult to generate. Instead of exact feed rates the experiments attempted to achieve a "low", "medium", and "high" feed rate within that regime. The data is separated based on test conditions with the same gas flow rates as seen in Table 1. For all the similar conditions except the "300 spout 100 sweep" condition there is a slight upward trend of increasing separation efficiency with increased mass feed rate. Since the mass feed rate shows negligible impact to the separation efficiency in this regime maximizing the feed rate is the goal for work moving forward with the current design.

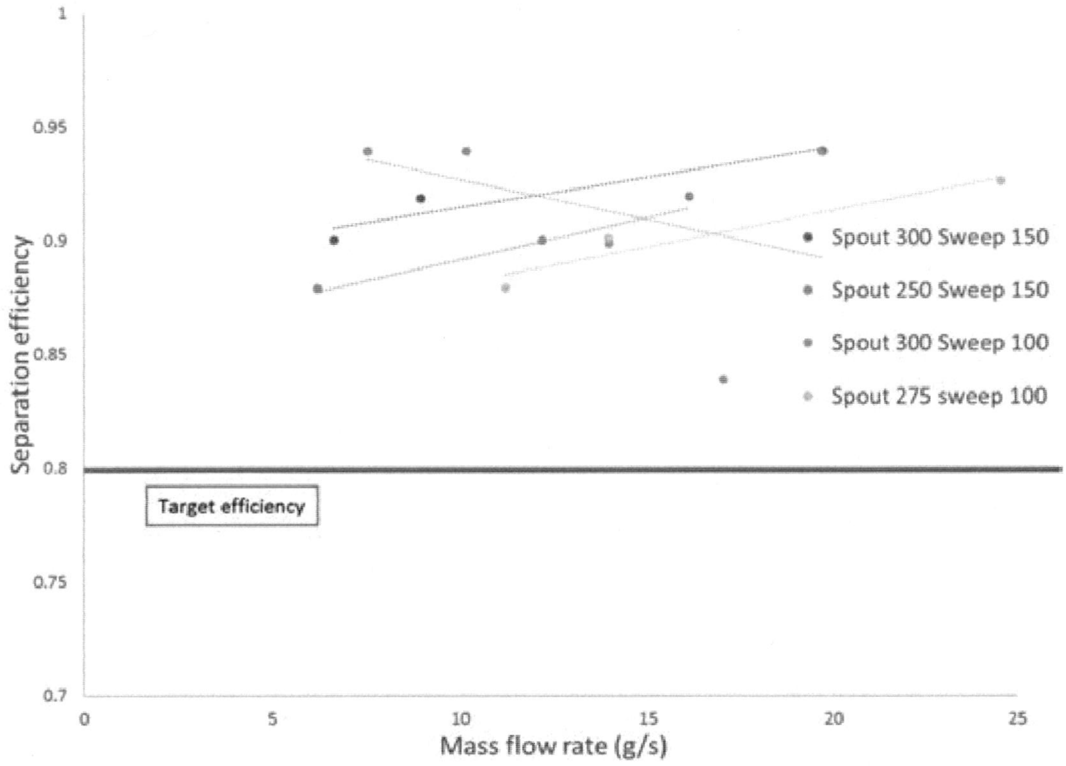

Figure 5:Mass feed rate vs separation efficiency

CONCLUSIONS

The target separation efficiency set in [1] of 80%, is well within reach based on the results from these tests so far. With the separation efficiency results being virtually independent of the test parameters, a normalized distribution of the results can be generated to determine the probability of achieving the 80% separation efficiency goal. Figure 6 shows the normalized distribution of the separation efficiency based off the average, 0.908, and standard deviation, 0.029, of the results. Overlaid on the normalized distribution is a histogram of the experimental results. From the normalized distribution the probability of reaching the 80% separation is 99.99%. While these are preliminary results of the total testing planned for the system, since all the parameters are showing minimal effect on the separation efficiency in the regimes tested, it is likely that efficiency of the system is effected more strongly by changes in the geometry. The future work planned so far includes changes to the spout, cone, and annular gap geometry, as well as expanding the gas flow and mass feed conditions. The goal moving forward with the flow conditions is to find the minimum gas flow rates that the system functions at along with maximizing the solids feed rate into the system.

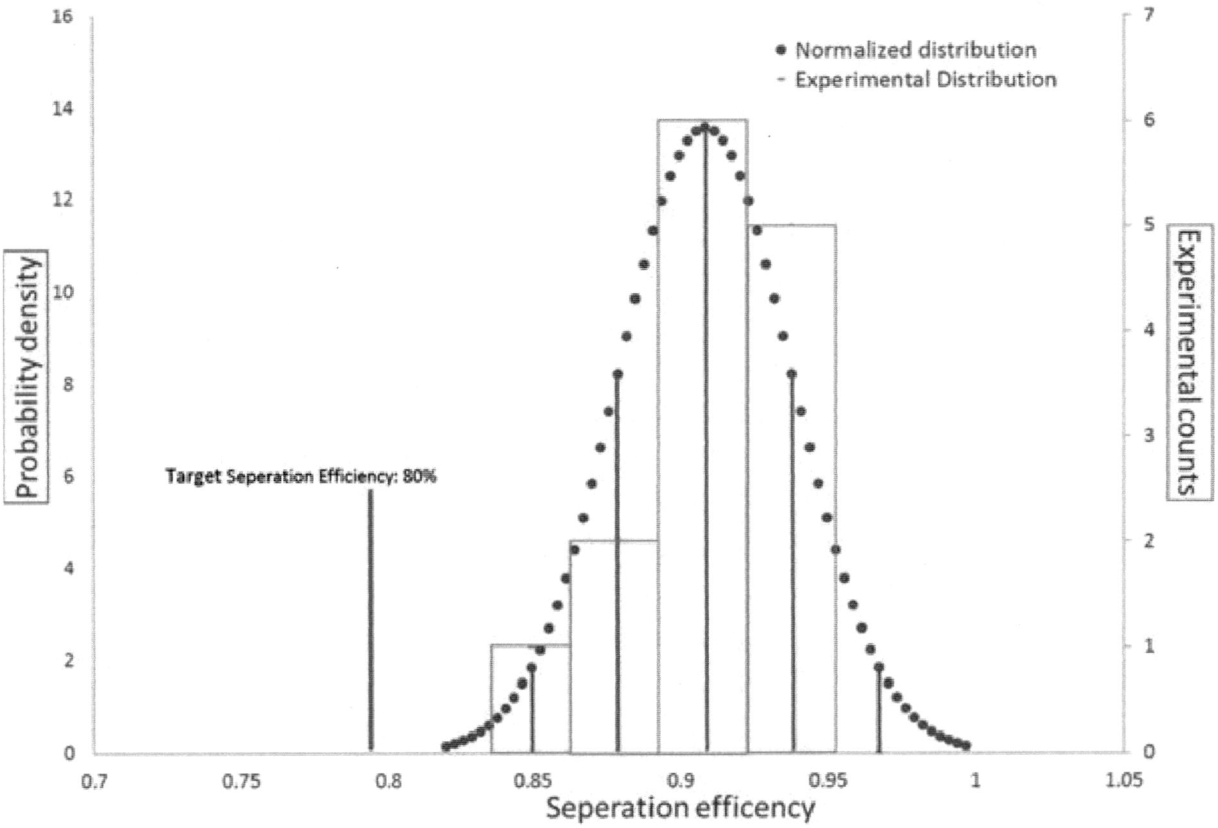

Figure 6: Separation efficiency probability function

[1] EIA, "U.S. Energy Information Administration," April 2018. [Online]. Available: https://www.eia.gov/outlooks/aeo/data/browser.

[2] A. Lynfelt and C. Linderholm, "Chemical Looping Combustion of Solid Fuels - status and recent progress," *Energy Procedia,* pp. 371-386, 2017.

[3] M. M. Hossain and H. I. de Lasa, "Chemical-Looping combustion (CLC) for inherent CO2 separations - a review," *Chemical Engineering Science,* pp. 4433-4451, 2008.

[4] M. Ishida, D. Zheng and T. Akehata, "Evaluation of a chemical-looping combustion power-generation system by graphic energy analysis," *Energy,* pp. 147-154, 1987.

[5] R. Stevens, R. Newby, V. Shah, N. Kuehn and D. Keairns, "Guidance for NETL's Oxycombustion R&D Program: Chemical Looping combustion Reference Plant Designs and Sensitivity Studies," DOE/NETL, 2014.

[6] L. Chen, S. Yong and A. Ghoniem, "Oxy-fuel combustion of pulverized coal: Characterization, fundamentals, stabilization and CFD modeling," *progress in Energy and Combustion Science,* pp. 156-214, 2012.

[7] A. Haider and O. Levenspiel, "Drag coefficient and terminal velocity of spherical and non-spherical parrticles," *Powder Technology,* pp. 63-70, 1989.

508596 Gravity-Based Percolation of Small Particles through an Assembly of Large Particles

Abstract:

The *Chemical-Looping Gasification* is one of several options being explored to design and operate a Chemical-Looping System using solid fuels (e.g. coal). There are many challenges for solid fuel chemical looping, such as the handling of the oxygen carrier, the continuous feeding of the solid fuel and its mixing with the oxygen carrier to maximize its conversion, and how to continuously remove of the solid wastes (e.g. ash) during the fuel consumption process to maintain efficient operations. One proposed configuration to address these challenges is a moving bed reducer (e.g. Stevens, Fan et. al., Velazquez-Vargas, Tong et. al.). The reactor consists of a slow downward moving bed composed of oxygen-carrier and larger solid fuel particles with a multiphase stream composed of gas and smaller particles moving through the pore network formed by these larger particles. The insight in the overall dynamics of this reactor configuration and its accurate prediction is of practical interest for the design and operation of these solid fuel-based moving bed reactors.

There are many interacting physical sub-processes within this system. In this work, we examine one relevant sub-processes: mono- and bi-disperse gravity-driven percolation of fine particles through a packed bed of larger particles. Using the results from extensive Discrete Element Method (DEM) simulations, we discuss percolation regimes through the analysis of the mean transit time and the radial variance of the positions of the percolating matter. The numerical results are also compared with experimental data (Lominé & Oger, 2006).

References

L. Velazquez-Vargas, “Final Report: Commercialization of an atmospheric iron-based CDCL process for power production. Phase I: Technoeconomic analysis”, DOE-FE009761 (2013).

https://www.osti.gov/scitech/servlets/purl/1222706/

L-S Fan., “Coal Direct Chemical Looping (CDCL) Retrofit to Pulverized Coal Power Plants for In-Situ CO2 Capture”, Project Review Meeting, DOE-NT0005289 (2014).

https://www.netl.doe.gov/publications/proceedings/10/co2capture/presentations/wednesday/Ray%20Kim-NT0005289.pdf

R. Stevens, R. Newby, D. Keairns, M. Woods, “Moving Bed Chemical Looping Combustion (MB-CLC) Reference Plant Design and Sensitivity Studies”, Clearwater Coal Conference (2016).

A. Tong, S. Bayham, M. V. Kathe, L. Zeng, S. Luo, L.-S. Fan. “Iron-based syngas chemical looping process and coal-direct chemical looping process development at Ohio State University”. Applied Energy 113, pp. 1836–1845 (2014)

F. Lominé, L. Oger. “Transport of small particles through a 3D packing of spheres: experimental and numerical approaches”. Journal of Statistical Mechanics: Theory and Experiment, pp. 1-13 (2006)

509590 On the Effect of Frictional Stress Modeling in Dense Fluidized Beds for Minimum Fluidization Conditions and in Bubbling Regime

Abstract:

The Euler-Euler approach commonly used to simulate bubbling fluidized applies the kinetic theory of granular flow to model the stress of solid phase. This theory, understood as an extension of the kinetic theory of gases, assumed that particle-particle collisions were instantaneous and binary. However, in dense regimes such as encountered in bubbling fluidized bed, the frictional contacts between neighboring particles may have a strong influence. Johnson and Jackson [1] proposed to sum the stress derived by the kinetic theory to the stress provided by frictional stress model. Several practical models derived from the soils mechanics were proposed for modeling frictional stress [2-3].

In recent works, some authors [4-6] investigated the performance of frictional models [2-3] for Euler-Euler simulations of dense fluidized beds. They studied the bubbles size and shape, pressure drop, bed expansion and particles circulation. However, these models were found to be very sensitive to the chosen activation threshold.

In this paper, according to experimental data and numerical simulations of the transition from fluidized beds to fixed bed, a new method was proposed to determine the activation threshold of frictional models. The 3D numerical simulations by NEPTUNE_CFD code [7-8] were carried out on a large number of particle types to predict the minimum fluidization velocity and the bed expansion in the transition regime. It was demonstrated that the threshold value is not sensitive to the particle size but is strongly dependent on the shape and surface properties of the particles.

Then, hybrid approach frictional-kinetic was applied to simulate bubbling fluidized beds of particle Geldart group A powder known to be very sensitive to mesh refinement [9]. Very fine meshes of several million cells were used to demonstrate that a frictional model was needed to achieve a mesh convergence. Finally, the influence of the introduction of this frictional stress were evaluated on the macroscopic flow variables such as drop pressure and time averaged solid volume fraction, but also on the particle velocity and solid volume fraction time-averaged variances.

References

[1] P.C. Johnson, R. Jackson, Frictional-collisional constitutive relations for granular materials, with application to plane shearing, J. Fluid Mech. 176 (1987) 67–93.

[2] D.G. Schaeffer, Instability in the evolution equations describing incompressible granular flow, J. Differ. Equ. 66 (1987) 61–74.

[3] A. Srivastava, S. Sundaresan, Analysis of a frictional–kinetic model for gas–particle flow, Powder Technology. 129 (2003) 72–85.

[4] N. Reuge, L. Cadoret, C. Coufort-Saudejau, S. Pannala, M. Syamlal, B. Caussat, Multifluid Eulerian modeling of dense gas–solids fluidized bed hydrodynamics: Influence of the dissipation parameters, Chem. Eng. Sci. 63 (2008) 5540–5551.

[5] A. Passalacqua, L.A. Marmo, A critical comparison of frictional stress models applied to the simulation of bubbling fluidized beds, Chem. Eng. Sci. 64 (2009) 2795–2806.

[6] Farzaneh, M., Almstedt, A.-E., Johnsson, F., Pallarès, D., Sasic, S., The crucial role of frictional stress models for simulation of bubbling fluidized beds. Powder Technology 270, (2015) 68–82.

[7] Neau, H., Fede, P., Laviéville, J., Simonin, O.. High performance computing (HPC) for the fluidization of particle-laden reactive flows. In : The 14th International Conference on Fluidization, From Fundamentals to Products, Noordwijkerhout, Netherlands (2013).

[8] R. Ansart, P. Garcia Triñanes, B. Boissière, H. Benoit, J.P.K. Seville, O. Simonin. Dense gas article suspension upward flow used as heat transfer fluid in solar receiver : PEPT experiments and 3D numerical simulations. Powder Technology, Volume 307, (2017) 25-36.

[9] Agrawal, K., Loezos, P., Syamlal, M., Sundaresan, S., The role of mesoscales structures in rapid gas-solid flows. Journal of Fluid Mechanics 445, (2001) 151–185.

507778 CFD-DEM Simulation of Fluidization in a 3D Spouted Bed and Flow Regime Prediction Using MFIX-DEM

Abstract:

The spout-fluidized bed offers several advantages in chemical looping combustion operation using solid fuels based on the multiphase behavior. The difference in size and weight between the large, spoutable oxygen carrier particles and the smaller coal and ash particles allows the oxygen carrier to be easily segregated for recirculation; the increased solids mixing due to dynamic flow pattern in the spout-fluidization regime prevents agglomeration. A 3D spouted bed simulation is set up in MFiX-DEM based on a previous flow regime mapping experiment using 4-mm-diameter glass beads. The minimum fluidization condition is determined based on the change in the relationship of the pressure drop with velocity as the bed transitions from a packed bed configuration to a fluidized one. The operating conditions in the CFD-DEM simulation are subsequently normalized by the minimum fluidization velocity.

At each operating condition, the flow is classified into the internal spout, spouting with aeration, spout-fluidization, jet in fluidized bed, or slugging bed regime based on the multiphase flow pattern. The flow regimes predicted by the MFiX-DEM simulation show excellent agreement with the experimental flow regime map at all operating conditions. Spectral analysis of the pressure fluctuations in the bed quantitatively corroborates the visual observations of the flow pattern based on the particle trajectories. The qualitative trends in the shape, frequency, and power of the dominant peak in the frequency spectra follow the expectations from theory and experiment. The results of this work confirm that the MFiX-DEM model can effectively predict the flow regime in a spouted bed with high accuracy.

A reduced value of the spring stiffness coefficient, required for the particle collision calculations, is used in this work as it would take months to complete the simulations with the exact value. A parametric study is conducted on the spring stiffness to rule out errors in the results due to the reduced value. It was found that the value used as a starting point in the simulation was sufficient to produce accurate results and increasing the value does not affect the flow regime prediction even in the worst-case scenario.

535406 Fully Coupled CFD-DEM Model for Dense Gas-Particle Flows in Cyclones

Abstract:

It is widely accepted that combined computational fluid dynamics (CFD) and discrete element method (DEM) (CFD-DEM) when fully coupled is the most accurate way to model the granular flows as it has the least amount of assumptions and resolves all particle-particle interactions. The downside being the heavy computational expense. To address this issue, coarse grain models are used to help with lowering down the computational requirements significantly. In coarse grain model, each simulated particle is actually a parcel that represents multiple smaller identical particles. This model treats each parcel as a single large particle and computes the contact between the parcels instead of individual particle therefore it can use a larger time-step and speed up the simulation at the cost of a small loss of fidelity in describing the details of individual contacts.

Here, we take advantage of the parcel contact coarse grain particle model, a newer type of coarse grain models that resolves parcel contacts without deriving the force from the inserted load on representative fine particle, to simulate the behavior of dense gas-particle flows in a Stairmand cyclone using STAR-CCM+. We compare the results and runtime with conventional approach and show the capability of this model for various industrial applications varying from coal processing to fluidized bed.

507522 Uncertainty Quantification and Industrial Applications of Coarse Grained Particle Method

Abstract:

Uncertainty quantification and industrial applications of coarse grained particle method

Liqiang Lu*, Sofiane Benyahia

National Energy Technology Laboratory, Morgantown, West Virginia 26507, United States

* Corresponding author: liqiang.lu@netl.doe.gov

In past decades, the continuum approach was the only practical technique to simulate large-scale fluidized bed reactors because discrete approaches suffer from the cost of tracking huge numbers of particles and their collisions. The coarse grained discrete particle method significantly improved the computation speed with controllable compromise in accuracy and has been extensively investigated in recent years. In this presentation, recent advances in this method and their verification and validations are discussed including extension to heat transfer, chemical reactions and coupling with time-driven hard-sphere (TDHS) method. The uncertainties of this method are also quantified in several simple homogeneous cooling systems and more complex fluidized beds. Finally, it is used to simulate several reactors including the scale-up of a rare earth element (REE) leaching reactor, a pilot scale methanol to olefins (MTO) reactor and an industrial FCC regenerator. The simulation results compared well with available experiment and industrial data and proved that this new approach can be used for efficient and reliable simulations of industrial-scale fluidized bed systems.

Refs:

1. **Lu, L**.; Li, T.; Benyahia, S., **2017.** An efficient and reliable predictive method for fluidized bed simulation. **AIChE Journal,** DOI: 10.1002/aic.15832
2. **Lu, L.**; Morris, A.; Li, T.; Benyahia, S., **2017.** Extension of a coarse grained particle method to simulate heat transfer in fluidized beds. ***Int. J. Heat Mass Transfer***, *111*, 723-735.
3. **Lu, L.,** Yoo, K., Benyahia, S., **2016**. Coarse-Grained-Particle Method for Simulation of Liquidâ€"Solids Reacting Flows. **Industrial & Engineering Chemistry Research** 55, 10477-10491.

507313 EMMS Application in 3D Full Loop Circulating Fluidized Bed Simulation

Abstract:

Flows inside circulating fluidized beds (CFBs) are inherently multiphase, involving at least one fluid phase and one solid phase. Accurate modeling is significant for the beds design, optimization, operation and scaling. Computational fluid dynamics (CFD) simulation is an efficient tool, since it is able to obtain flow dynamics in detail. The simulation accuracy is determined by many factors, drag model is a crucial one, which dominates the particle transport and internal recirculation in the system. Most drag models are conventional drag models, only true for homogeneous flow, but in the CFB system, especially in the dense region, the flow is heterogeneous, including bubbles, clusters etc., such meso-scale structures. To resolve these, new models are needed. Energy minimization multi-scale (EMMS) drag model has been found successful in correcting the drag and predicting the flow distribution in CFBs.

In this study, EMMS drag models, obtained based on the specific operating conditions and geometries, taking into account both local slip velocity and voidage, are coupled with CPFD (computational particle fluid dynamics) software Barracuda, to simulate two pilot scaling full loop three-dimensional CFBs, one with rectangular, the other with annular combustion chamber, to investigate the details inside respectively. The simulation results are analyzed in terms of full loop pressure drop, particle volume fraction distribution and particle circulation flux, etc., and compared with available experimental data. Through these, it can give guidance to the beds design and optimization.

506934 Hydrogen Transport through Consolidated Particulate Suspensions of Nuclear Corrosion Products; Insights from X-Ray Tomography

Abstract:

The UK and USA have accumulated substantial inventories of highly heterogeneous radioactive particulate sludge wastes comprised of consolidated fuel cladding corrosion products and by-products from legacy reprocessing activities respectively. These consolidated sediments experience in situ hydrogen generation from metallic fuel corrosion and radiolysis of water due to water soluble fission products such as caesium and strontium, while hydrogen bubbles can be retained within the settled sludge matrix over long periods. Subsequently, significant interest has arisen in understanding gas retention in order to mitigate against periodic acute releases of flammable gas during decommissioning of legacy waste silos, compounding the existing research interest in biogenic methane release from natural aquatic sediments, which represents a significant contributor towards global greenhouse gas emissions.

High resolution x-ray computed tomography (CT) has been employed to visualise bubbles retained by two predominantly brucite, $Mg(OH)_2$, test materials, which represents the major corrosion product from Magnox fuel cladding used in first generation reactors in the UK. A bespoke *corroded magnesium sludge* test material differed from a commercial $Mg(OH)_2$ test material as it contained a fraction of coarser particles, in the 100-1000 μm range, and achieves around double the yield strength at comparable solids concentrations within the 30-40 % w/w range. Low field nuclear magnetic resonance (LFNMR) relaxation time distributions also revealed that the microstructure of CMgS is characterised by a more constricted pore size distribution, with a median pore throat radius of 152 nm within a 42 % w/w sample. These subtle distinctions in sediment microstructure and mechanical properties manifested in considerable differences in the retained bubble populations observed under x-ray CT. A 30 Pa yield stress sample of CMgS supports a coarser population of 0.9-8 mm bubbles as opposed to 0.1-2 mm bubbles within an equivalent strength commercial $Mg(OH)_2$ sample and the coarser bubble population exhibits substantially lower connectivity between adjacent bubbles. Over 98 % of bubbles within the commercial $Mg(OH)_2$ soft sediment were found to *partially coalesce* and form a continuous void-network, spanning the 71 mm test vessel, thereby enabling chronic gas release from the bed by diffusion through the void-space. Conversely, the coarser bubbles within CMgS demonstrated fewer coalescence events, with distinct chains of up to 40 bubbles comprising less than 10 % of the total voidage and spanning less than 30 mm in length. Consequently, the coarse bubbles within the CMgS test material lack the buoyant stress to rise through the sediment and are insufficiently networked to promote release by diffusion through the gas phase; in the absence of an adequent mechanism for chronic gas release, the sediment experiences substantial expansion to a voidage approaching 40 % of the bed. Void fractions of this magnitude would have a substantial impact on decommissioning the legacy waste silos at Sellafield, with bulk waste swell requiring the provision of excess interim storage capacity at £10s millions in expense, while high voidage, low density sediments also present an enhanced risk of acute flammable gas release from Rayleigh-Taylor instabilities.

Investigation of the Complex Gas-Solids Flow Characteristics in a Fluidized Bed with multi-cyclone separators system By Process Tomography and CPFD Simulation

Haigang Wang[1], Hanqiao Che[1], Qiuya Tu[1], Jiamin Ye[1] and Wuqiang Yang[2],

(1)Institute of Engineering Thermophysics, Chinese Academy of Sciences, Beijing, China
(2)The School of Electrical and Electronic Engineering The University of Manchester, Manchester, United Kingdom

Circulating fluidized bed (CFB) is one kind of clean coal combustion technologies which plays an important role in the power generation and coal gasification industry. Large thermal capacity and high steam pressure is a tendency for the development of CFB boilers. To meet the demands for high steam parameter and large thermal capacity, high efficiency of gas-solid separation is a key to achieving high combustion efficiency, reducing limestone consumption and NOx emission. Over the past two decades computational fluid dynamics (CFD) and electrical capacitance tomographic (ECT) techniques have been developed to see inside particle processes, and the knowledge gained from such studies has improved energy efficiency and product quality. Examples from industrial research with different scale of CFB system are presented to illustrate the use of CFD and ECT in challenging particle process measurements. Further, a new approach based on computational particle fluid dynamic (CPFD) method combined with electrical capacitance tomography (ECT) is presented to investigate the hydrodynamic behavior of gas-solid flow in a CFB with multip-cyclone separators in order to improve the design and performance of a large scale CFB boiler.

To verify and optimize the measurement result, computational fluid dynamic (CFD) based on the kinetic theory of granular flow (KTGF) is used to investigate the flow characteristics in the same bed. Different drag models, i.e. the energy-minimization multi-scale (EMMS) model, Wen-Yu and Gidaspow model, are used to investigate the gas-particle interface effect on the flow characteristics. Key process parameters, including the solids concentration, solids velocity and pressure drop, are analyzed and verified by the measured results. Both the CFD simulation and tomography measurement results provide valuable information to reveal the hydrodynamic characteristics in the Wurster type fluidized bed. With the lab-scale fluidized bed, the results of Wen-Yu and Gidaspow drag models have a good agreement with the ECT measurement. Based on the CFD simulation and ECT measurement results, optimized parameters are chosen to improve the solids circulation efficiency and coating quality.

Figure 1 shows the CFB geometry scheme used in the simulation. The model includes a rectangular combustion chamber/riser whose height and main cross section are 6 m and 0.4 m × 1.2 m, respectively, six axis symmetric cyclone separators with diameter of 0.31 m, and six naturally balanced U - loop seal, through which the solid particles from cyclones recirculate to the riser to balance the solid inventory. The primary gas inlet is at the bottom of the whole system. Figure 1 gives the CPFD simulation result for the pressure and velocity distribution among different cyclones.

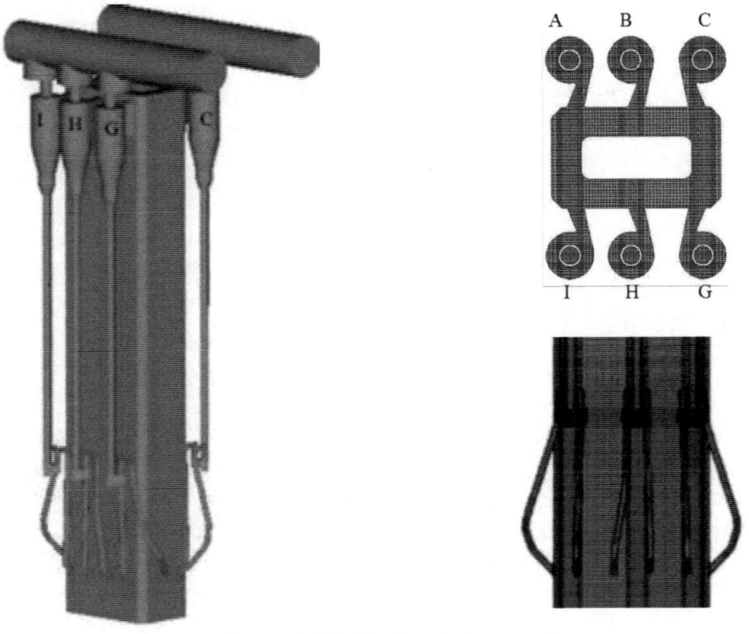

Figure 1 CPFD simulation mesh

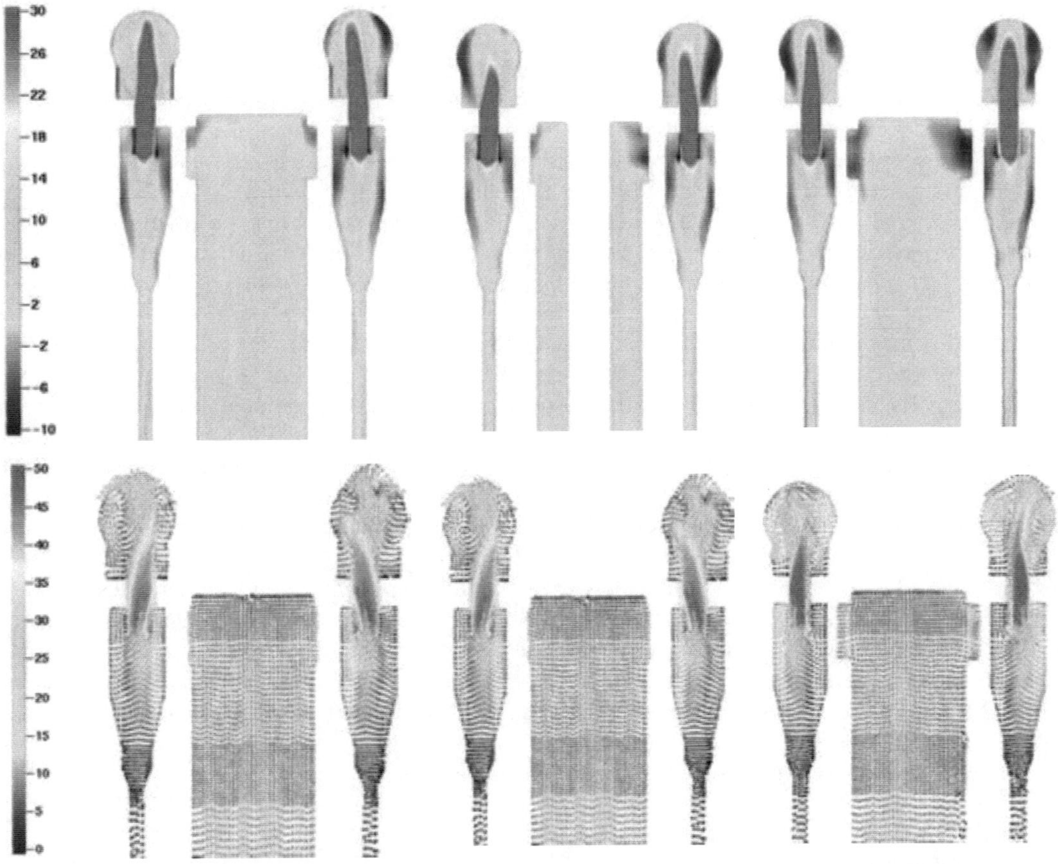

Cyclone-I and A Cyclone H and B Cyclone G and C

Figure 2 Flow velocity distribution

Experimental study on flow characterization in a rectangular spouted bed by image processing

Jingsi Yang [a], Ronald W. Breault [b*], Steven L. Rowan [a], Justin M. Weber [b]

[a] *Oak Ridge Institute for Science and Education, Morgantown, WV, United States*
[b] *National Energy Technology Laboratory, U.S. Department of Energy, Morgantown, WV, United States*

Abstract

In this study, systematic image processing methods are applied to characterize gas-solid flow under different operating conditions in a 30.2 mm × 101.6 mm rectangular spouted bed. Images are obtained by a high-speed video camera with Geldart D particles as the bed materials. Superficial gas velocity, initial bed height, and nozzle size are investigated as influence factors. By applying image adaptive thresholding and logical operation, the original color images are transformed into binary images exhibiting detailed boundary information. To analyze the binary image using a self-developed algorithm in combination with mask operations, the bed height is successfully determined, through which the variation of the bed expansion ratio with superficial gas velocity can be obtained. The application of image morphological operations, i.e. dilation and erosion on the binarized images ensures the accurate identification of the contours of the jet and fountain so that the jet and fountain heights can be determined. Based on the image processing methods, the variation of bed expansion ratio, jet height and fountain height with the three influence factors are discussed. The current study highlights the application of image analysis in investigating the flow characterization and obtaining more hydrodynamic information in a rectangular spouted bed.

Keywords: Image processing; Bed expansion ratio; Jet height; Fountain height; Rectangular spouted bed

Introduction

A spouted bed system is a composite of a dilute central core with particles transported upward by a concurrent gas flow and a dense annulus of downward particle flow. Hydrodynamic characterization of the gas-solid flow in spouted beds is a significant objective due to its close relationship with the wide applications in drying,[1] coating,[2] pyrolysis,[3] combustion,[4] and gasification.[5] In spouted beds, the jet and the fountain are considered as two distinctive formations that have remarkably effect on the hydrodynamic and efficiency of the system. The jet is inherently responsible for gas bypassing and erosion of bed internals,[6, 7] while the fountain plays an important part in spouted bed reactor design and particle segregation.[8, 9] Extensive experimental and simulation studies about the jet and fountain heights have highlighted their great potential support to the industrial application.[6-15] It's worth mentioning that the bed expansion ratio, as an important hydrodynamic characteristic reported in different kinds of gas-solids systems, has been rarely reported in spouted beds possibly due to the intrinsic complication of gas-solids interaction and the limitation of the measurement technique. With the development of modern video cameras and imaging analysis, it is expected that more hydrodynamic information can be provided.

In gas-solids two phase study, both intrusive and non-intrusive techniques have been applied to investigate the hydrodynamic characteristics. The former is expedient in providing and revealing localized information and particulate behavior,[6, 16, 17] but the inevitable unquantified flow disturbance has been a debatable topic. As a comparison, the latter is superior notably for none disturbance to the flow hydrodynamics, whereas the more complex data interpretation relies on complicated programming and analysis.[18-20] In spite of the pros and cons of each technique, the non-intrusive technique has become more popular promoted by the development of digital image acquisition system and the improvement of image processing methods. The combination of high-speed video and digital image analysis has been applied in different gas-solids systems such as circulating fluidized beds,[21, 22] fluidized beds,[23, 24] packed beds,[25, 26] and spouted beds.[27, 28] To fully broaden and utilize the superiority of the imaging technique, different image processing algorithms and operations need to be explored and extended further so as to enhance the accuracy, objectivity and automation capability of the measurement.

In this paper, systematic image processing methods are proposed to study the hydrodynamic characteristics i.e. bed expansion ratio, jet height, and fountain height in a rectangular spouted bed. One objective is to verify the feasibility and effectiveness of the image processing methods in assessing the solids behavior in the spouted beds. Another objective is to establish a basis for specific hydrodynamic studies in the future.

Experimental Apparatus

Rectangular spouted bed and operating conditions

The spouted bed used in the current study consists of a 609.6 mm high acrylic column with a rectangular cross section of 30.2 mm×101.6 mm, a 317.5 mm height of cone with a cone angle of 75° (i.e. the included angle of 30°), and two filters

* Corresponding author: E-mail address: ronald.breault@netl.doe.gov (R. W. Breault).

connected to the top of the rectangular column. An interchangeable slotted nozzle is assembly installed at the bottom of the cone, through which the air enters into the bed. Two different sizes of rectangular slotted nozzle are used in the experiment with inner cross section of 3.8 mm×19.1 mm and 6.6 mm×19.1 mm, respectively. The corresponding equivalent cross-sectional area diameters (D_{eqn}) are 9.6 mm and 12.7 mm.

The bed materials are nylon and alumina particles classified as Geldart Group D. The particle properties are listed in Table 1. Increasing the superficial gas velocity (U_g) from 0 to 2.5 m/s, experimental investigation are conducted under three initial bed heights (H_0): 0.0762 m (3 in), 0.102 m (4 in), and 0.127 m (5 in) with two different nozzle sizes respectively.

Table 1 Particle properties

Particle	Sauter mean diameter (μm)	Particle density (kg/m^3)	Minimum fluidization velocity (m/s)	Particle sphericity (-)	Geldart Classification
Nylon	3190	1130	1.088	0.95	D
Alumina	1709	3150	1.101	0.95	D

Visualization system

A Sony RX10II high speed video camera is applied with a Dracast LED500 light used as the backup illumination light. To study the flow characteristics in the spouted bed, the frame speed of 500 frame per second (fps) with shutter time of 0.001 s is set, under which the resolution is 1920 pixel × 1080 pixel. A diffusion panel is employed in front of the LED light to enable the uniform illumination of recorded area and eliminate the undesirable shadows and intensity gradients.

With the acquired color images transformed into the grayscale, noise reduction and adaptive threholding are applied as preliminary processing to prepare the images for the following specific analysis. Depending on the different investigated characteristics, different image processing methods including image logical and mask operations, and morphological operations viz. dilation and erosion are adopted to "extract" the target information, the procedure of which are introduced in the corresponding sections. The frame-by-frame image processing is achieved through PYTHON, utilizing OpenCV and Numpy libraries.

The choosing of adaptive thresholding rather than the conventional global threholding is because of its regional-thresholding function which uses local thresholds determined by the pixel values in the neighborhood of each pixel in the image. Therefore, the adaptive thresholding utilizes multiple thresholds instead of one global value to binarize the image, which shows prominent advantage in discriminating the phase boundary (solid and gas phases). Under the current visualization setup, the pre-calibrated meter-to-pixel ratio is 0.00026 m/pixel, i.e.12 pixels/particle. Thus, the optimum neighborhood size of 7×7 with the lowest standard deviation value of 0.00015 is chosen for the adaptive thresholding.

Results and discussion

Bed expansion ratio

The bed expansion ratio is obtained on the basis of the determination of the bed height, i.e. the height difference between the bed surface and the bed bottom. The precise boundary detection of the bed surface and the bed bottom can be achieved by the following image processing procedure: 1) Converting the original color image into grayscale image, after which Gaussian filter is applied to remove the Gaussian white noise; 2) Using adaptive thresholding to transform the de-noised image into binary image; 3) Performing image logical and mask operations to eliminate interference information such as background shades and jet/bubble geometry. After the processing procedure, the extracted image with interfaces of the bed surface and bottom is obtained, as shown in Fig. 1. By calculating the average intensity of each row in the extracted image through a self-developed algorithm, both the bed surface and bottom are indicated by two dominant peaks in the generated average intensity profile, i.e. red line in Fig. 1. Plotting the average intensity with the image size, the location of the bed surface and bottom can be determined using pixel as the unit. With the pre-calibrated meter-to-pixel ratio, the bed height is obtained from the height difference (Fig. 1).

Fig. 2 shows the profiles of bed expansion ratio with U_g under different H_0. For both nylon and alumina particles in this study, the bed expands with increasing U_g till the onset of the external spouting, i.e. the initial formation of the fountain. It is clearly shown that before the external spouting, the bed expands with the U_g and a lower H_0 leads to higher bed expansion. After the external spouting, i.e. the jet breaks through the bed surface, increasing U_g results in a decrease of the bed expansion ratio because more particles are carried above the bed surface by the increasing gas drag. Compared to thicker bed (i.e. larger H_0), it is easier for the gas to penetrate the bed surface with more particles brought into the freeboard, which causes the sharper decrease of the bed expansion ratio curve. It is noted that the bed reaches a maximum expansion at the onset of the external spouting for both particles. Considering the maximum bed expansion ratio, the bed seems to expand the same ratio independent of H_0 and D_{eqn} for the same particle (as shown in Fig. 2 (a) and (b)). However, the maximum value varies with different particle properties, as indicated in Fig. 2 (a) and (c), the constant value of 1.4 for nylon, while that of 1.16 for alumina. Different kinds of particles will be involved to investigate the effect of particle properties on the maximum bed expansion ratio.

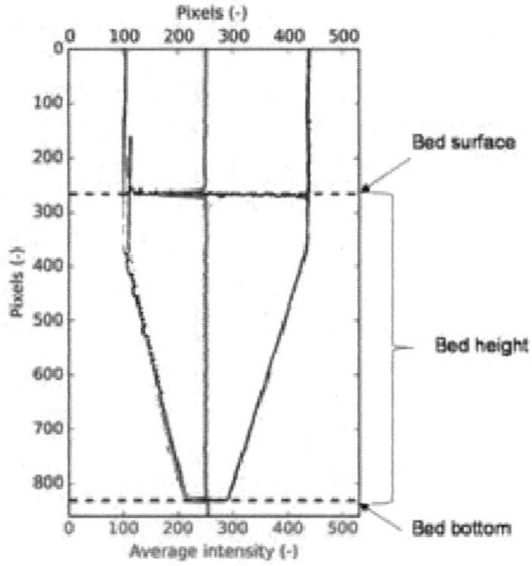

Fig. 1 Interfaces detection

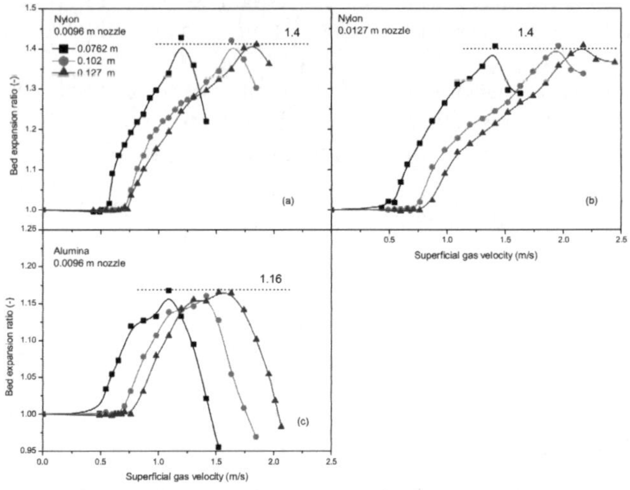

Fig. 2 Profiles of bed expansion ratio with U_g and H_0

In combination of profiles of the pressure drop across the column and the bed expansion ratio, the determination of different flow regimes can be achieved. As shown in Fig. 3, from the intersection point between the linear fitting lines of pressure drop decreasing section and the platform section, an internal spouting velocity (U_{is}), i.e. the superficial gas velocity at which the internal spout originates, can be determined. Similarly, the intersection point between the liner fitting lines of bed expansion ratio increasing and decreasing sections defines an external spouting velocity (U_{es}), the superficial gas velocity that the internal spout breaks through the bed surface leading to the fountain formation.

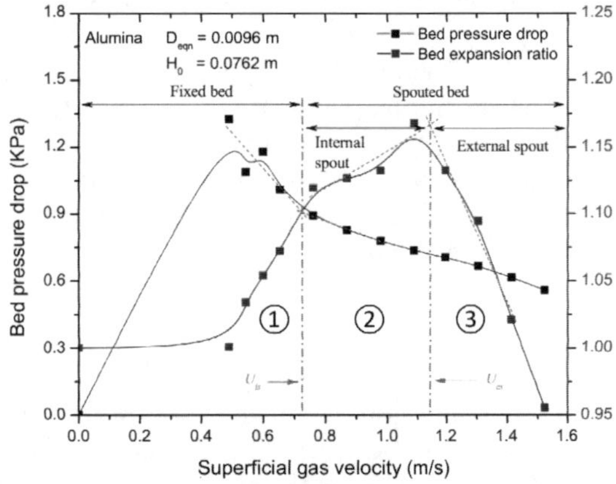

Fig. 3 Flow regime discrimination

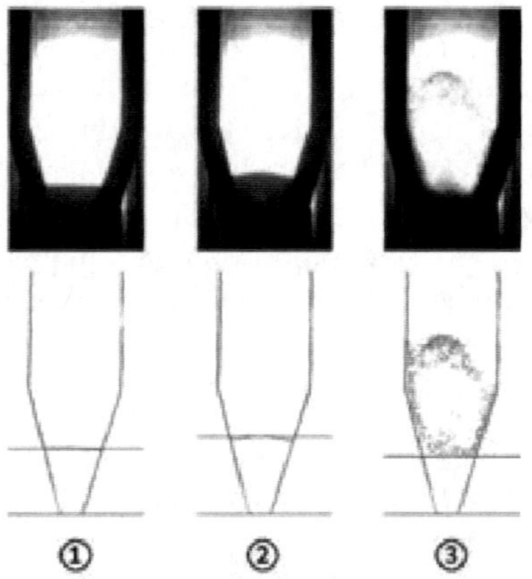

Fig. 4 Typical bed height variation in different flow regimes

The U_{is} and the U_{es} represent the initiation velocities of internal cavity and the fountain respectively. The former can be used as the discrimination velocity between the fixed bed and spouted bed, while the latter as the one between the internal and the external spout. The discriminated flow regimes are shown in Fig. 3 with the U_{is} and the U_{es} of 0.72 and 1.14 m/s respectively. Typical images of bed height in different flow regimes are shown in Fig. 4, which clear demonstrates the increase and decrease of the bed height under different U_g.

Jet height

The jet height has been determined by various methods such as visual observation,[11, 12] photographic,[10] Pitot tube technique,[7] and multi-channel pressure signal analysis.[29] In addition, it was reported that image analysis was applied to determine the jet height by frame-by-frame analysis of instantaneous images represented the end of a cycle of jet development,[29] but without detailed image processing introduction. In this study, an image-based jet contour detection is utilized to determine the jet height, which aims to provide a novel method in jet height determination using image processing methods.

Detailed image processing procedures to detect and recognize the well-defined jet contour are demonstrated in Fig. 5 with the following steps: 1) Converting original image (Fig. 5 (a)) into grayscale image and applying Gaussian filter; 2) Transforming the grayscale image (Fig. 5 (b)) into binary image through adaptive thresholding and retaining only the target information, i.e. the jet, using mask operation, see Fig. 5 (c); 3) Extracting the jet contour by morphological closing operation, dilation followed by erosion, and applying labelling operation to eliminate discrete regions caused by uneven solids holdup distribution around jet. The distinctive jet contour in white foreground is shown in Fig. 5 (d). The jet bounding box can be obtained accurately with the well-defined jet contour, as shown in Fig. 5 (e). Fig. 5 (f) shows the recognized jet bounding box in the color image, which indicates that the current image processing guarantees a precise jet contour recognition. With the detected jet contour, the jet height can be determined precisely with pre-calibrated meter-to- pixel ratio of 0.00026 m/pixel.

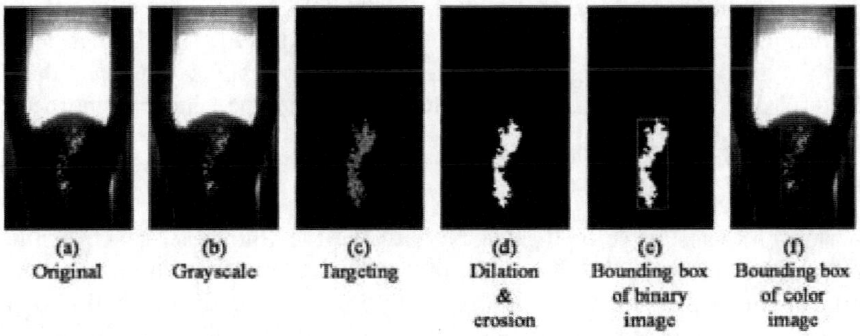

Fig. 5 Demonstration of jet height determination

In this study, the term of "jet height" is used from the angle of image processing so as to conform to contour and morphology operations. Through the observation of the jet evolution, it is found that the internal jet develops from a cavity state with a constant height to a dynamic state with a periodical changing of jet height due to the periodic bubble detaching at the top of the jet. A dynamic equilibrium seems to be established between the jet shrink and the bubble generation. Based on the jet height determination by the image processing method, Fig. 6 shows the variation of the instantaneous jet height in an image sequence of 960 images under different U_g with typical images. Considering the intrinsically-unsteady character of the jet, the effective jet height is a time-averaged value. As shown in Fig. 6, an average jet height (H_j) of 0.0609 m with a standard deviation (σ) of 0.00975 at lower U_g (0.98 m/s) reflects the stable vertical internal cavity. A higher H_j of 0.116 m with σ of 0.0237 obtained at higher U_g (1.64 m/s) corresponds to a periodic manner of the jet collapsing and the bubble generating with a certain frequency. Fig. 6 also illustrates that the image-based determination method not only records the instantaneous jet height, which can illustrate the dynamic behavior of the jet, but also generates a more meaningful and representative average jet height. Profiles of determined average jet height under different operating conditions are shown in Fig. 7 (a). For both nozzle sizes, the H_j increases with the increase of U_g. Under the same U_g, smaller nozzle size (D_{eqn} = 0.0096 m) leads to larger H_j. For the same D_{eqn}, a higher H_0 requires higher U_g to initiate the internal jet. Once the jet is established, the H_0 seems to have no effect on the jet height illustrated by the overlapping profiles of jet height under different H_0. The effect of H_0 on H_j is consistent with the conclusions of previous studies,[6, 7] which verifies the effectiveness and usefulness of the image processing method developed in this study.

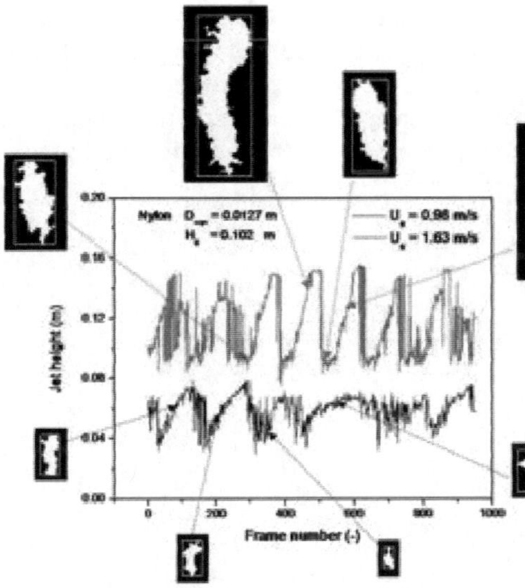

Fig. 6 Variation of instantaneous jet height under different U_g

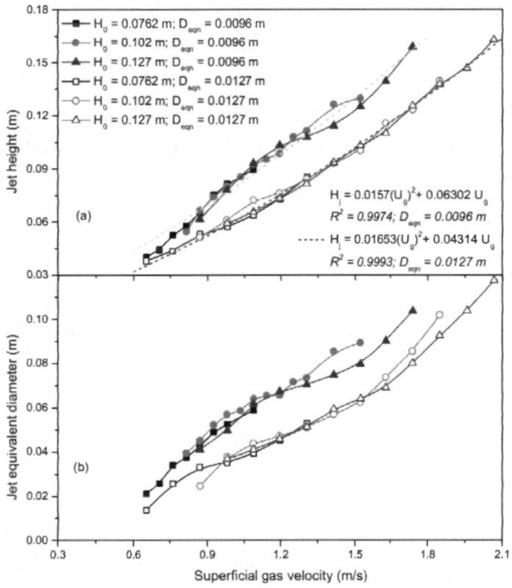

Fig. 7 Variation of jet height and equivalent diameter with U_g

To express the relationship between the H_j and U_g, a polynomial fitting with higher adjusted R^2 is applied for the two D_{eqn} respectively. The generated quadratic polynomial expressions for both nozzle sizes are shown in the following:

$$H_j = 0.0157\, {U_g}^2 + 0.06302 U_g \quad (1)$$

$$H_j = 0.01653\, {U_g}^2 + 0.04314 U_g \quad (2)$$

Eq. (1) and (2) correspond to the nozzle sizes of 0.0096 and 0.0127 m with R^2 values of 0.9974 and 0.9993, respectively. Both expressions suggest that the jet height is related to the jet kinetic energy (U_g^2) and momentum (U_g) under a certain nozzle size. Comparing the quadratic and linear coefficients of the two expressions indicates that the variation of nozzle size has more effect on the jet momentum than the kinetic energy. The higher jet momentum of the smaller nozzle leads to the higher H_j. In addition, when the jet width is taken into account, a jet equivalent diameter can be obtained from the area of the jet bounding box. The variation profiles of the jet equivalent diameter with U_g are shown in Fig. 7 (b), which indicates that not only the jet height, but also the jet equivalent diameter is independent of the H_0. This consistency suggests that the jet formation may be related closely to the voidage/solids holdup of the jet influenced by the nozzle inlet gas momentum and deserves further investigation.

Fountain height

The fountain height is determined as the vertical distance between the fountain surface to the dense bed surface. The determination process is demonstrated in Fig. 8. Similar to the jet height determination, a binary image exhibiting the fountain geometry is obtained by application of the adaptive threholding after transforming the original into grayscale image, as shown in Fig. 8 (a) to (c). The detection of fountain contour is achieved by the morphological closing operation with a neighborhood size of 12×12, i.e. dilation followed by erosion. With the dilation filling the voids and the erosion deleting isolated particles to restore the fountain boundary, the fountain contour is delineated clearly, as shown in Fig. 8 (d). Fig. 8 (e) shows the determined top surface of the fountain contour. With the determined bed height introduced above, the fountain height can be obtained, as shown in Fig. 8 (f).

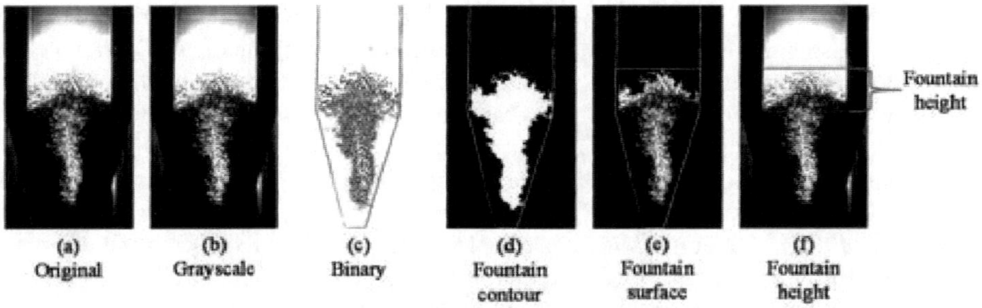

Fig. 8 Demonstration of fountain height determination

Fig. 9 shows the variation of the fountain height (H_f) with the U_g. For both nylon and alumina particles, the H_f increases with the increasing U_g because the particles can reach higher level above the bed surface with higher momentum caused by the higher gas drag. Compared to the lower H_0, the higher H_0 requires higher U_g to reach the same H_f due to the larger resistance that the gas encounters when passing through the bed to form the fountain. Thus, higher U_g is needed to balance the additional gas momentum consumption. Fig. 10 shows the effect of nozzle size on the H_f. It is indicated that after penetrating the same H_0, smaller nozzle leads to higher H_f, which may be attributed to the higher inlet gas velocity under the same gas flow rate so that the larger gas momentum leads to a higher reachable height of the spouted particles. These results are consistent with previous studies,[15, 30] which, on the other hand, proves that the proposed image processing method is effective and useful.

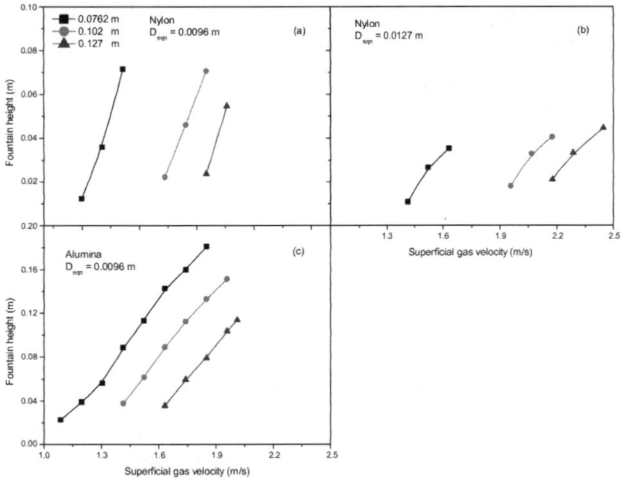

Fig. 9 Profiles of fountain height with U_g

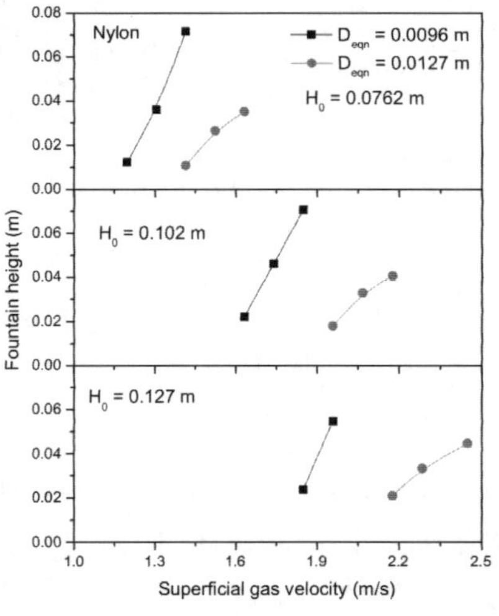

Fig. 10 Effect of nozzle size on the fountain height

Furthermore, it is found that under the same D_{eqn} and H_0, the H_f of Alumina particles is higher than that of nylon particles, as shown in Fig. 11. The H_f difference between the two particles seems to increase with the increase of H_0. The effect of particle properties on the H_f is a focus of future study. The image processing methods applied in this paper are summarized in a flow chart provided by Fig. 12, which have been proved to be feasible and functional in studying flow characteristics in the spouted beds.

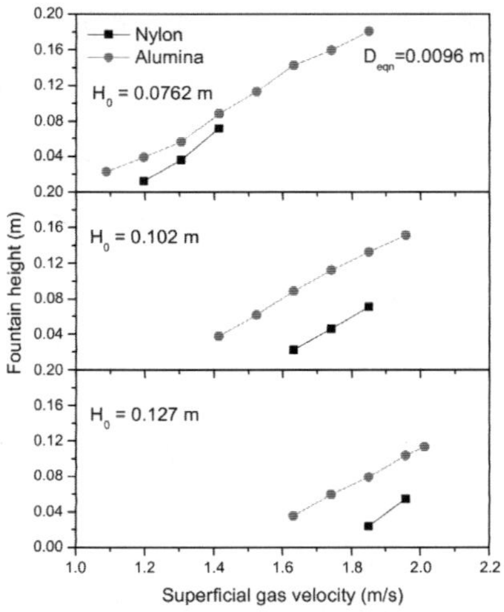

Fig. 11 Profiles of fountain height for different particles

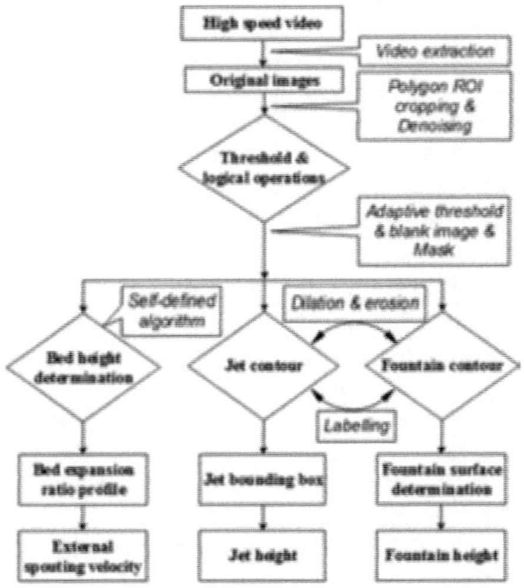

Fig. 12 Flow chart of applied image processing methods

Conclusions

Systematic image processing methods have been proposed to study the flow characteristics in a spouted bed. Using a high speed video camera, the experiment is conducted in a rectangular spouted bed with a cross section of 30.2 mm×101.6 mm under three initial bed heights (H_0 = 0.0762, 0.102 and 0.127 m) for two different nozzle sizes (D_{eqn} = 0.0096 and 0.0127 m). Nylon and alumina particles, belonging to Geldart Group D, are used bed materials.

Preliminary image processing including noise reduction and thresholding is applied on the acquired original image to benefit the follow-up specific processing corresponding to different characteristics. Based on the adaptive thresholding, logical and mask operations, an algorithm is applied to determine the bed expansion ratio. Through contour detection achieved by the morphological closing operation and the labelling operation, the jet and fountain heights are determined effectively. Both heights are time-averaged values of series of instantaneous values.

The feasibility and effectiveness of the proposed image processing methods are verified by the variation profiles of bed expansion ratio, jet and fountain heights with influence factors of superficial gas velocity (U_g), initial bed height (H_0) and nozzle size (D_{eqn}). Conclusions are shown in the following.

1) The bed expansion ratio increases with the increase of U_g before the initiation of the external spouting, beyond which the bed expansion ratio decreases. At the onset of the external spouting, the bed reaches a maximum expansion ratio, which varies from particles, but is independent of the H_0 and D_{eqn}. The ratios of nylon and alumina are 1.4 and 1.16 respectively in this study. In addition, different flow regimes can be determined through profiles of the pressure drop across the column and the bed expansion ratio.

2) The jet height (H_j) increases with U_g. Under the same U_g, smaller nozzle size (D_{eqn} = 0.0096 m) leads to larger H_j. The H_0 has almost no effect on the H_j but the incipient internal jet velocity. The relationship between the H_j and U_g suggests jet height is related to the jet kinetic energy and momentum under a certain nozzle size. The nozzle size has more effect on the jet momentum than the kinetic energy. The jet formation may be related closely to the voidage/solids holdup of the jet influenced by the nozzle inlet gas momentum, which deserves further study.

3) The fountain height (H_f) increases with the increase of U_g. To reach the same H_f, the higher H_0 needs higher U_g to balance more resistance compared to the lower H_0. Under the same H_0, the smaller nozzle leads to higher H_f, which is attributed to the higher inlet gas velocity under the same gas flow rate. Furthermore, under the same D_{eqn} and H_0, the H_f of Alumina particles is higher than that of nylon particles. The effect of particle properties on the H_f is a focus of future study.

Acknowledgements

The authors acknowledge the Department of Energy for funding the research through the office of Fossil Energy Commercialization Technology Fund and the Advanced Gasification Program. Additionally, this research was supported in part by an appointment to the National Energy Technology Laboratory Research Participation Program, sponsored by the U.S. Department of Energy and administered by the Oak Ridge Institute for Science and Education.

References

1. S. Wiriyaumpaiwong, S. Soponronnarit, S. Prachayawarakorn (2004), "Drying and urease inactivation models of soybean using two-dimensional spouted bed technique", Drying Technol. 24, pp. 1673–1681.
2. G.S. da Rosa, S.C. dos Santos Rocha (2010), "Effect of process conditions on particle growth for spouted bed coating of urea", Chem. Eng. Process: Process Intensif. 49, pp. 836-842.
3. J. Alvarez, M. Amutio, G. Lopez, I. Barbarias, J. Bilbao, M. Olazar (2015), "Sewage sludge valorization by flash pyrolysis in a conical spouted bed reactor", Chem. Eng. J. 273, pp. 173-183.
4. R.K Konduri, E.R. Altwicker, M.H. Morgan III (1999), "Design and scale-up of a spouted-bed combustor", Chem. Eng. Sci. 54, pp. 185–204.
5. G. Lopez, A. Erkiaga, M. Amutio, J. Bilbao, M. Olazar (2015), "Effect of polyethylene co-feeding in the steam gasification of biomass in a conical spouted bed reactor", Fuel 153, pp. 393-401.
6. R. Hong, H. Li, M. Cheng, J. Zhang (1996), "Numerical simulation and verification of a gas-solid jet fluidized bed", Powder Technol. 87, pp. 73-81.
7 R.Y. Hong, Q.J. Guo, G.H. Luo, J.-Y. Zhang, J. Ding (2003), "On the jet penetration height in fluidized beds with two vertical jets", Powder Technol. 113, pp. 216-227.
8. J.R. Grace, K.B. Mathur (1978), "Height and structure of the fountain region above spouted beds", Can. J. Chem. Eng. 56, pp. 533-537.
9. M. Olazar, M.J. San Jose′, M.A. Izquierdo, S. Alvarez, J. Bilbao (2004), "Fountain geometry in shallow spouted beds", Ind. Eng. Chem. Res. 43, pp. 1163-1168.
10. V.I. Markhevka, V.A. Basov, T.Kh. Melik-Akhnazarov, D.I. Orochko (1971), "The flow of a gas jet into a fluidized bed", Theor. Found. Chem. Eng. 5, pp. 80-85.
11. J.M.D. Merry (1975), "Penetration of vertical jets into fluidized beds", AIChE J. 21, pp. 507-510.
12. T.M. Knowlton, I. Hirsan (1980), "The effect of pressure on jet penetration in semi-cylinderical gasfluidized beds", In: Grace J.R., Matsen J.M. (eds) Fluidization. Springer, Boston, MA, pp. 315-324.
13. J. Y. Day (1990), "The fountain height and particle circulation rate in a spouted bed", Chem. Eng. Sci. 45, pp. 2987-2990.
14. Y. L. He, C.J. Lim, J.R. Grace (1997), "Scale-up studies of spouted beds", Chem. Eng. Sci. 52, pp. 329-339.
15. W. Zhong, Q. Li, M. Zhang, B. Jin, R. Xiao, Y. Huang (2008), "Spout characteristics of a cylindrical spout-fluid bed with elevated pressure", Chem. Eng. J. 139, pp. 42–47.
16. Y.L. He, S.Z. Qin, C.J. Lim, J.R. Grace (1994), "Particle velocity profiles and solid flow patterns in spouted beds", Can. J. Chem. Eng. 72, pp. 561 – 568.
17. M. Olazar, M.J. San José, S. Alvarez, A. Morales, J. Bilbao (1998), "Measurement of particle velocities in conical spouted beds using an optical fibre probe", Ind. Eng. Chem. Res. 37, pp. 4520–4527.
18. G.Q. Liu, S.Q. Li, X.L. Zhao, Q. Yao (2008), "Experimental studies of particle flow dynamics in a two-dimensional spouted bed", Chem. Eng. Sci. 63, pp. 1131–1141.
19. V.S. Sutkar, N.G. Deen, V. Salikov, S. Antonyuk, S. Heinrich, J.A.M. Kuipers (2015), "Experimental and numerical investigations of a pseudo-2D spout fluidized bed with draft plates", Powder Technol. 270, pp. 537–547.
20. H. Zhang, M. Liu, T. Li, Z. Huang, X. Sun, H. Bo, Y. Dong (2017), "Experimental investigation on gas-solid hydrodynamics of coarse particles in a two-dimensional spouted bed", Powder Technol. 307, pp. 175–183.
21. J. Yang, J.X. Zhu (2014), "A novel method based on image processing to visualize clusters in a rectangular circulating fluidized bed riser", Powder Technol. 254, pp. 407–415.
22. J. Yang, J.X. Zhu (2015), "Visualization of solids phase separation in a rectangular CFB riser using a novel image calibration method", Powder Technol. 273, pp. 76–82.
23. A. Busciglio, G. Vella, G. Micale, L. Rizzuti (2008), "Analysis of the bubbling behaviour of 2D gas solid fluidized beds: Part I. Digital image analysis technique", Chem. Eng. J. 140, pp. 398-413.
24. J.A. Laverman, I. Roghair, M. van Sint Annaland, H. Kuipers (2008), "Investigation into the hydrodynamics of gas–solid fluidized beds using particle image velocimetry coupled with digital image analysis", Can. J. Chem. Eng. 86, pp. 523 – 535.
25. M.H. Köhl, G. Lu, J.R. Third, M. Häberlin, L. Kasper, K.P. Prüssmann, C.R. Müller (2013), "Magnetic resonance imaging (MRI) study of jet formation in packed beds", Chem. Eng. Sci. 97, pp. 406-412.
26. M.H. Köhl, G. Lu, J.R. Third, K.P. Prüssmann, C.R., Müller (2014), "Magnetic resonance imaging (MRI) of jet height hysteresis in packed beds", Chem. Eng. Sci. 109, pp. 276-283.
27. M. Kiani, M.R. Rahimi, S.H. Hosseini, G. Ahmadi (2017), "Mixing and segregation of solid particles in a conical spouted bed: Effect of particle size and density", Particuology 32, pp. 132-140.
28 Y. Zhang, W. Zhong, X. Rui, B. Jin, H.Liu (2017), "Evaluation of mixing and mixing rate in a multiple spouted bed by image processing technique", Int. J. Chem. Reac. Eng. 15, pp. 103-115.
29. W. Zhong, M. Zhang (2005), "Jet penetration depth in a two-dimensional spout-fluid bed", Chem. Eng. Sci. 60, pp. 315-327.
30. O.M. Dogan, L.A.P. Freitas, C.j. Lim, J.R. Grace, B. Luo (2000), "Hydrodynamics and stability of slot-rectangular spouted beds", Part I: Thin bed, Chem. Eng. Comm. 181, pp. 225-242.

509597 Atomic Layer Deposition and Other Applications of the Littleford Plow Mixer As a Mechanically Fluidized Bed

Abstract:

The Littleford plow mixer is a well-known workhorse in commercial operations ranging from pharmaceuticals to iron ore. The unit operations involved include blending, dissolution, spray coating, vacuum drying, agglomeration, direct-contact cooling, and interfacial chemical reactions. Each of these applications is enhanced by the **use of the plow mixer as a mechanically-fluidized bed.**

The geometry and operating parameters of the mixer will be briefly reviewed, and compared to other types of fluidized beds. In particular, the ability of the unit to operate under high vacuum will be explained.

Several operations using the Littleford plow mixer will be discussed to illustrate the utility of the unit as a fluidized bed. One area of interfacial chemical reactions of particular interest is **atomic layer deposition**.

A typical procedure is as follows: The substrate was first loaded into the 22-liter mixer. The plow speed was then set, and the hot-oil unit was started and recirculated to heat the product. Once the material reached the target temperature, it was held at temperature to completely remove any moisture. The active chemical cylinders were weighed and then attached to the vapor header. The nitrogen supply was used to check for leakage, assuring that the flow was zero when the appropriate valves were closed. Vacuum was then applied to the unit, and an initial nitrogen purge (300 sccm) was put through each addition line for several minutes. A rotary lobe blower was connected in series with the 2-stage liquid-ring vacuum pump to achieve pressures below 25 torr. Dosing time was determined experimentally by periodically stopping the dosing and re-weighing the cylinder(s) to determine rate. Once the desired dosing cycles were completed, the material was cooled using the recirculating system. The vacuum was shut off and the nitrogen purge was continued until the vessel reached atmospheric pressure. The material was discharged through the bottom valve, with the plows running, and weighed. Degree of coverage was determined by analytical means.

Other unit operations will be discussed, as time and interest allow.

510390 The Divide between Academia and Industry in Modeling Gas-Solids Reacting Flows

Abstract:

Developing new and efficient commercial scale multiphase reactors is extremely difficult as one needs to consider the complex interactions over a wide range of both temporal and spatial scales encountered in these systems (from molecules to macroscale). Therefore, it is important to codify and automate the knowledge collectively acquired so far, reserving human resources for the creative solutions that build upon codifying past learnings. Computational science (algorithms, theory and modeling, computer science, etc.) combined with exponential growth in computing hardware has great potential to provide us valuable tools for improving the effectiveness of existing plants but also design new reactors. Industry is in desperate need of fast simulation tools that are comparable to the accuracy of the single-phase reacting flows to advance the adoption of multiphase flow reactors and primarily to reduce the risk of scale-up. On the other hand, academia and national laboratories are focused on improving numerical methods, submodels or certain physics to push the state-of-the-art and predictability of the models. However, there is tremendous need for rapid integration of various capabilities that exploit latest computer hardware improvements, robust model reduction techniques, etc. that would improve the computational tools available to the industry to deliver the next generation of efficient and cost-effective reactors. These developments typically fall into the valley of death where industry and software vendors feel these efforts are far-fetched while the academic world considers it incremental. One way to address this divide is to develop mechanistic upscaling framework to coarse-grain information from more detailed modeling approaches to fast models accessible to the industry with relatively minimal information loss.

I will present such an approach using a gas-phase fluidized bed polymerization reactor as an example. First, I will provide an overview of the various models currently used at the different scales in gas-solids reacting flows. The coupling across the scales will be introduced through few different approaches:

1. Discrete-Continuum Coupling using Discrete Element Method for particles
2. Upscaling data from Discrete Element Modeling results to continuum based Computational Fluid Dynamics (CFD)
3. Upscaling information from CFD to simplified reactor network models

In conclusion, the presentation will elucidate the approaches and opportunities in bridging the gap between academic advances and industry needs to model and advance our understanding of gas-solids reacting flow reactors.

507029 A CFD-DEM Analysis of Particle Attrition in a Cyclone

Abstract:

Cyclones are used to separate entrained particles from the flue gas streams. Particles entering a cyclone are subjected to a radial centrifugal force field, driving them to the cyclone walls, where they experience collisional and rapid shearing stresses. For frangible particles this could lead to attrition and dust generation, whilst for hard particles wear of cyclone wall could occur.

In this work we address the issue of attrition of manganese di-oxide particles used in the CLC process. A CFD-DEM analysis of particle motion and interactions is used to calculate the frequency of collisions, impact velocities and shear stresses. These are then coupled with a model of particle breakage due to impact and of surface wear, in order to predict attrition in a cyclone. The outcome of the work is presented, providing a correlation of the attrition extent as a function of particle properties and operating conditions.

DEM Simulation of Cylinders and Capsules in a Fluidized Bed

*Oleh Baran**
Thomas Eppinger
Kuanjin Han

Material presented at World Congress on Particle Technology in Orlando on April 23, 2018

Abstract

Fluidization of non-spherical particles of cylindrical or close to cylindrical shape is part of many industrial processes, especially in relation to biomass fuel production, fluidized bed gasifiers and pharmaceutical tablet coating processes. In these applications accurately accounting for the details of particle shape is important for safe and efficient design of industrial equipment. In this work the numerical model based on the Discrete Element Method (DEM) fully coupled with CFD is used to predict the effect of particle aspect ratio and presence of sharp edges in the particle's shape on the fluidized bed hydrodynamics. The model is implemented within commercial software Simcenter STAR-CCM+ version 12.06 developed by Siemens PLM. The model assumes particles are cylinders or spherocylinders (capsules). It accurately resolves the collisions between cylindrical particles or between capsules without employing conventional method of "spheres glued together" for non-spherical particles.

Cylindrical Particles

Capsules

The model is then used to simulate cylindrical particles or capsules of the same volume, mass, aspect ratio, and frictional properties in a cylindrical container with air inlet at the bottom and pressure outlet boundary at the top with the range of inlet velocities both below and above minimum fluidization velocity. The results for pressure drop, particle packing and particle preferred orientation are analyzed for both cylinders and capsules and compared to the correlations reported in the literature. The results for the percentage of particles aligned with fluid flow direction in bubbling flow regime indicate the presence of surprisingly longer timescale for establishing steady state preferred orientation of particles. The effect of the absence of the sharp edges in capsules compared to cylinders on the hydrodynamics of the fluidized bed is discussed in details.

Introduction

The computational models for the flow of elongated particles in fluidized bed can bring valuable insight in understanding the effect of particle shape on fluidization state inside industrial reactors and ultimately help designing cost effective fluidization processes and equipment. Notably Discrete Element Method (DEM) is recognized as promising computational approach for modeling the flow of non-spherical particles. For example, (Borzsonyi and Stannarius 2013), (Guo, et al. 2013) (Abreu, Tavares and Castier 2003) (Williams and Philipse 2003) use DEM to study interesting aspects on packing and shear resistance of rods without coupling to fluid flow and (Mema, et al. 2017), (Zhong, et al. 2009), (Mahajan, et al. 2017), (Cai, et al. 2016) use DEM coupled with CFD to investigate the flow of elongated particles in fluidized bed. Experimental data for validating DEM-CFD simulations of above systems are also becoming available recently, see, for example, (Buist, et al. 2017), (Liao 2013). However, DEM by itself and DEM-CFD coupling for non-spherical particles are known to be computationally too expensive to be applied to even smaller scale industrial scale problems.

In this work new DEM-CFD model is introduced for simulating cylinders or capsules in a fluidized bed which offers the computational speed advantages due to particular choice of modeling the contacts between non-spherical particles and applicability to complex geometries advantage due to special cell clustering approach in 2-way coupling between solid and fluid phases.

Contact model for arbitrarily shaped discrete elements

Resolving collisions between particles in the distinct feature of DEM, see Fig 1 below.

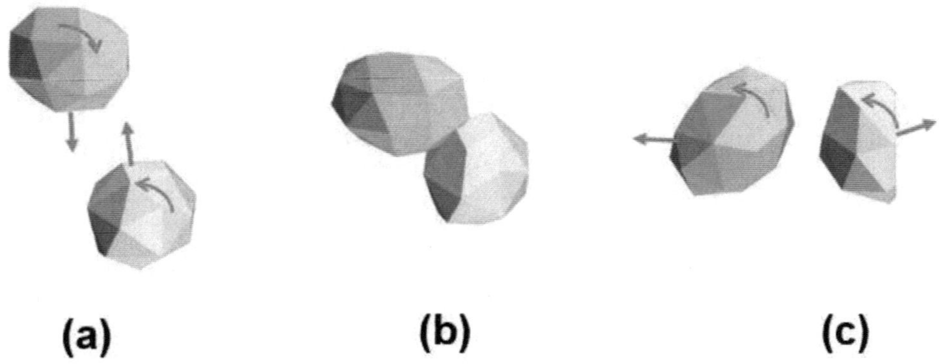

Figure 1. Collision between non-spherical particles. (a) Before collision. (b) During collision. (c) After collision. Blue arrows show particle's linear velocities and green arrows show particle's rotational velocities

Generalizing well validated and tested soft sphere approach for DEM on non-spherical particles is described next. Within this approach particle collision (see Fig 1 (b)) takes finite time or many DEM timesteps and can be described by the physics of spring-dashpot system as shown on schematics in Fig 2.

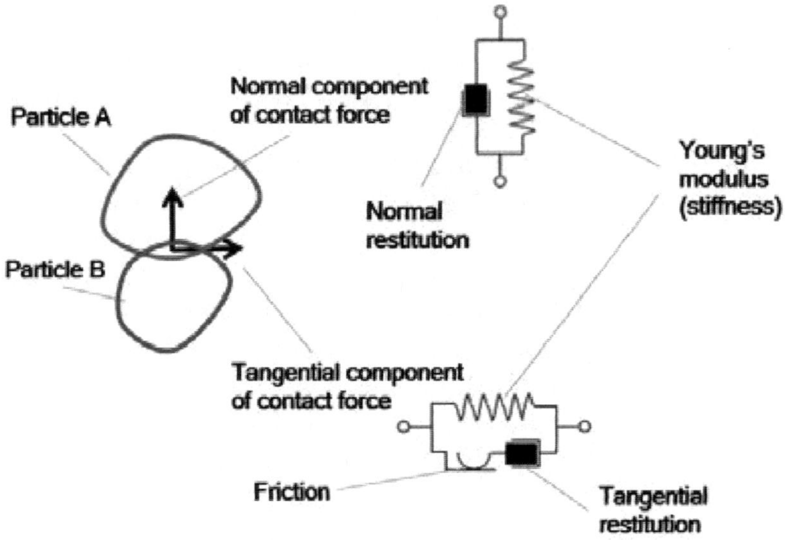

Figure 2. Schematics for soft collision approach in DEM

Within this new soft collision approach colliding Particle A and Particle B are rigid bodies that do not deform. During collision the overlap between these two arbitrary shape objects is monitored and contact force which depends on this overlap is calculated. Total contact force is a sum of normal and tangential components. According to (Feng, et al., 2012) the direction of the normal component can be uniquely determined and it provides the maximum rate of slowing down two impacting particles.

As shown in (Feng, et al., 2012) the elastic component of the normal contact force must conserve mechanical energy and, in order to do so, it must satisfy several requirements described next.

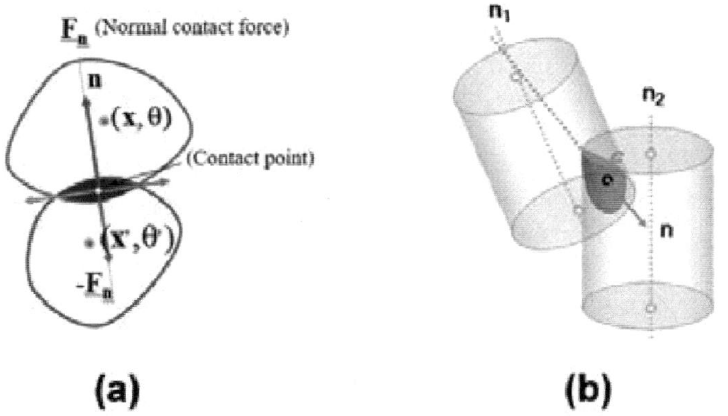

Figure 3. (a) Two colliding bodies with reference points. (b) Collision between axi-symmetric objects

In a fixed global coordinate system two points on the two colliding bodies have coordinates $\boldsymbol{x}$ and $\boldsymbol{x}'$ and the orientations of these two bodies are described in terms of variables Θ and Θ', see Fig. 3 (a). The normal elastic component of the contact force is the function of position and orientation of two bodies, $\boldsymbol{F_n} = \boldsymbol{F_n}(\boldsymbol{x}, \boldsymbol{x}', \Theta, \Theta', \ldots)$ and it must obey the following "Energy Conservation Theorem". Contact model will be energy conserved if there exists a scalar potential function $\phi(\boldsymbol{x}, \Theta)$ such that

$$F_n = -\nabla_x \phi(x, \Theta) = -\frac{\partial \phi(x, \Theta)}{\partial x}$$

For more specific cases of axi-symmetric objects such as cylinders or capsules, see Fig. 3 (b), the following "Normal Contact Theorem" applies. For a pair of axi-symmetric bodies in contact with contact point c, the normal contact line n , which passes the two central lines, can be uniquely determined. In particular,

$$c = x + \frac{n \times M}{|F_n|} + \alpha n$$

Here $M = -\nabla_\Theta \phi(x, \Theta) = -\frac{\partial \phi(x,\Theta)}{\partial \Theta}$ is a contact moment.

With above described approach the DEM simulations of various shape particles were conducted and validated first under conditions that do not require coupling to fluid. One such validation example involves piling up particles of different shape. The final state of the simulated process is shown in Fig 4 below.

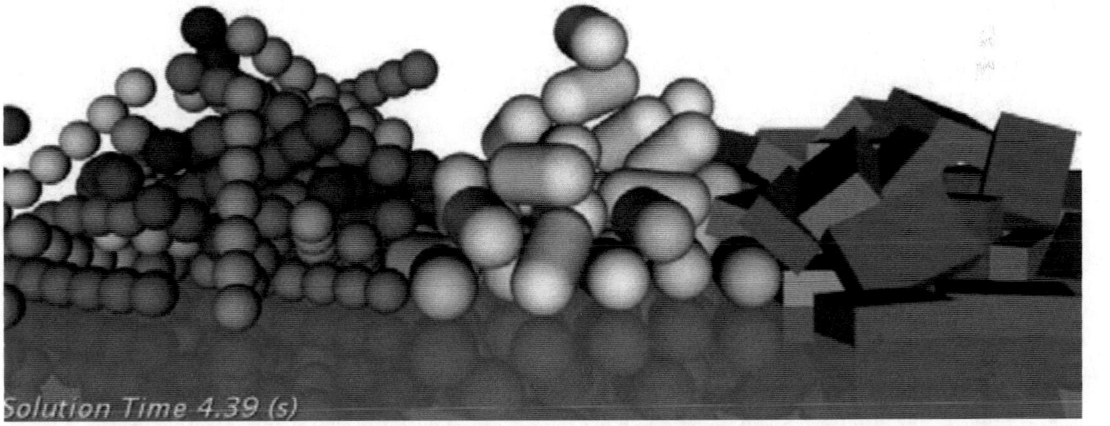

Figure 4. Interaction between different particle types. Left: blue clumped particles (flexible rods). Middle: Yellow capsules (spherocylinders). Right: Green polyhedral particles (boxes)

DEM-CFD two-way coupling for non-spherical particles

The novel source smoothing using cell clustering method for 2-way coupling between DEM and CFD is used in all simulations discussed next. This source smoothing algorithm considerably reduces the mesh size dependence in DEM-CFD coupling. This algorithm was first introduced in (Baran, Greening and Kodl, Numerical Simulations of Fluidized Beds in Fine-Grid Geometries Using Source Smoothing with Cell Clustering Method for DEM-CFD Coupling 2016) and (Baran, Eppinger, et al. 2016).

The drag force on capsules or cylinders is set to be a combination of Wen-Yu model and Ergun based model

$$\overrightarrow{F_{drag}} = \frac{1}{2}\rho_f C_d A_d |\vec{u} - \vec{v_p}|(\vec{u} - \vec{v_p}); \qquad C_d \text{ is a function of } Re_p = \frac{\rho_f |\vec{u} - \vec{v_p}| d}{\mu}$$

$$C_d = \begin{cases} \frac{4}{3}\left(\frac{150(1-\epsilon)^2}{\epsilon Re_p} + 1.75\right); & \text{if } \epsilon < \epsilon_{\min};\ Ergun \\ \frac{24 + 3.6Re_p^{0.687}}{Re_p}\epsilon^{-3.65}; & \text{if } \epsilon \geq \epsilon_{min};\ Wen\,Yu \end{cases}$$

ϵ_{min} is the cutoff void fraction (=0.8)

A_d is a projected area of particle. For cylinders and capsules projected area is recalculated every DEM timestep as the orientation of particle relative to the fluid flow is continuously changing, see Fig. 5.

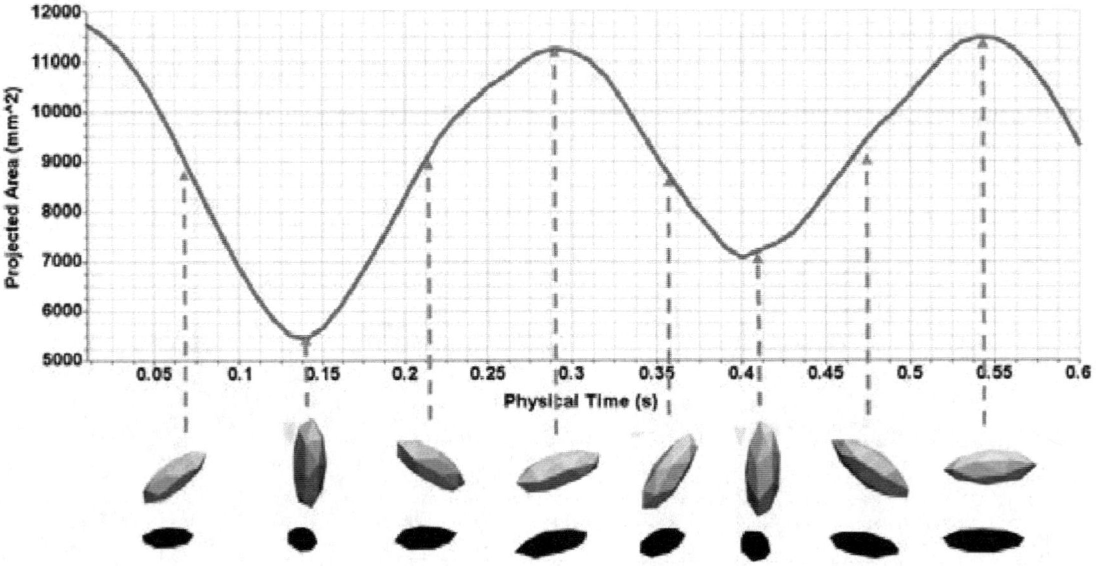

Figure 5. Projected area of elongated particle rotating in constant fluid flow coming from bottom to top. Orientation of particle for selected points in time are shown below the bottom axis along with the shape of projected area colored in black.

Fluidized bed setup

The simulations of capsules and cylinders of various aspect ratio in fluidized bed are discussed next. Geometry includes simple cylindrical container of diameter 0.8 m and height 2 m with velocity inlet at the bottom and pressure outlet at the top with gravity pointing downwards, see Fig. 6. Total simulated time is 20 seconds. The inlet velocity is zero for the first second during settling phase, then it is linearly ramped over the next four seconds to the final maximum inlet velocity $v_{inlet} = 4\ m/s$ and stays at this value until the end of simulation. In all simulated cases the number of particles is always 28320. The density of material particles are made from and their equivalent sphere diameter are fixed to have values $2200\ kg/m^3$ and $20\ mm$ respectively. This means regardless the shape of particle the total amount of solids is always the same. The shape and dimension of particles used in this study are given in Fig. 6. For the base case of capsules with aspect ratio 2 (short capsules) the maximum inlet velocity is 1.22 times minimum fluidization velocity as this study focuses mostly on the effect of particle shape on the hydrodynamics in a fluidized state regime.

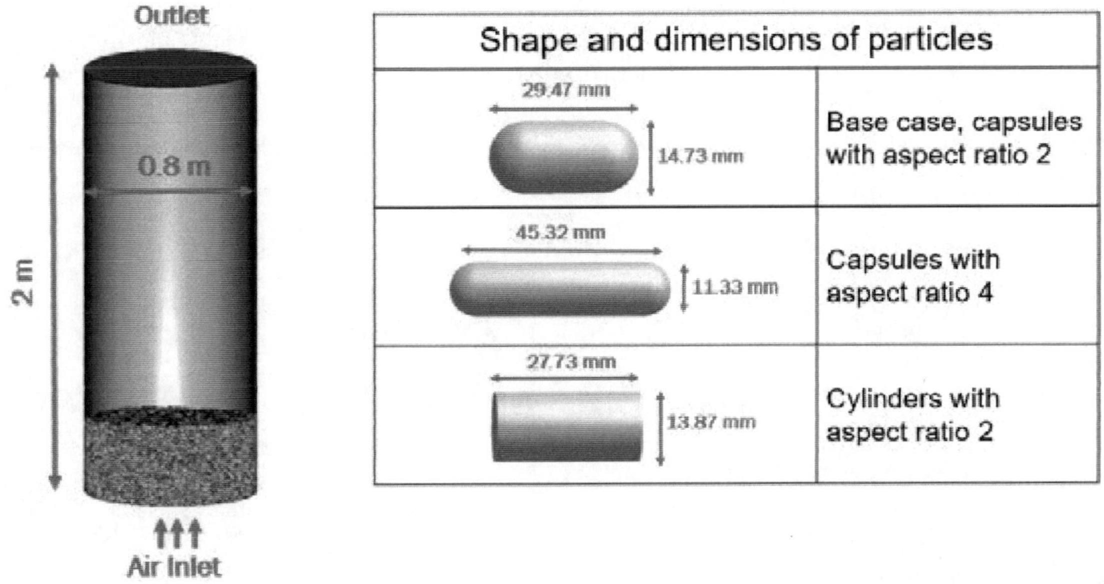

Figure 6. Fluidized bed setup and types of particles used in simulation.

All simulations are set to monitor the following quantities.

- Pressure Drop between inlet and outlet is calculated and compared with Ergun prediction for inlet velocities below fluidization
- Bed height is calculated as the vertical distance between inlet and center of mass of all particles times 2
- Void fraction is obtained by averaging in volume cells below fill level. It is used for evaluating the Ergun prediction for pressure drop for the inlet velocities below minimum fluidization velocity
- Alignment of cylinder's or capsule's orientation with the direction of fluid flow, see Fig. 7

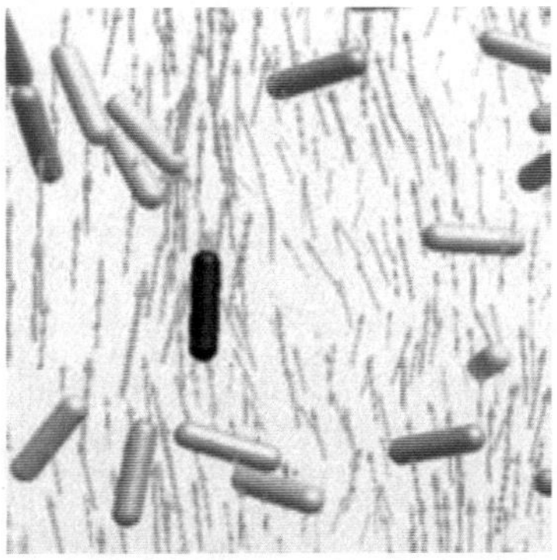

Figure 7. The fluid direction is shown by vectors. Particle aligned with fluid direction is shown in dark green

In particular, the angle between particle's main symmetry axis and local direction of fluid, angle of attack, is evaluated for every particle at the end of each fluid timestep and the status of "aligned" is set for each particle with angle of attack smaller than 25 degrees. Number of aligned particles or fraction of particles aligned with flow as a function of time is then available for analysis.

Simulation results

The snapshots for particle-fluid states in fluidized state for three shapes of particles are shown in Fig. 8

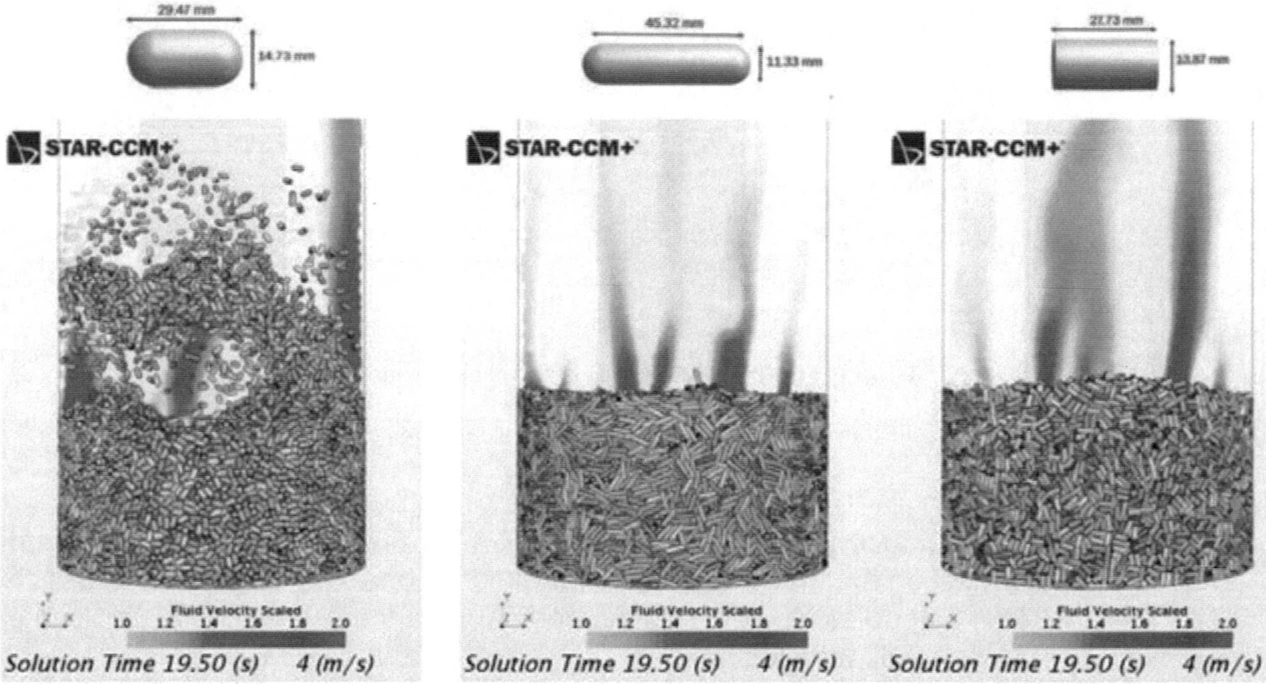

Figure 8. Rendering of particles in 3D (yellow) and fluid velocity magnitude (blue) on cross-section plane for cases of capsules with aspect ratio 2 (left), capsules of aspect ratio 4 (middle), and cylinders of aspect ratio 2 (right).

3D visualization of the magnitude of fluid velocity using resampled volume technology is shown in Fig. 9.

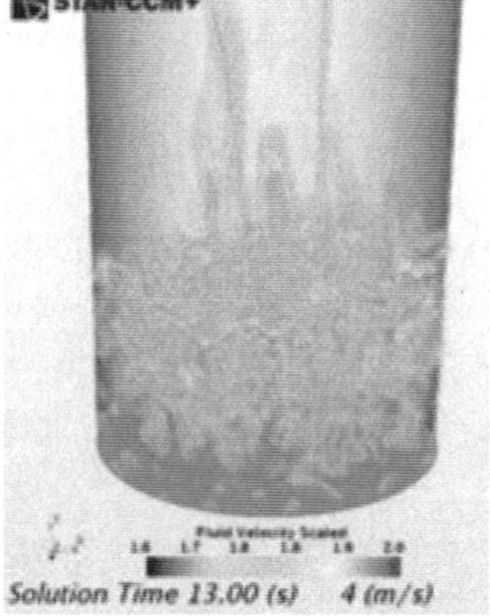

Figure 9. 3D rendering of fluid velocity for the case with capsules of aspect ratio 2

The pressure drop as a function of time for each considered case is shown in Fig. 10.

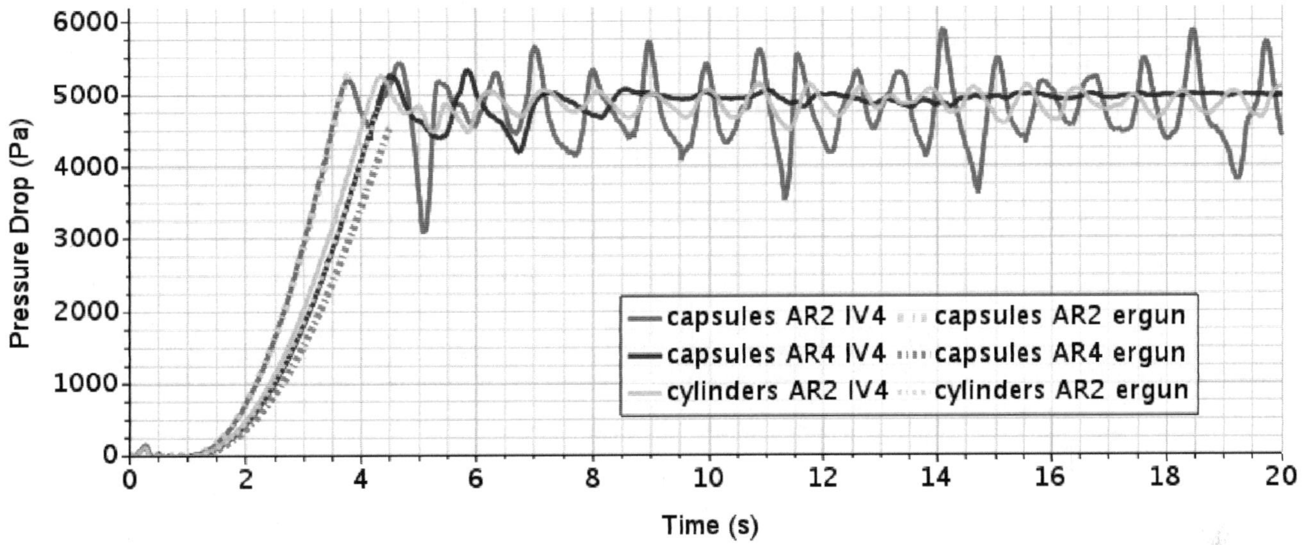

Figure 10. Pressure drop as a function of time

The results show good agreement with Ergun prediction for the inlet velocities below minimum fluidization velocities (which correspond to time interval between 1 s and 3.8 seconds). They also show much higher fluctuations of pressure for the case capsules with aspect ratio 2 compared to both other cases. This is consistent with visual analysis of the animations for each case where bubbling flow regime was prevailing for short capsules and channeling flow regime was prevailing for both long capsules and short cylinders.

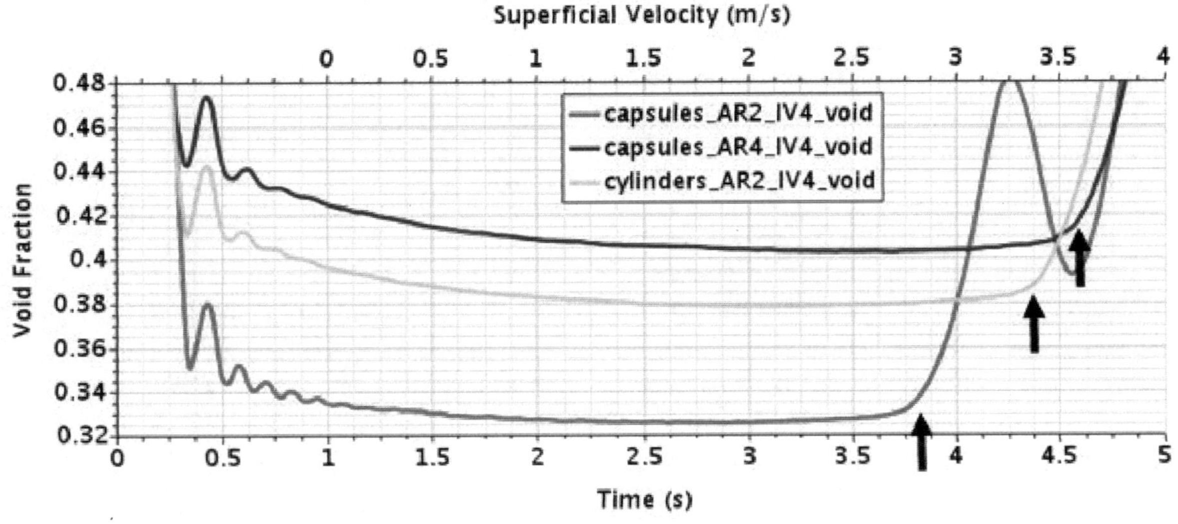

Figure 11. Void fraction below fill level as a function of time (bottom axis) and superficial velocity (top axis). Black arrows indicate the state of the onset of fluidization.

Fig. 11 shows the results for void fraction of packed particles. These results are used to estimate the minimum fluidization velocity as the threshold inlet velocity above which void fraction starts rapidly increasing. In particular, the minimum fluidization velocity is 2.7 m/s for short capsules, 3.5 m/s for long capsules and 3.2 m/s for short cylinders. The values of void fraction were also used in calculating the Ergun prediction for pressure drop plotted in Fig. 10.

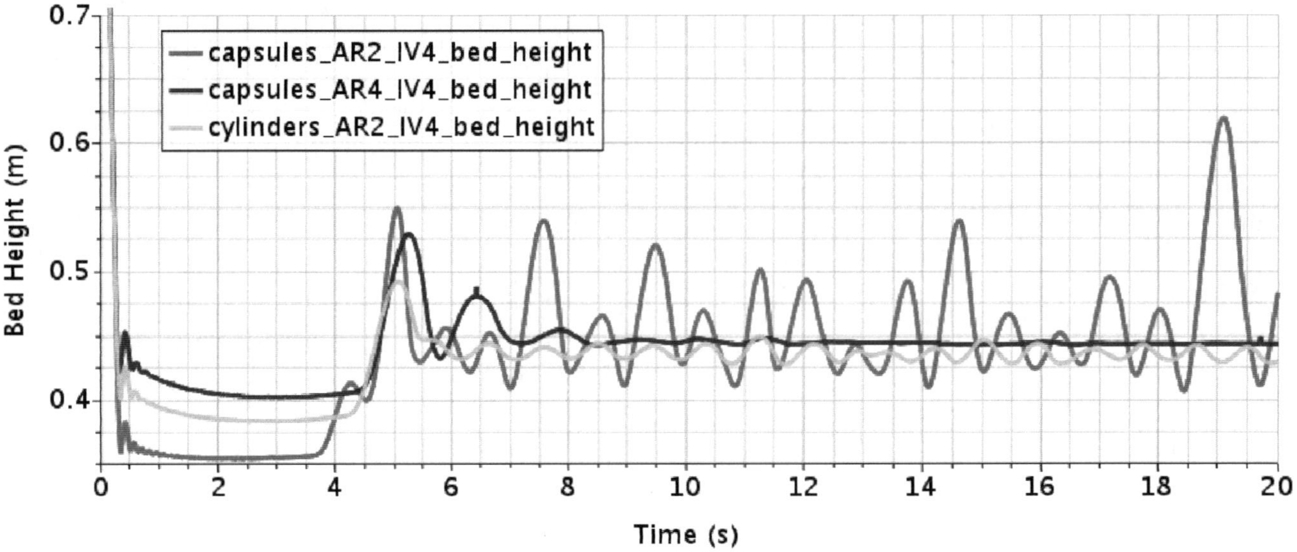

Figure 12. Pressure drop as a function of time

Bed height as a function of time is shown in Fig 12. It is consistent with previous results indicating more dense packing of short capsules compared to long capsules and short cylinders. It also shows much higher fluctuation in bed height in fluidized regime for short capsules compared to other two cases.

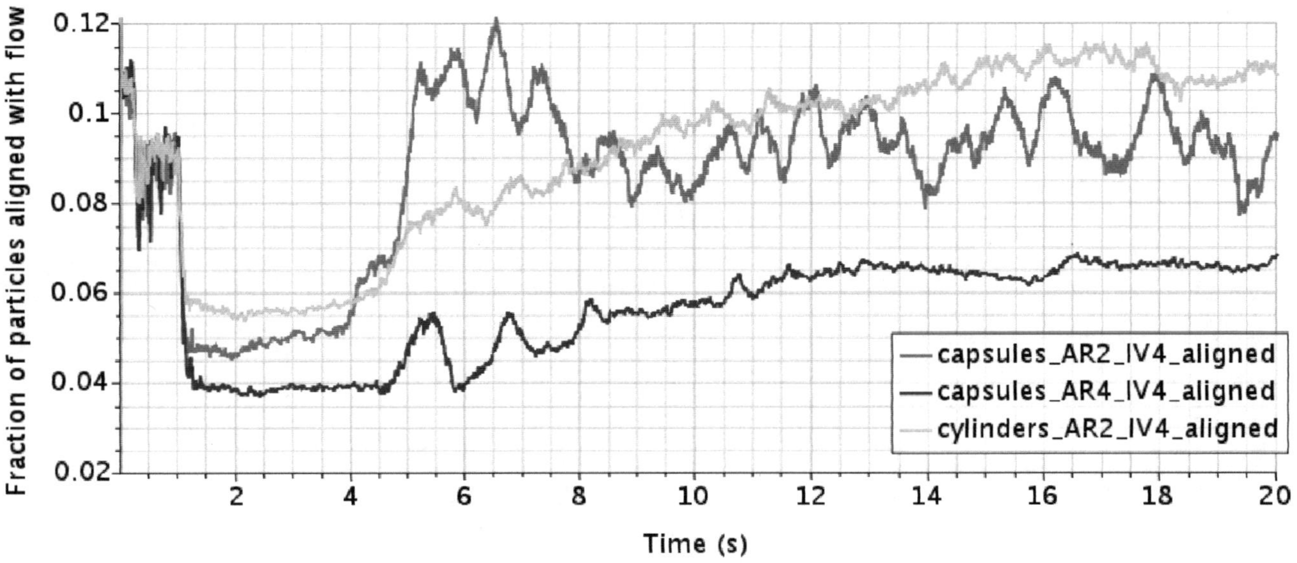

Figure 13. Ratio of aligned with the flow particles to all particles as a function of time

Finally, Fig. 13 shows the results for the number of aligned with the flow particles as a function of time. Unlike previous results for pressure, bed height, and void fraction, the steady state orientation of particle has much higher dependence on the aspect ratio of particles. In particular both short cylinders and short capsules end up with more particles aligned with flow direction in steady state compared to long capsules. These results will be discussed in more details next.

Discussion

A fully coupled DEM-CFD method for simulating capsules or cylinders in fluidized bed is introduced and used to study the effect of particle shape on fluidization in fluidized bed. This algorithm is implemented within commercial software Simcenter STAR-CCM+ which allows running simulations massively in parallel. One of the values of this algorithm is computational cost being bigger but still comparable with the cost of running the equivalent simulation with spherical particles. In particular, it takes less than 2.5 hours to simulate 20 seconds of physical time on 36 cores for any of the considered in this study shape of particles.

The results confirmed that with the increase of particle's aspect ratio

- Packing density decreases
- Initial bed height increases
- Minimum fluidization velocity increases
- Channeling flow regime is replacing bubbling flow regime for the same inlet velocity
- Takes longer to get steady state in particle's orientation

Similar changes but of lesser degree are observed when changing shape from capsule to cylinder while keeping the same aspect ratio. All recorded observation are consistent with the similar simulation or experimental results reported in the literature (Williams and Philipse 2003), (Buist, et al. 2017), (Cai, et al. 2016), (Mahajan, et al. 2017).

References

Abreu, Charlles R.A., Frederico W. Tavares, and Marcelo Castier. 2003. "Influence of particle shape on the packing and on the segregation of spherocylinders via Monte Carlo simulations." *Powder Technology* 134: 167– 180.

Baran, Oleh, Doran Greening, and Petr Kodl. 2016. "Numerical Simulations of Fluidized Beds in Fine-Grid Geometries Using Source Smoothing with Cell Clustering Method for DEM-CFD Coupling." *AIChE.* San-Francisco.

Baran, Oleh, Thomas Eppinger, Ravindra Aglave, and Simon Lo. 2016. "A fully coupled CFD-DEM algorithm for S-L mixing in stirred vessels." *AIChE.* San-Francisco.

Borzsonyi, Tamas, and Ralf Stannarius. 2013. "Granular materials composed of shape-anisotropic grains." *Soft Matter* 9: 7401.

Buist, Kay A, Pavithra Jayaprakash, J.A.M. Kuipers, Niels G. Deen, and Johan T. Padding. 2017. "Magnetic Particle Tracking for Nonspherical Particles in a Cylindrical Fluidized Bed." (AIChE Journal) 63: 5335.

Cai, Jie, Zhengbiao Peng, Charley Wu, Xiaobao Zhao, Zhulin Yuan, Behdad Moghtaderi, and Elham Doroodchi. 2016. "Numerical Study of the Orientation of Cylindrical Particles in a Circulating Fluidized Bed." *Ind. Eng. Chem. Res.* 55: 12806-12817.

Feng, Y.T., K. Han, and D.R.J. Owen. 2012. "Energy-conserving contact interaction models for arbitrarily shaped discrete elements." *Comput. Methods Appl. Mech. Engrg.* 205: 169-177.

Guo, Y., C. Wassgren, B. Hancock, and J. Curtis. 2013. "Granular shear flows of flat disks and elongated rods without and with friction." *Physics of Fluids* 25: 063304.

Liao, Lingzhi. 2013. "Influence of particle shape and bed height on fluidization." thesis, University of Florida.

Mahajan, Vinay, Tim M.J. Nijssen, Barry W. Fitzgerald, Jeroen Hofman, Hans Kuipers, and Johan T. Padding. 2017. "Fluidization of spherocylindrical particles." *Powders & Grains* 140: 06019.

Mema, Ivan, Vinay V Mahajan, Barry W Fitzgerald, Hans Kuipers, and Johan T Padding. 2017. "Effect of lift force on dense gas-fluidized beds of non-spherical particles." 12th International Conference on CFD in Oil & Gas, Metallurgical and Process Industries.

Williams, S. R., and A. P. Philipse. 2003. "Random packings of spheres and spherocylinders simulated by mechanical contraction." *Phys. Rev. E* 67: 051301.

Zhong, Wenqi, Yong Zhang, Baosheng Jin, and Mingyao Zhang. 2009. "Discrete Element Method Simulation of Cylinder-Shaped Particle Flow in a Gas-Solid Fluidized Bed." *Chem. Eng. Technol.* 32: 386-391.

APPLICATION OF CPFD METHOD IN THE SIMULATION OF VERTICAL DENSE PHASE PNEUMATIC CONVEYING OF PULVERIZED COAL

Yong Jin[†], *East China University of Science and Technology, Shanghai, PRChina*
Haifeng Lu, East China University of Science and Technology, Shanghai, PRChina
Xiaolei Guo, East China University of Science and Technology, Shanghai, PRChina
Xin Gong[‡], *East China University of Science and Technology, Shanghai, PRChina*

Abstract

Flow stability is a key issue to be addressed for the development of dense phase pneumatic conveying. Flow patterns are closely related with flow stability. From the view of experimental materials, angle of repose, packing density and tapped density were determined by PT-X powder tester. And flowability was determined by FT4 Powder Rheometer. The experimental results shew that the tested pulverized coal was a kind of Geldart's group C powder with strong cohesion. This study contributes with a computational particle fluid dynamic (CPFD) simulation which is based on the numerical solution of continuity and momentum balance equations in a three-dimensional (3-D) horizontal pipeline. A novel correction model was developed, describing interaction among particles and between fluid and particles decided by the experiment mentioned before. CPFD approach predicted the particle velocity and pressure drop were generally below 5% for the stable cases. What's more, the twelve different flow patterns were predicted by the CPFD approach embedded in the correction model, which was confirmed by the electrical capacitance tomography (ECT) in the experiment facility set up by our own. With the help of our correction model, it could reveal the mechanism of the change of flow patterns.

Keywords: CPFD; flow patterns; flow stability; ECT; correction model;dense

Highlights

Multiphase particle-in-cell model is used to predict dense phase pneumatic conveying.
Pressure drop and velocity profiles show good agreement with experiments.
A improved model was proposed to give the right simulation results.

1. Introduction

The practical application of the pneumatic conveying technology of particulate materials are extremely extensive, especially for the field of pulverized coal pressurization and gasification process. High pressure dense phase pneumatic conveying of pulverized coal is one of the key technologies of entrained flow gasification technology. The stability of transportation directly affects the gasification temperature, gasification products and the safety of the gasifier. Depending on the materials to be conveyed, operational conditions and facility geometries, transportation can

[†] Present author: Yong Jin, PhD student, E-mail: mhzxjinyong@126.com
[‡] Corresponding author: Prof. Xin Gong, E-mail: gongxin@ecust.edu.cn

occur in different flow regimes, which usually can be categorized as dilute or dense. Despite wide applications, efficient pneumatic system design remains challenging and it is complicated by interactions between the particle phase and the turbulent carrier flow as well as particle-particle and particle-wall interactions. Because of these complications, a fundamental understanding of the complex physics, involved in gas-solids flows is crucial for the proper design, control and optimization of any pneumatic conveying systems.

For the dense phase pneumatic conveying system, transportation spped is low, the solid concentration is high and the flow patterns are complex, where the transportation requirements and conditions of the system are quite different from the general conveying system. Nowadays, many researches focused on the pressure drop analysis [1], flow patterns [2], signal analysis [3], where the research methods are experiment. However, the complex interactions among the particle phase and fluid phase are difficult for people to explore by experiments, like particle-particle and particle-fluid micro interactions, resulting the macro particle flow. Numerical simulation emerges at a historic moment. The computational particle fluid dynamics (CPFD) numerical scheme incorporated with the multiphase-particle-in-cell (MP-PIC) method is a recently developed model for gas-solids flows. To date, this method has been used to predict certain processes; however, to the authors' best knowledge, published information is unavailable on the application of this model in dense pneumatic conveying. Previous applications of the MP-PIC method include bubbling and circulating fluidized beds (CFBs), spouting beds, fluidized bed gasifiers (FBG), fluidized beds (FB) for carbon capture, gas/liquid/solid fluidized beds, Rayleigh-Taylor mixing layers and etc.

As in all other Eulerian-Lagrangian approaches, the MP-PIC method treats the gas as a continuous phase and the particles in a discrete manner. Solving a transport equation called a Liouville equation for the particle distribution function predicts the particle dynamics. The particle distribution function contains particle properties such as particle spatial location, particle velocity, particle mass, and time. The MP-PIC method uses the concept of parcels (numerical particles) in order to reduce the number of particles involved in the computations; this provides a significant acceleration of simulation speed. A parcel is a collection of several real particles with the same properties (species, size, density, temperature, etc.). Unlike DEM models that calculate particle-to-particle force with a spring–damper model and direct particle contact, the MP-PIC methodology models particle collision force for each particle as a spatial gradient. A particle normal stress model is developed from this concept to describe the particle collisions.

Therefore, this work uses the MP-PIC method to perform three-dimensional simulations of the dilute-phase pneumatic conveying of plastic pellets in a horizontal circular pipe.. In addition to investigating the validity of the method for pneumatic conveying applications, we also develop a model, describing interaction among particles and between fluid and particles decided by the experiment for the accurate prediction of pressure drops and flow patterns in dense phase pneumatic conveying. Moreover, some of the major model parameters deemed important in predicting pneumatic conveying characteristics, namely particle-particle and particle-wall collision parameters. The CPFD numerical methodology incorporated in the commercially available Barracuda® 15 code is the platform used for the modelling and simulations in this study.

2. Model description

2.1. Governing equations

The governing equations of the models for describing gas and solid phase are presented below. The gas phase equations consist of mass, momentum and energy conservation equation.

The gas phase continuity equation without interphase transfer is given by.

$$\frac{\partial \theta_f}{\partial t} + \nabla \bullet (\theta_f \mathbf{u}_f) = 0 \qquad (1)$$

Where $\mathbf{u}_f$ is gas velocity and θ_f is gas volume fraction.

The gas phase velocity components in the *x*, *y* and *z* directions are given by time-averaged Navier-Stokes equations.

$$\frac{\partial(\theta_f \mathbf{u}_f)}{\partial t} + \nabla \cdot (\theta_f \mathbf{u}_f \mathbf{u}_f) = -\frac{1}{\rho_f}\nabla \mathbf{P} - \frac{1}{\rho_f}\mathbf{F} + \theta_f \mathbf{g} \qquad (2)$$

Where ρ_f is fluid density, **P** is fluid pressure, **g** is gravity, **F** is the rate of momentum exchange per volume between the fluid and particles phases.

The constitutive equation for the non-hydrostatic part of the gas stress tensor is given by

$$\tau_g = \mu_{eff}(\nabla \mathbf{u}_g + \nabla \mathbf{u}_g^T) - \frac{2}{3}\mu_{eff}\nabla \cdot \mathbf{u}_g \mathbf{I} \qquad (3)$$

where μ_{eff} is the shear viscosity, which is the sum of the laminar shear viscosity and the turbulent viscosity ($\mu_{t,g}$). The large eddy simulation (LES) model is used for calculating the gas phase turbulence. In LES, large eddies are calculated from the flow equations and the subgrid turbulence is captured with a Smagorinsky subgrid scale model.

$$\mu_{t,g} = C_S \rho_g \Delta^2 \left|\nabla \mathbf{u}_g + \nabla \mathbf{u}_g^T\right| \qquad (4)$$

where Δ, $\Delta=(\delta x \delta y \delta z)^{1/3}$ is the subgrid length scale. Smagorinsky constant = 0.01 is used for the current simulations.

The interphase momentum transfer is given through an interphase viscous drag force ***F***.

Gidaspow [4,5] proposed a drag function by blending Wen-Yu and Ergun functions; this model is used as the drag function in the present study. A switching function is used in order to prevent possible numerical problems due to sharp transitions or discontinuities in the blended model. In the present study, $\theta_{cp} = 0.6$ is used.

$$D = \begin{cases} D_1 & \theta_p < 0.75\theta_{cp} \\ (D_2 - D_1)\left(\dfrac{\theta_p - 0.75\theta_{cp}}{0.85\theta_{cp} - 0.75\theta_{cp}}\right) + D_1 & 0.85\theta_{cp} \geq \theta_p \geq 0.75\theta_{cp} \\ D_2 & 0.85\theta_{cp} < \theta_p \end{cases} \qquad (5)$$

Where, $D_1 = \dfrac{3}{8} C_d \dfrac{\rho_f \left|\mathbf{u}_f - \mathbf{u}_p\right|}{\rho_p r_p}$, $D_2 = 0.5\left(\dfrac{C_1 \theta_p}{\theta_f Re} + C_2\right)\dfrac{\rho_f \left|\mathbf{u}_f - \mathbf{u}_p\right|}{r_p \rho_p}$

The dynamics of the particle phase is predicted by solving a transport equation for the particle distribution function (PDF), $\varphi(\boldsymbol{x}, \mathbf{u}_p, m_p, \Omega_p, t)$ (Eq. (12)). Since the particles are non-vapourizing and chemically inert, the dependency of φ on particle temperature and material composition is ignored in this study.

The PDF $\varphi(\boldsymbol{x}, \mathbf{u}_p, m_p, \Omega_p, t)$ is calculated from the Liouville equation

$$\frac{\partial\varphi}{\partial t}+\nabla(\varphi\mathbf{u}_p)+\nabla(\varphi A)=0 \tag{6}$$

Where $\boldsymbol{x}$ is the particle position, $\mathbf{u}_p$ is the particle velocity, m_p is the particle mass, Ω_p is the particle volume and t is the time

The trajectory of a given particle is solved by Newton's second law. The particle acceleration in the original MP-PIC method is

$$A = D_p(\mathbf{u}_f - \mathbf{u}_p) - \frac{1}{\rho_p}\nabla\mathbf{P} + \mathbf{g} - \frac{1}{\theta_p\rho_p}\nabla\tau \tag{7}$$

The particle volume fraction in the above equations is given by

$$\theta_p = \int\int\int \varphi\Omega_p d\Omega_p d\rho_p d\mathbf{u}_p \tag{8}$$

Moreover, θ_p and θ_f are related by

$$\theta_p + \theta_f = 1 \tag{9}$$

τ in Eq. (7) is the particle stress function (also called particle normal stress or isotropic solid stress), which is used to model particle-particle interactions. Due to the difficulties in calculating the particle stress gradient for each particle in a dense flow, the particle-particle interaction forces are calculated as a spatial gradient (as a gradient on the grid) that is mapped back to the particles. The particle stress model used here is an extension of the model from Harris and Crighton [6].

$$\tau = \frac{P_s\theta_p^\beta}{\max[(\theta_{cp} - \theta_p), \varepsilon(1-\theta_p)]} \tag{10}$$

where P_s is constant with units of pressure, β is supposed to lie between 2 and 5 [7], and ε is a small number on the order of 10^{-7} to remove singularity. The values used in the current simulations are 1, 3 and 10^{-8}, respectively. However, the particle stress model has a negligible effect on the particle behaviour in the dense phase. The stress model will begin to affect particles as the particle volume fraction of a region begins to approach the close-packed level.

2.2. Model modification

For the particle force term, the external force term was modified by the the polydisperse system contact model of Liu et al [8], which was based on the Rumpf [9] equation, a model with elastic-plastic deformation was considered for the granular system:

$$\sigma_{t,iso} = \frac{1-\varepsilon}{\pi}\sum_{j=1}^{n} f_j d_j \sum_{i=1}^{n} k_j F_{ji} \Big/ \sum_{j=1}^{n} f_j d_j^3 \tag{11}$$

Here, the interaction force between particle i and particle j is F_{ji} :

$$F_{ji} = F_{H0}(j,i) + h(R_{ji}^*)[F_{H0}(j,i) + F_N] \tag{12}$$

3 Computational details

The geometry, particle properties and operating conditions of the present simulations are based on self-established experimental facilities. work done by Tsuji and Morikawa [64]. These details including solution procedure are explained in the Subsections; 3.1–3.6. The fluid flow is isothermal, compressible and no reactions are involved. The effect of gravity is included.

3.1. Geometry and grid generation

The closed-loop experimental system designed to investigate the dense phase flow of pulverized coal is illustrated in Figure 1. The mass flow rates of three gaseous streams feeding the conveying system were individually controlled to achieve the desired operating conditions. The ECT sensor was installed on the 50 mm I.D. vertical pipe to monitor flow patterns. Two pressure transducers were mounted at the top of the horizontal pipe to obtain the wall static pressure signals at a sampling frequency of 1 Hz for a total of more than 200 sample points. The ECT and pressure signals were both recorded in a computer data acquisition system. The test section with the ECT sensor was located near the pressure transducers. Dry compressed air was used as the conveying gas.

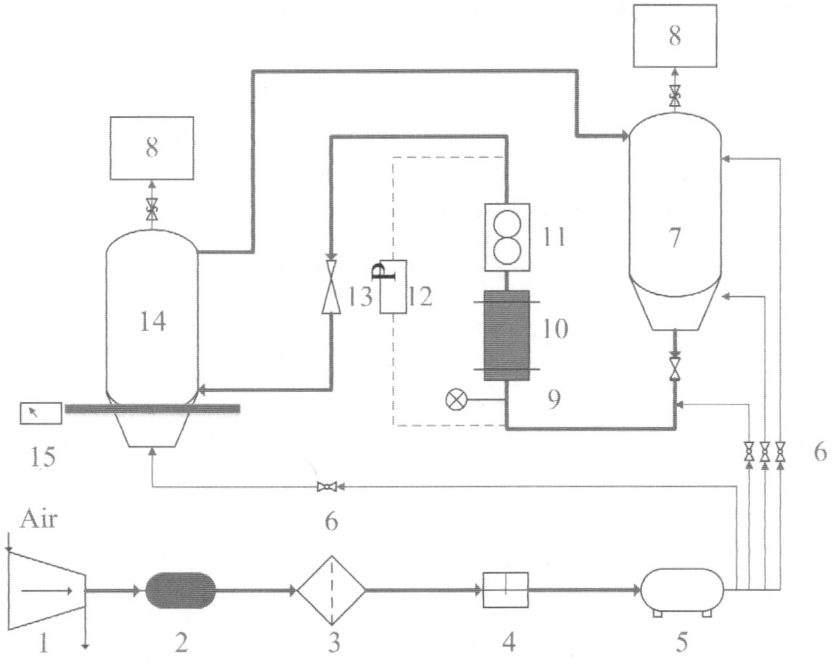

Figure 1. Schematic diagram of the vertical experimental facility.

1-air compressor; 2-buffer tank; 3-refrigeration compressed air drier; 4-filter; 5-gas distributor; 6-gas flowmeter; 7-feeding tank; 8-fabric filter; 9-pressure sensor; 10-solid mass flowmeter; 11-ECT sensor; 12-differential pressure sensor; 13-venturi; 14-receving tank; 15-weighing module.

A three-dimensional 4.2 m vertical circular pipe with an internal diameter of 50 mm was created (Figure 2 (left)). The geometry was meshed by a built-in module in Barracuda software. The mesh of the pipe cross section is shown in Figure 2 (right). The mesh had a total number of 89000 cells (5 × 5 × 5) and these were cube. The computational time was approximately 1–3 days for a high performance computer (Intel® Xeon® Processor X5650) characterized as 12 M Cache, 2.66 GHz, RAM 24 GB and 2 TB.

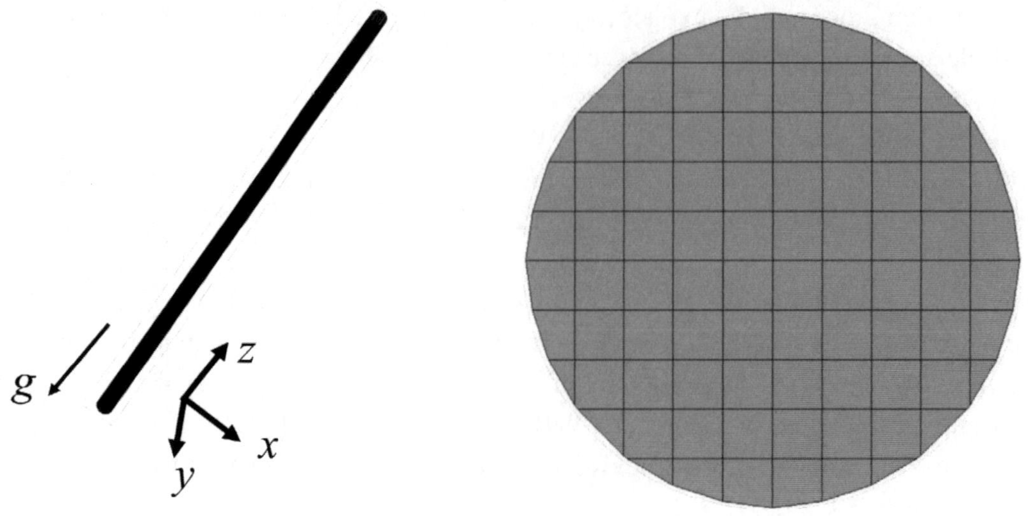

Figure 2. The computational domain (left) and grid of pipe cross section (right).

3.2. Material properties

The particles were lignite pulverized coal with a particle density of 1530 kg·m^{-3}. Particles were considered to be spherical. The used particle size distribution and characteristic particle size are shown in Fig. 2. The air viscosity was 1.84469 × 10^{-5} kg·m^{-3}.

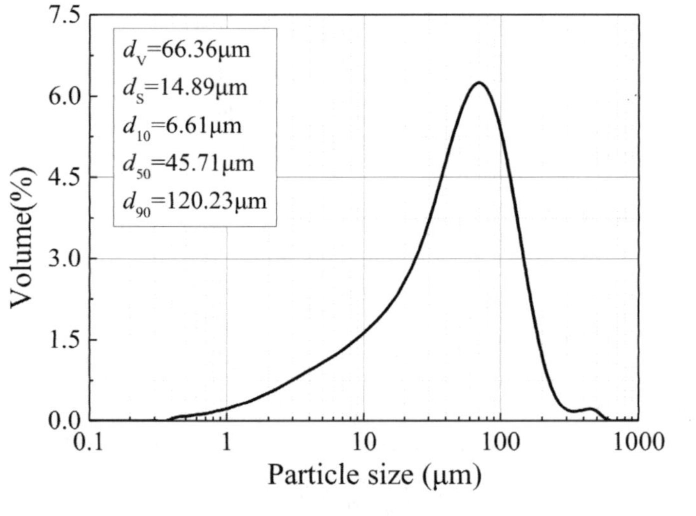

Figure 3. Particle size distribution of plastic pellets.

Table 1 Material properties of pulverized coal

ω(%)	θ_r(°)	θ_f(°)	θ_d(°)	θ_s(°)	ρ_b(g·cm^{-3})	ρ_t(g·cm^{-3})	ρ_p(g·cm^{-3})
3.35	50.09	28.09	22.00	60.70	0.4475	0.8593	1.490

According to the Carr's flowability index chart[10], the flowability index of our tested pulverized coal is 37, between 20 and 39, which are classified as the poor flowability powder, like Geldart's Group C powder. [11] Usually, this kind of powder shows great cohesive and there is a necessity for people to use detailed force equations to describe the motion of particles. However, in an MP-PIC scheme, each cell contains a number of numerical particles (parcels). Particle properties are mapped to and from the Eulerian grid to obtain grid-based properties for the particles; fluid properties are, in turn, mapped to discrete particle locations.

3.3. Initial and boundary conditions

The pipe is initially filled with 1 atm air, no particles, and an air velocity of zero. The mass flow rates are calculated according to the superficial air velocities and solids loadings used in the experiments. As for the mass flow rate calculation, the air density at the inlet is 1.2 $kg \cdot m^{-3}$, which corresponds to 1 atm pressure at a temperature of 298 K, as specified at the inlet in the Barracuda setup. The solids have the same velocity to the air at the inlet. In Barracuda setup, it is possible to specify "the numerical particle feed per average volume" which controls the feed rate of parcels (numerical particles) into the domain. In the present study, it is specified as 1,000,000 which gives around 2–4 real particles/parcel for all the cases. At the outlet, a pressure boundary condition is adopted with a specified pressure of 1 atm. No slippage of the gas phase and a partial slip of the particles are assumed as the wall boundary conditions. A bounce-back condition is used to model the particle-wall interactions. It is assumed that 95% of the normal and tangential momentum of the particles are retained after the collision with the pipe wall, i.e., a normal-to-wall momentum retention = 0.95 and a tangent-to-wall momentum retention = 0.95 in the Barracuda setup.

3.4. Solver numeric and convergence criterion

The equations are solved in a transient state. An initial time step of 0.0001 s is applied to the CPFD simulation. However, there is a built-in-model in Barracuda that uses a Courant-Friedrichs-Lewy (CFL) algorithm to automatically determine the proper time step based on the simulated meshes and parcels in order to maintain stability, accuracy and calculation speed [13]. The lower and upper limits for CFL are 0.8 and 1.5, respectively. Each simulation was run over 6 s (in real time) to achieve a quasi-steady state. No major changes were observed in the variables after 1 s. As for the level of accuracy, the limited residual values for convergence were 10^{-7}, 10^{-8} and 10^{-7} for volume, pressure and velocity, respectively.

4. Results and discussion

The simulations are carried out with a certain set of model parameters, which will be published after the acceptance of our paper.

4.2. Prediction of pressure drop

Having determined the importance of the correction models, the results from the simulations carried out with the new collision terms are compared with experimental data.

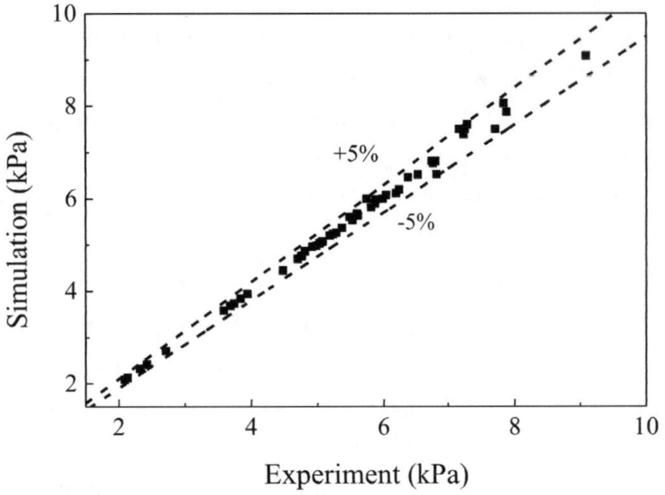

Figure 4. Comparison of 52 groups pf experimental data and simulation results

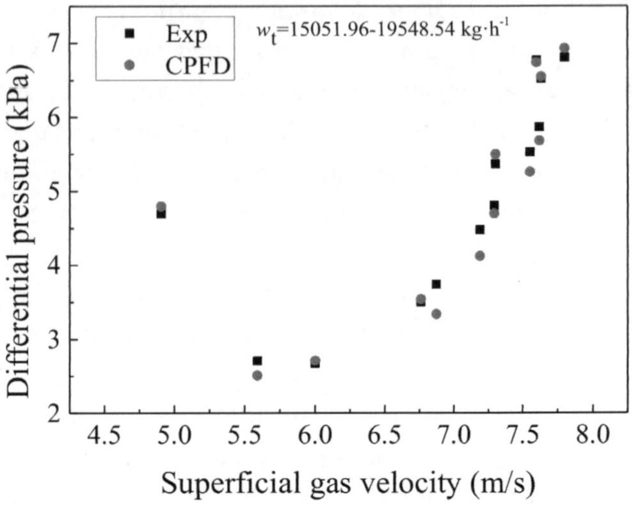

Figure 5. Phase diagram simulation (14 groups)

As shown in Figure 4, the predicted pressure drop results that the model provide better agreement with the experimental results. The predictions deviate from the experimental data is approximately 20%. Figure 5 gives the simulation of phase diagram.

4.2. Prediction of particle velocity

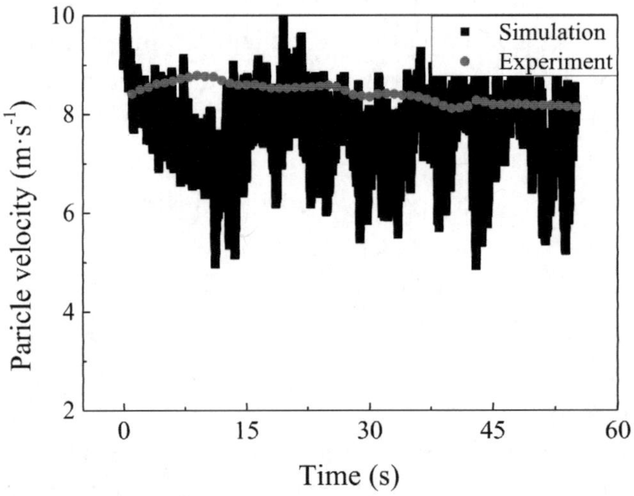

Figure 6. Comparison of simulated particle velocity and experiment velocity

Figure 6 shows that the deviation is about 10%.

4.3. Prediction of flow patterns

Table 2 give the simulation results and experiment results which shows great agreement.

Table 2 The relationship between superficial gas velocity and different flow patterns.

a	b	c	d	e	f	g	h	i	j	k	l

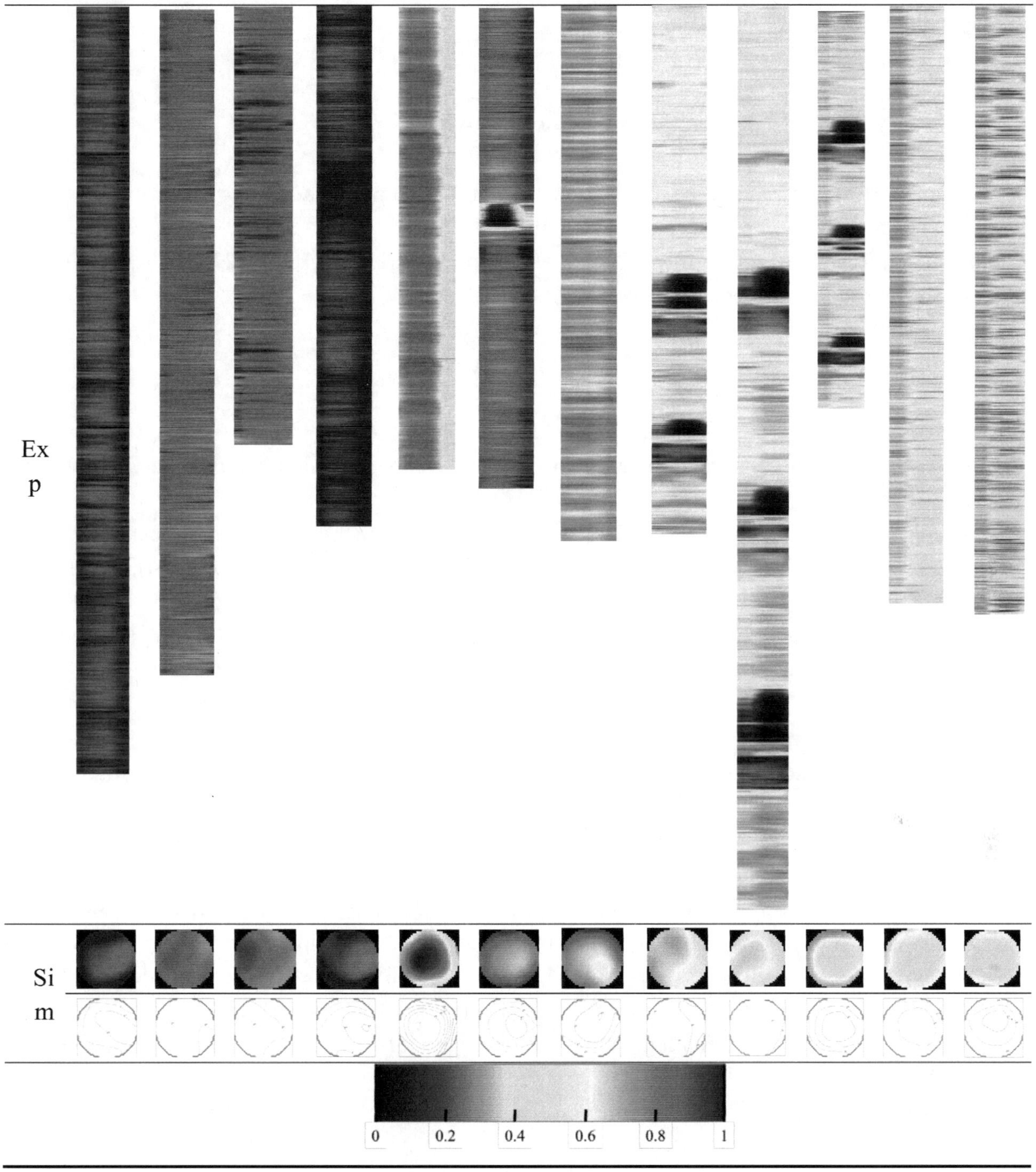

5. Conclusions

In this paper, a novel correction model was proposed to describe interaction among particles and between fluid and particles decided by the experiment mentioned before. CPFD approach predicted the particle velocity and pressure drop were generally below 5% for the stable cases. What's more, the twelve different flow patterns were predicted by the CPFD approach embedded in the correction model, which was confirmed by the electrical capacitance tomography (ECT) in the experiment facility set up by our own. With the help of our correction model, it may help people understand the change of flow patterns.

Acknowledgements

This work was supported by the National Natural Science Foundation of China (21206041), and Shanghai Science and Technology Commission Project (14DZ1200100, 15DZ1200802), the Fundamental Research Funds for the Central Universities (222201717004, 222201514338) and the key scientific research project in Shanxi (MH2014-01).

References

1. CW A, JH H, Hp S. Pressure Drop in Flow of Dense Coal-Air Mixtures. *Ind.Eng.Chem.* 1951;43(8):1837-1840.

2. Cong X, Guo X, Lu H, et al. Flow patterns of pulverized coal pneumatic conveying and time-series analysis of pressure fluctuations. *Chemical Engineering Science.* 2013;101:303-314.

3. Xingliang Cong. Study on the relationship between the flow type and the pressure signal in the pipeline. Shanghai, East China University of Science and Technology, 2012. (doctoral dissertation)

4. Wen CY, Yu YH. Mechanics of fluidization. *Chem.Eng.Progr.Symp.* 1966;62:100-110.

5. D.Gidaspow. *Multiphase Flow and Fluidization, Continuum and Kinetic Theory Description.* Boston: Academic Press; 1994.

6. Harris SE, Crighton DG. Solitons, solitary waves, and voidage disturbances in gas-fluidized beds. *J. Fluid Mech.* 1994;266:243 - 276.

7. Auzerals FM, Jackson R, Russel WB. The resolution of shocks and the effects of compressible sediments in transient settling. *Journal Fluid Mesh.* 1988;195:437-462.

8. Liu Y, Lu H, Poletto M, Guo X, Gong X. Bulk flow properties of pulverised coal systems and the relationship between inter-particle forces and particle contacts. *Powder Technology.* 2017.

9. Rumpf DIH. Grundlagen und Methoden des Granulierens. *Chemie Ingenieur Technik.* 1958;30(3):144-158.

10. Carr.R.L, Div.B. Evaluating flow properties of solids. *Chemical Engineering.* 1965;18(1):163-168.

11. D.Deldart. Types of gas fluidization. *Powder Technology.* 1973;7:285-292.

DIFREX Reactor and Technology Solutions for Many Particle Types and Sizes in Catalytic and Non-Catalytic Reaction Systems

Subhash Dutta, Stephen Arnold, Aashish Gaurav, Jim Brenner
DIFREX LLC., 8900 Astronaut Blvd. Cape Canaveral, FL 32920 USA

Full understanding of particle flow and hydrodynamic behavior, characterization of the properties of these materials, and of their effects on the operation and performance of industrial reactors, is essential for successful industrial reactor design and scale-up. This discussion will focus on gas-solid reactor systems and will use the term "particle" or "solid" to represent both particulates and particles. Where the terms are distinguished, "particulates" are defined as fine particles usually suspended in gas and in dynamic and/or flow mode with a bulk gas flow stream, and are the solid types used in fluidized bed reactors of various kinds and in pneumatic transport lines. The term "particles", on the other hand, is used for relatively larger particles that are usually utilized in fixed or packed bed reactors and remain static or in flow mode in moving bed reactors. Schematics of the four basic industrial solid-gas reactor types are shown in **Figure 1**: packed bed, tubular fixed bed, bubbling/turbulent fluid bed, and circulating fluid bed.

Reactor Types and Particle Technology

Common industrial solid-gas reactors are of two types: a packed/fixed bed reactor (PKB/FXB) or a fluidized bed reactor, with variations of each. A fluidized bed reactor uses solid particulates. Two prominent versions of the fluidized bed are a bubbling/turbulent fluidized bed (BFB) and a circulating fluidized bed (CFB). Impacts of solid-gas interactions, and effects of particulate flow and hydrodynamic behaviors of fluidized beds, on heat and mass transfer characteristics of reaction systems are discussed with examples, including designs and performances of the reactors.

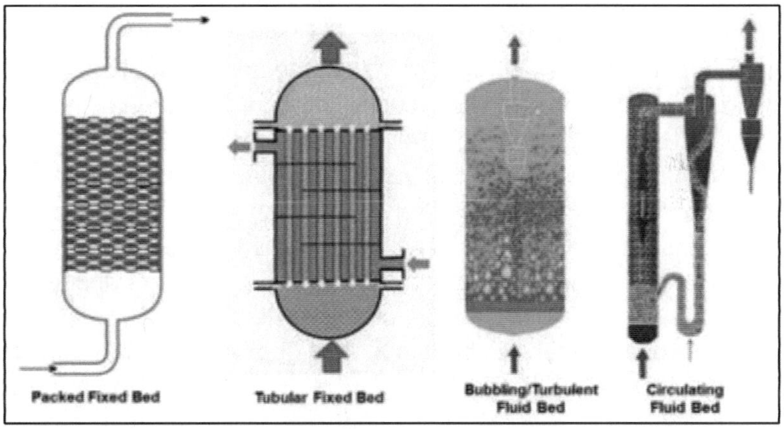

Figure 1: Basic Commercial solid-gas reactor types

Packed/Fixed Bed Reactors, which employ larger particles in a static bed, are traditionally viewed as the most obvious choices in industry due to easy design and scale-up. For a multi-tubular FXB reactor (TFXB) development and testing, testing in a full-size single tube is ordinarily sufficient for the design of an industrial size reactor of virtually any capacity – the industrial size simply having multiple tubes (hundreds or thousands of the single tube). An FXB reactor also eliminates many potential issues typically associated with a fluidized bed reactor, like solid attrition, entrainment loss and dust emission. However, a multi-tubular FXB reactor suffers from many serious drawbacks that include the following:

1. High heat and mass transfer resistances leading to a steep axial temperature profile in the reactor.
2. Poor catalyst utilization or low catalyst effectiveness factor, due to its use of relatively large particles.
3. Possible maldistribution of gas between all tubes (including impact of catalyst not loaded equally).
4. No continuous solid (catalyst or material) regeneration and replacement capability.
5. No load-following capability.
6. Hot spot formations possibly leading to catalyst activity/selectivity loss and even catalyst/material vitrification.
7. Higher pressure drop.
8. Restricted window of allowable feed composition, temperature and/or pressure to avoid possible temperature runaway for highly exothermic reactions.

Fluidized Bed Reactors, both Bubbling/Turbulent (BFB) and Circulating (CFB), do not suffer from the drawbacks of PKB/FXB reactors indicated above. The smaller particle sizes in these particulate systems lead to better heat and mass transfer rates, allowing operation at uniform bed temperature, which is normally preferred. Close to 100% catalyst utilization can be achieved, with no gas maldistribution. These reactors also provide continuous catalyst/material regeneration and replacement capability due to fluidizability of the chosen particulates that can be continuously discharged and recharged after regeneration and/or feeding a make-up stream. These reactors also have load-following capability, allowing operation across a broad window of production capacity as desired. They can be operated with lower pressure drops and are usually also free of hot spots. The windows of allowable feed composition, pressure and temperature are much wider compared with a fixed bed reactor, with less chance of reactor temperature runaway. On the other hand, the key drawbacks of a fluidized bed reactor include the following:

1. Potential dust emission with possible environmental concern.
2. Loss of expensive catalyst by entrainment, to the extent that fines are not fully captured and returned.
3. Particulate attrition causing excessive fines buildup, leading to excessive solid loss or depletion of fines that are needed for good fluidization, and other potential problems like line plugging and agglomeration.
4. Potential erosion of lines or internal heat transfer tubes.
5. Poorer solid-gas mass and heat transfer rates for BFB through the interface films of gas bubbles or voids.
6. Potential neglect of safety concerns (even unacceptable bypassing of safety instrumentation and procedures) if operators do not fully understand, and become complacent with respect to the high flame-arresting safety benefits of a BFB, e.g., for partial oxidation (POX) reactions, including knowing when or where they do not apply.

Circulating Fluidized Bed Reactors (CFBs) are usually designed for fast reactions and for reactor-regenerator dual-bed systems. In the dual-bed system, the solid particulates, catalyst or non-catalytic material, can be regenerated in the regenerator bed, or sometimes in a transfer line, and be returned back to the reactor in a continuous circulation of the solid. (This is also done with some BFBs.) The primary section of a CFB system operates without gas bubbles or voids, and has minimal solid-gas heat and mass transfer resistances. However, it suffers from the following drawbacks compared with the other systems:

1. The bed voidage throughout these reactors is usually very high (in the range of 88-98%), thus, solid concentration per unit bed volume very low, resulting in much lower reaction rate per unit bed volume.
2. These reactors operate at relatively much higher gas velocity and therefore are applicable to fast reactions only.
3. At the high gas velocities, which includes the catalyst movement, they demand much tougher or more attrition-resistant catalyst or material, unless the material is desired to be broken down (or reacted away) and to pass once-through as in gasifiers or combustors.
4. The reactors need to be much taller (in the order of 30 meters or even more) for adequate reaction residence time to achieve satisfactory conversions at their high gas velocities.

DIFREX Reactor Solutions with Particle Technology

DIFREX software, services and solutions integrate chemical reaction engineering with particle technology to provide the right solutions for design, development, scale-up and successful commercialization of gas-solid reactors, enabling one to promptly establish the design and performance of virtually any particle or solid-gas reactor type for reaction systems of different complexities. Since it has fundamental basis in the two engineering sciences (reaction and particle) and avoids much of the empiricism of traditional process development, especially in the HC processing industry, it is better suited for new designs and also revamp or upgrade of an existing technology/ reactor/ process, particularly for a new feed or product stream. This refers to both catalytic and non-catalytic solid-gas reactions, including the use of a catalyst (active components with or without carrier) as the solid for catalytic processes or a non-catalytic solid particle like coal, biomass or a mineral (FeO or NiO, for example) for combustors, gasifiers and metal extraction/roasting processes, and also solid particles as adsorbents to transfer atoms and other molecular species from one stream or vessel to another.

Figure 2 illustrates the extensive and convenient framework and easy use of the DIFREX software as represented as a ***dial-a-reactor***. It incorporates detailed flow and hydrodynamic behavior for many different reactor types, allowing prompt and easy calculation switching (dialing) from one reactor type to another, for the same chemistry and kinetics of the reaction system. Better, faster initial determination of the best conceptual/ preliminary design for a reaction system, catalyst or process can therefore ***now*** be achieved by comparing the design and performance details projected by the software at the earliest project stages. The best choice could be a packed bed (PKB), a tubular fixed bed (TFXB), a bubbling/turbulent fluid bed (BFB), a circulating fluid bed (CFB), all with or without quench and/or recycle, or a hybrid reactor (CSTR-CFB-BFB) as in a typical FCC reactor or other system. The best **Reactor Of Choice (ROC)** can now be determined along with the **Catalyst or Material Of Choice (COC/MOC)** and much earlier – even while the solid catalyst or material is being developed. Final determination of the best ROC, however, would be determined by **consideration of other criteria** that include cost, particulate handling/recycle, strength and attrition resistance of solid catalyst, and miscellaneous other characteristics like particulate fluidizability and agglomeration problems for fluidized bed systems. However, the software allows one to reach each **GO/NO-GO decision gate** much sooner, leading to substantial savings in both cost and time for catalyst, process or technology development.

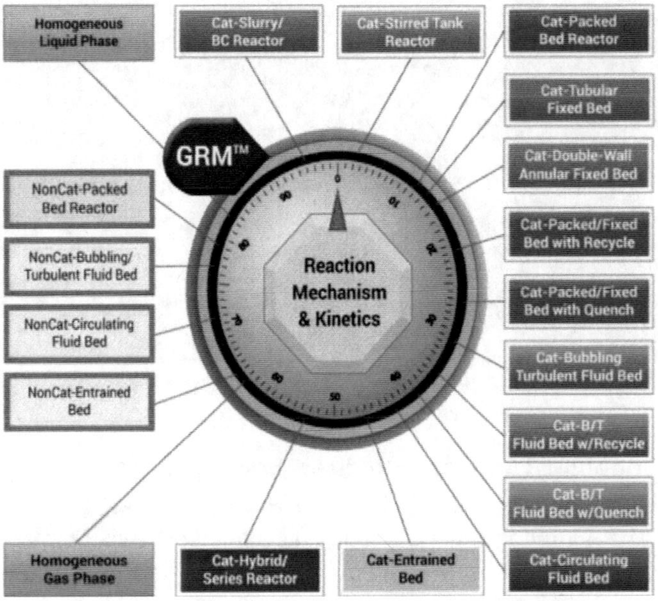

Figure 2: DIFREX: Selection for Reactor of Choice

Current DIFREX team members were instrumental in successful commercialization of several new process technologies, involving all four basic solid-gas reactor types: PKB, TFXB, BFB and CFB reactors and their variations. A prime example is the ALMA process, a fluidized bed POX process for production of maleic anhydride from n-butane and air. These designs were products of seamless integration of particle technology (including particulate flow and hydrodynamic behaviors) with chemical reaction engineering principles. Results of some of these works are available in a series of publications by Dutta, Arnold and other authors[1-20]. Many of the theories and procedures used in these works can be found in the two classical books on Fluidization Engineering and Chemical Reaction Engineering by Kunii and Levenspiel[21-22]. The following discussion provides information on some of these reactor designs and the results achieved.

Packed/Fixed Bed Reactors

The Fischer-Tropsch process is of great industrial significance. Projected axial profiles for temperature, conversions and pressure drop of a typical Fischer-Tropsch reaction system are shown in **Figure 3**, for syngas conversion to paraffins, olefins and alcohols, carried out in a steam-cooled Tubular Fixed Bed (TFXB) reactor system. PR and SR refer to primary reactant CO and secondary reactant H_2. There is a steep temperature rise near the reactor entrance, due to high reaction rate at the concentrated inlet composition, plus relatively high heat transfer resistance typical of this solid-gas fixed-bed reactor system. Temperature runaway with inadequate temperature control, hot spot formation, catalyst selectivity loss, deactivation and possibly even vitrification are potential problems of this system. **Figure 3** also shows the profiles of primary, secondary and undesirable product (PP, SP, UDP) selectivity, per-tube productivity, catalyst productivity (kg/day of PP per kg catalyst), and catalyst effectiveness factor (% utilization) for the primary reaction.

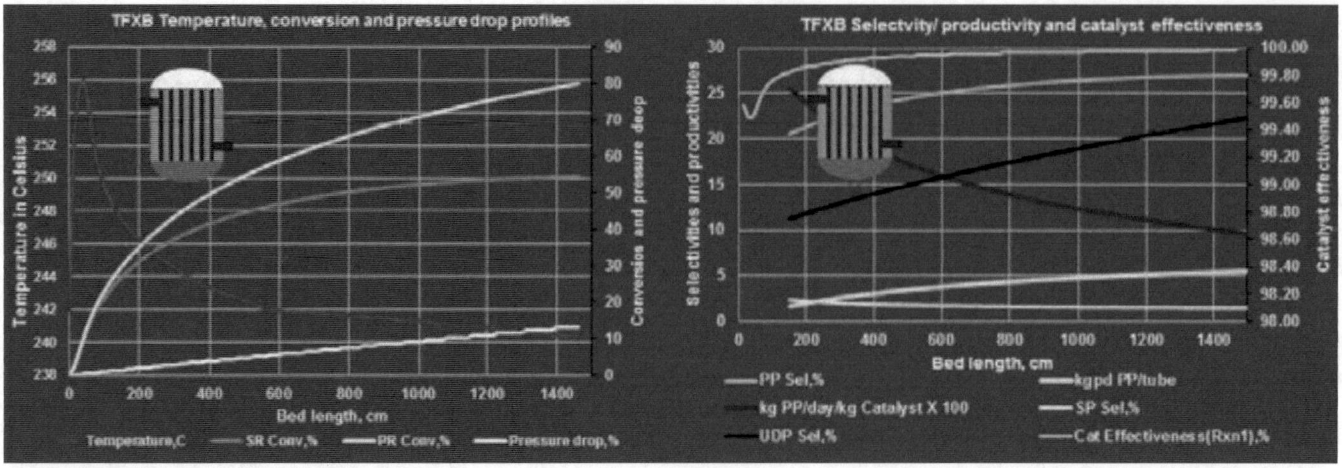

Figure 3: DIFREX capabilities in detailed reactor design studies: Case of FT synthesis in a TFXB reactor

Bubbling/Turbulent Fluidized Bed Reactors

The conversion of n-butane to maleic anhydride (B→M) presents a unique example where extensive performance tests simultaneously in both tubular fixed bed (TFXB) and bubbling or turbulent fluidized bed (BFB) reactors, accompanied by modeling and scale-up, led to successful commercialization of the world's leading fluidized bed process for this reaction system, called the ALMA Process. Unification of particulate technology principles, flow and hydrodynamic behavior with chemical reaction engineering formed the core of the reactor model that was developed. The model was also applied to simulation and scale-up analysis of two other catalytic reaction systems (N→P and P→A) that vary considerably in mechanism and kinetics. The three reaction systems are demonstrated in **Figure 4**. The data for the N→P system were obtained from published data of a commercial reactor.

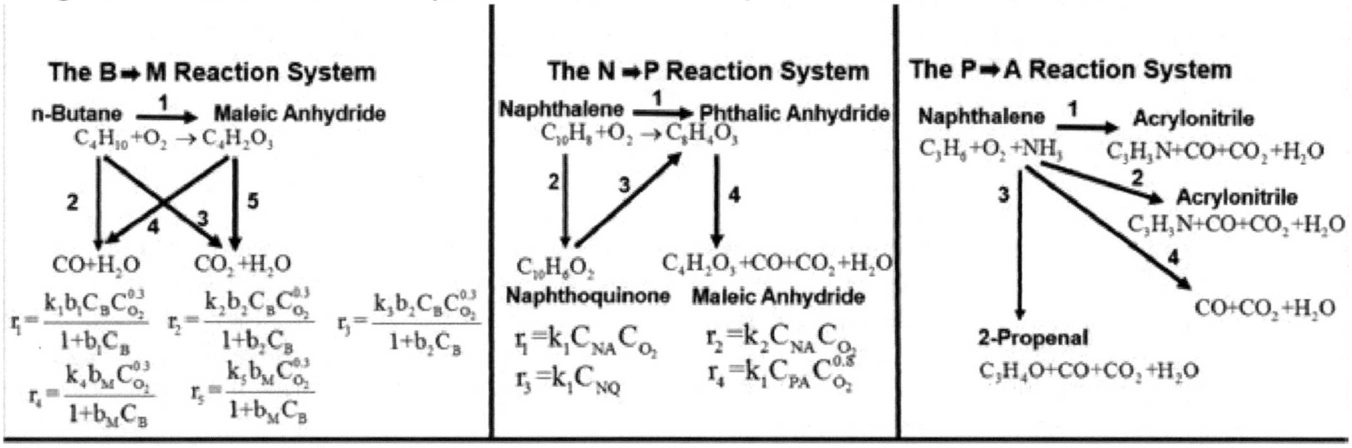

Figure 4: Catalytic reaction systems modelled and studied in TFXB and BFB reactors

In the unified model, the entire height of the BFB is viewed to consist of 5 possible zones: 1) ***grid zone*** where bubbles form and grow continuously until they become slugs or reach a constant average size, 2) ***bubble zone*** where bubbles of a constant average size (maximum stable size) rise up through surrounding solid-gas emulsion phase, 3) ***slug zone*** where slugs of constant size rise, filling up the vessel diameter (mostly in smaller-diameter lab or bench-scale reactors), 4) ***splash zone*** where the bubbles burst and splash the particles above the dense bed below and then the catalyst density decreases with height in the dilute phase, and 5) ***post-TDH zone*** where the density remains constant in the dilute phase ("freeboard") above the Transport Disengagement Height. The DIFREX model accounts for particulate flow and dynamic behavior in each zone, and how it affects the reaction system and reactor performance, including conversions and selectivities of each reaction. Details of these treatments are available in the publications of Dutta et al.[14-15].

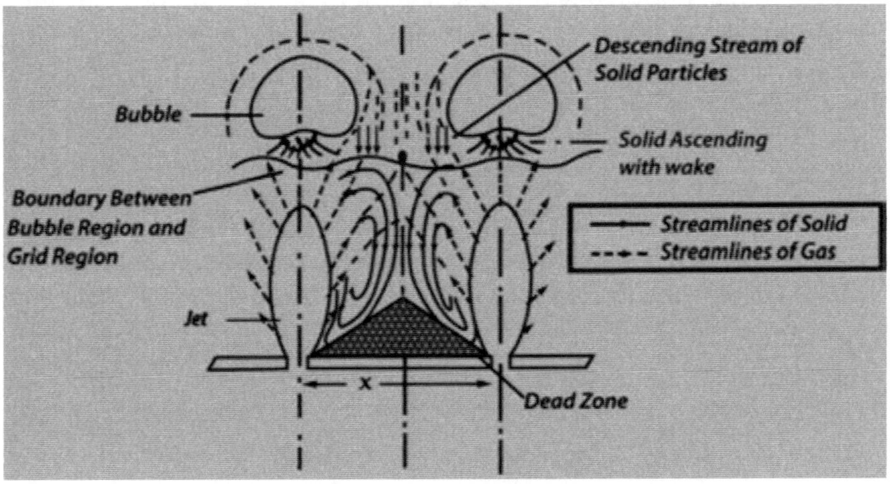

Figure 5: Schematic of gas and solids flow patterns and dead zone formation in the grid region

Unlike fixed bed (FXB), the particulate system behavior is quite complex in BFB and may change dramatically from reactor bottom to the top. Proper account of these changes and their effect on the reaction system is necessary for developing a successful BFB scale-up model. A schematic representation is shown in **Figure 5** of gas and solid flow patterns immediately above the grid at the very bottom of a commercial BFB. Due to intense solid-gas mixing, turbulence and smaller bubbles in this zone, plus the higher feed concentration, as much as 60% of the total conversion of a commercial BFB can happen in the ***grid zone***.

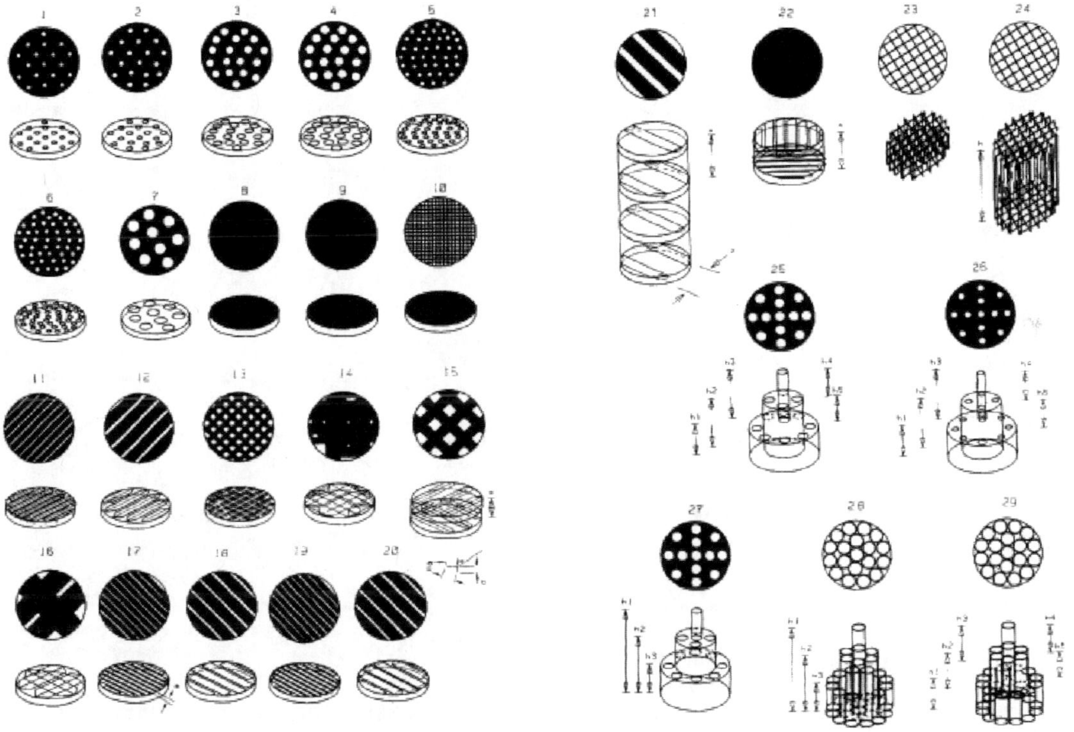

Figure 6: Top views of experimental baffles [Adapted from Dutta and Suciu[10]]

Grid design of a BFB is important and a challenging task since several objectives need to be achieved in this design. These include support strength, optimal pressure drop, minimum dead zone, minimum jet attrition of particles, and avoidance of particle slippage through the grid openings. Similar considerations affect the design of internal baffles, etc. A study of designs for the grids and other internals, with different open areas, layers, slot angles, egg-crate shapes, etc., was conducted for the ALMA process as exemplified in **Figure 6**. DIFREX team members have extensive experience in design and commissioning of such internals.

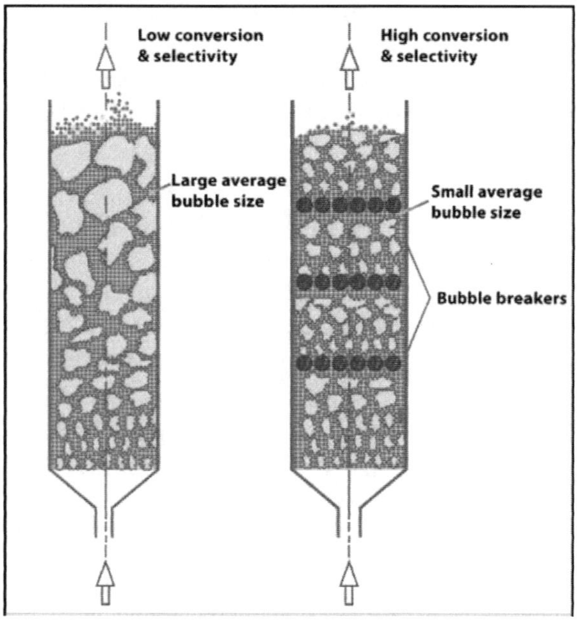

Figure 7: Bubble size control in BFB

Bubble size control above the grid zone is also an important consideration for higher reactor conversion, because smaller bubbles mean both higher and more selective conversion, while bigger bubbles mean lower conversion and poor selectivity. This is due to higher mass transfer resistance for the latter case, which slows the exchange (or mass transfer) rate between the rising bubbles with the bulk of the gas and the surrounding emulsion phase containing most of the catalyst. Proper design of bubble breakers therefore can also be an important part of a BFB design. DIFREX can provide help in this area as well. The concept of bubble size control is shown schematically in **Figure 7**.

Applicability of three BFB models were examined and compared during the ALMA process development, as described in a publication by Dutta, Arnold, Suciu and Verde[16]. Schematics of these models[21, 23, 24] are shown in **Figure 8**. Modification of the Kunii-Levenspiel (K-L) model[21] to account for growing bubbles and also the ***splash zone*** as proposed by Hatano et al.[25] are three important features of the DIFREX BFB reactor model. More details are available in Dutta and Suciu[14].

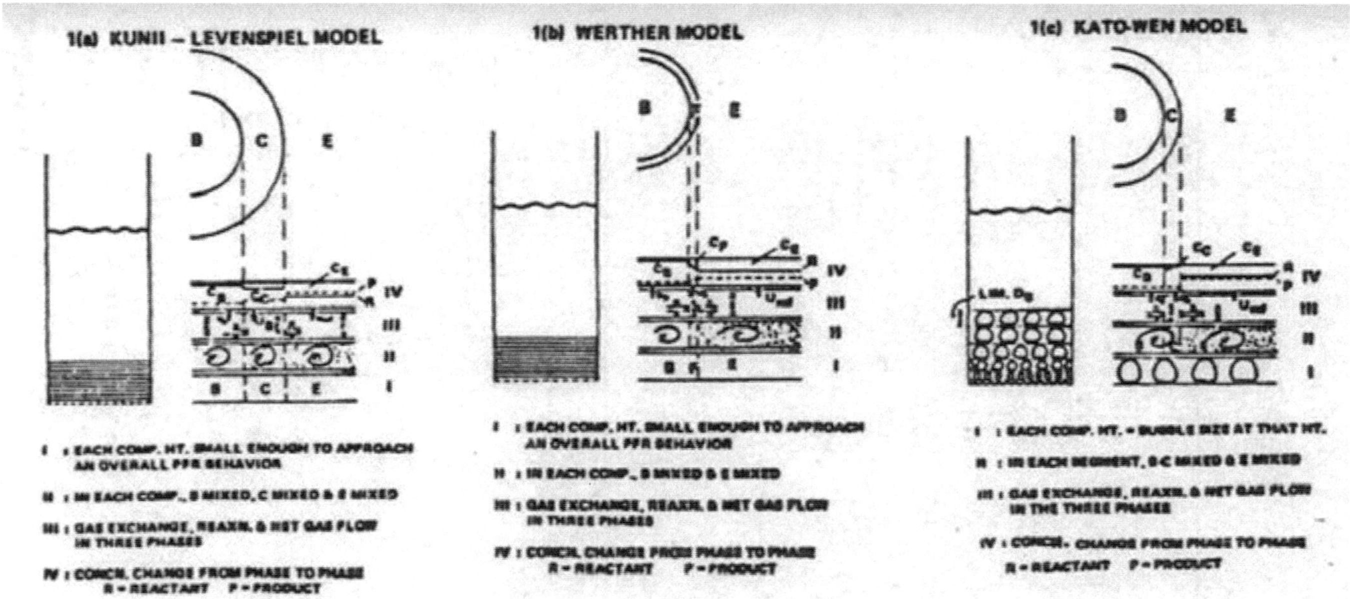

Figure 8: Schematic representation of three BFB models [Adapted from Dutta et.al [16]]

Projected BFB reactor performances and the key particulate hydrodynamic behaviors affecting these for the same FT reaction system used in the FXB reactor (those projections shown in Figure 3) are shown in **Figure 9**. The figure shows bubble size growth and expanded ***dense phase*** - the two critical properties of the particulate system that primarily influence the performance of a BFB reactor. The case shown is for a reactor with bubble size control to a maximum size of 10 cm. This figure also shows solid loadings and dilute phase voidage up to the TDH in the dilute phase region, where the declining solid loading (shown by the dotted curve) characterizes the ***splash zone***. The model estimates the TDH, the information necessary to decide the minimum height necessary for the dilute phase of the reactor before the exit gas nozzle location (or entry into internal cyclones, if used). The BFB temperature is ordinarily constant within the bed, and so the temperature profile is not shown in the figure. No significant reaction is noticed in the ***dilute phase*** for this particular case and reaction system used. However, depending on the design and operating conditions, reaction system, relative volume and temperature of the two zones (dilute phase vs. dense phase), significant reaction may occur in the dilute phase, too.

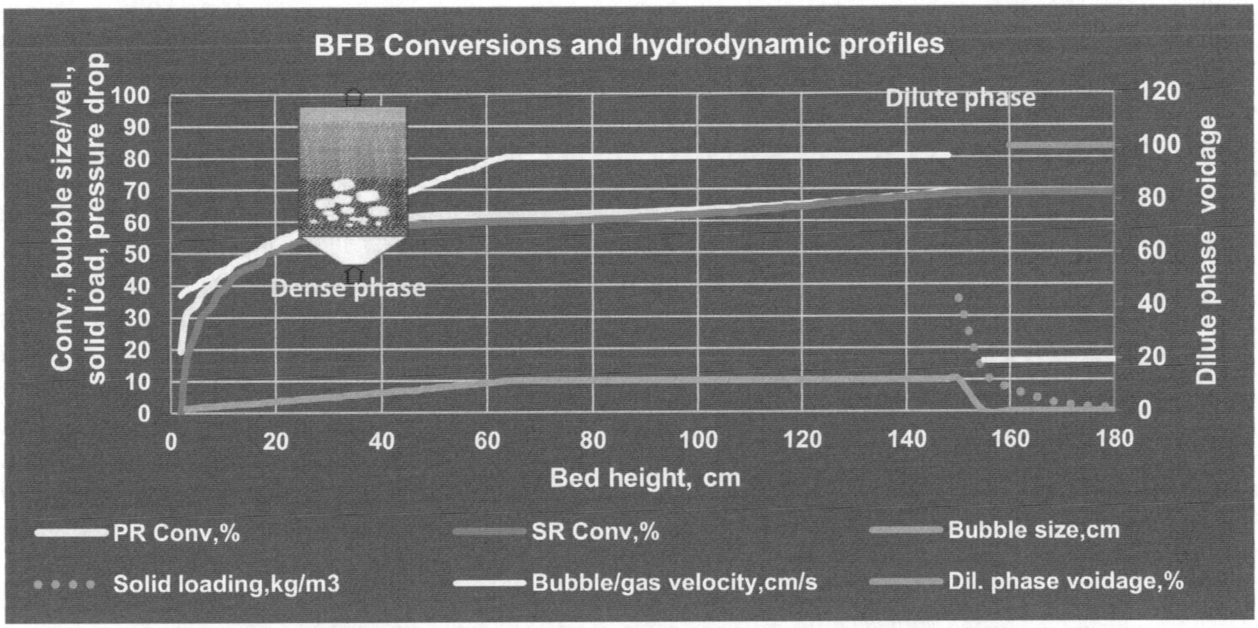

Figure 9: Projected BFB reactor performances and key particulate hydrodynamic effects

The DIFREX reactor design and model package GRM^{TM} enables quick generation of the complete design and operating conditions and performance comparison of alternative reactors (FXB, BFB or CFB) if the information required for complete definition of the reaction system involved can be provided or is generated with DIFREX help. **Figure 10** shows one such comparison of complete design and operating conditions of three different solid-gas reactor types - a TFXB (Tubular Fixed Bed), a BFB and a QTFXB (Quenched Tubular Fixed Bed), for 100 TPD production of the primary product (PPX) or product group (like alcohols, paraffins or olefins) from syngas for an FT (Fischer-Tropsch) reaction system (as in Figures 3 and 9). **Figure 11** shows the performance comparison of the three reactors. The best Reactor Of Choice (ROC) can be quickly determined based on the criteria of the user that usually include cost, mechanical and other considerations. In **Figure 11**, the best choice shown is based solely on the highest yield.

Uncertainty/ fear of scale-up of a BFB reactor from bench to commercial scale can now be drastically reduced if proper account of advances made in particulate technology is incorporated in the design and scale-up of this reactor system.

Circulating Fluidized Bed Reactors

Successful design and operation of a circulating fluidized bed (CFB) reactor also require proper interplay of particulate flow and hydrodynamics behavior with reaction chemistry and kinetics and heat and mass transfers. The two key advantages of this reactor are: 1) It offers an elegant means of continuous regeneration of spent solid (catalyst or non-catalytic/ reactive particles) in a recirculation loop that removes the solid after only a short exposure (residence) time in the reactor, and subsequently returns the regenerated material back to near the reactor bottom. It is therefore particularly suited for fast-deactivating catalyst and for regenerable catalysts as in an FCC reactor. It is also a prime candidate for applications in **coal/ biomass/ tar sand combustion and gasification and minerals processing**, fast reactions where particle ***attrition*** is not an issue of concern. 2) CFB is also a ***bubble-less*** reactor above the bottom mixing zone, unlike BFB, thereby enjoying minimum heat and mass transfer resistances when compared with both FXB and BFB reactors.

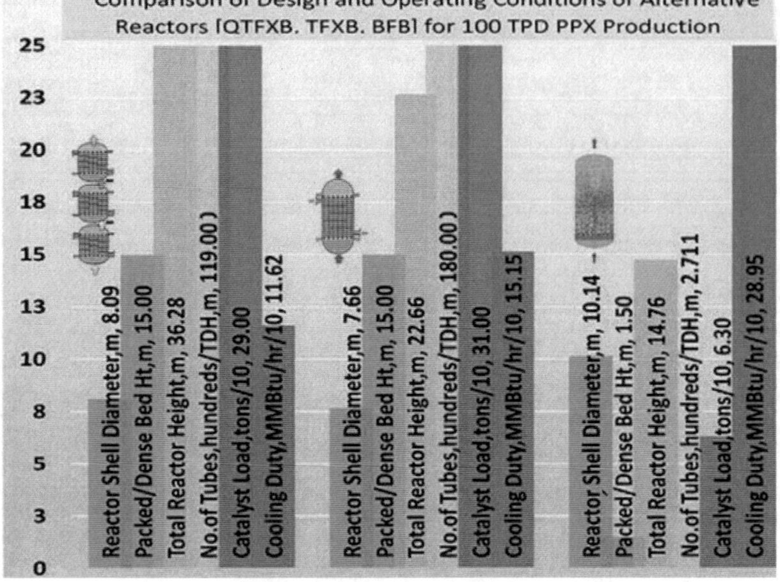

Figure 10: Reactor design and operating conditions for 3 different solid-gas reactor types

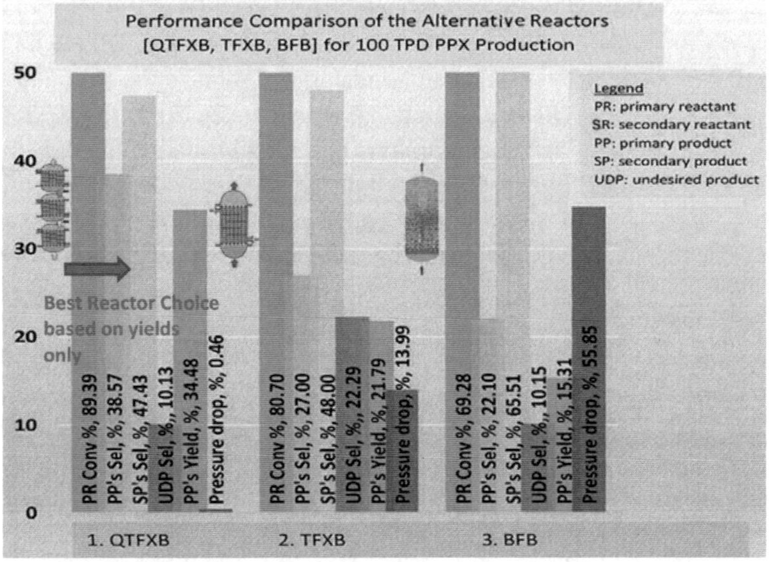

Figure 11: Performance Comparison for 3 solid-gas reactors

The two most important particulate system parameters used in the design of a CFB are reactor voidage (**ε**) profile and solid loading or "flux" of the up-flow solid-gas stream. Voidage profile correlations proposed by Kwauk et al.[26] used in the CFB reactor design and model have been first tested for valid simulation of voidage profiles of common solids of commercial interest and found to be satisfactory, as shown in **Figure 12.** For the CFB, reactor pressure drop was calculated by using four major pressure drop components: bed hydrostatic pressure, solids acceleration, solids friction and gas friction, as proposed by Yang and Keairns[27]. More details of these treatments are available in Dutta et al[13]. The DIFREX CFB design and modeling tool has been applied to both catalytic and non-catalytic reactor designs that include ***acrylonitrile production from propylene***, ***FCC catalyst regeneration***[13], ***high-grade crude extraction from oil shale***, ***iron ore reduction*** **and** ***hot gas desulfurization by zinc titanate***[6]. Application of particulate technology and principles is integral to the design of these reactors.

Temperature, conversions and bed/ particulate hydrodynamics for a propylene-to-acrylonitrile CFB system is shown in **Figure 13**, where PR and LR denote primary reactant propylene (C_3H_6) and limiting reactant ammonia (NH_3). In this reactor design, the bed bottom gas-solid mixing zone has a smaller diameter than the top/main bed. Profiles of temperature, reactor voidage, local gas velocity and solid loading of the up-flowing gas are shown.

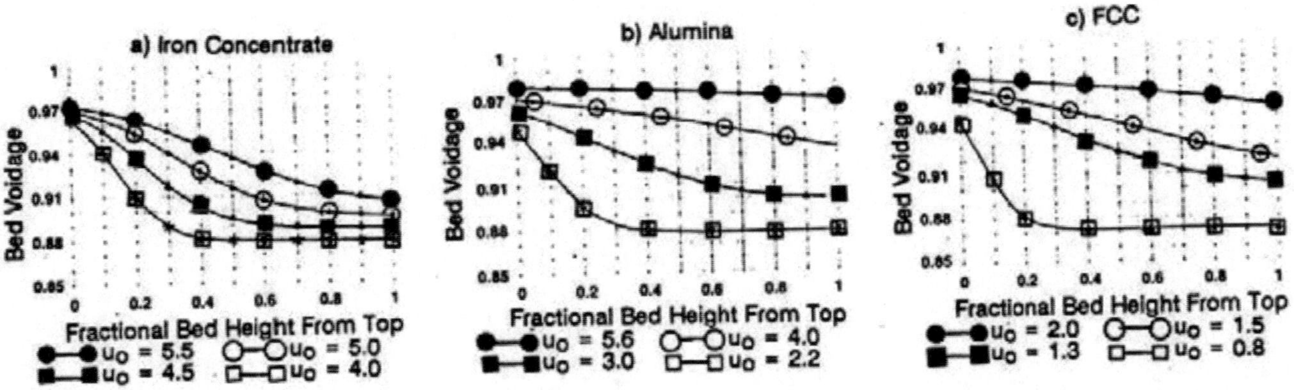

Figure 12: Simulation of CFB model based on voidage profile correlations proposed by Kwauk et al.[26]

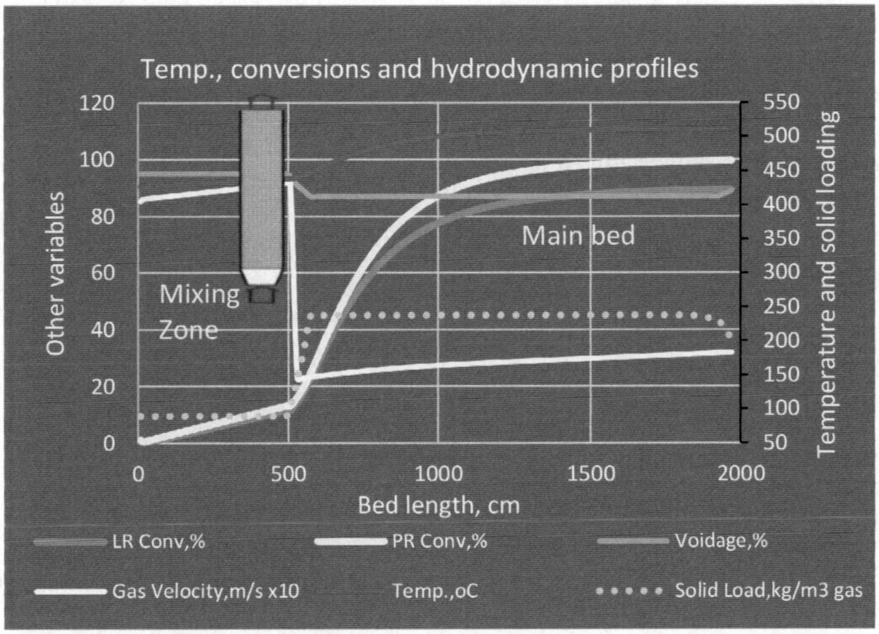

Figure 13: Profiles (Temperature, conversion and hydrodynamics) for a CFB system

An Interesting Observation

An interesting observation while studying the effect of CFB hydrodynamics on a reaction system is worth mentioning. A *CFB* can be viewed as a suspended and moving fixed bed (*FXB*) of particulates (fine particles) but with a very high bed voidage and with significant solids slippage (falling backward in the annulus along the wall). However, due to the intimate best heat and mass transfer rates possible, the bed temperature (rise) profile (for an exothermic reaction) in an adiabatic CFB reactor does not show as steep a temperature rise or spike as can be problematic in an FXB. Also, the bed temperature rise can be controlled by both gas flow and solid circulation rate.

Due the various impacts, the upper portion of a CFB that has a higher average bed voidage (a CFB "suspended bed" with lower solid/catalyst concentration per unit volume) may achieve higher conversion than a CFB "suspended bed" with lower bed voidage. Lower solid loading in this more diluted suspended bed allows a higher temperature rise due to lower volumetric heat capacity because of its lower solid content. The effect of rising temperature in the CFB may then more than compensate for the effect of lower solid/catalyst concentration. This is illustrated in **Figure 14**. Higher temperature and higher conversions are observed for the case with higher voidage and lower solid loading (kg/m^3 gas) in spite of the low solid concentration. On the other hand, since the inlet temperature is the same for both cases and the average and upper zone temperatures are higher for the case with higher bed voidage (due to extra temperature rise), the catalyst deactivation is likely to be faster and the selectivity is also likely to be different (for several reasons).

It is important to understand that, unlike the dense bed of a BFB, a CFB is not isothermal even though it is more isothermal than an FXB.

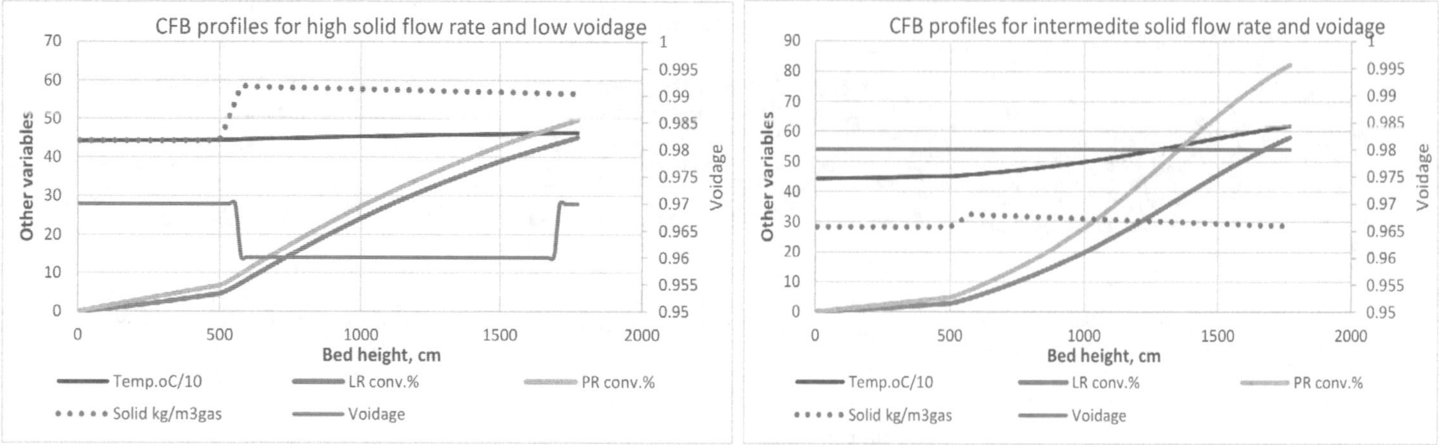

Figure 14: Effect of solid flow rate and bed voidage in a CFB case study

Hybrid Particulate Systems

A prime application of particulate technology is in FCC reactors. There are a variety of configurations of these reactors. In many installations, the FCC reactor is a "riser" type, with high gas velocity carrying all the catalyst upward in a dilute phase, with minimal residence time, etc. For this riser kind of FCC reactor, it can be represented as a hybrid/ combination of three particulate-gas contact types: a bottom regenerated solid-feed gas mixing (CSTR or BFB) zone, a middle which is the main CFB (riser) section, and a gas-solid disengaging zone at the top (which may create a further bed that can be simulated as a BFB). This representation is shown in **Figure 15.**

Using 1) a 5-lump reaction scheme with gasoil feed as shown in **Figure 16**, 2) surrogate compounds to represent the components involved, and 3) particulate system behavior in the three zones, a simple FCC reactor model was developed by Dutta and Gualy[8] that simulated the observed performance data reasonably well. The simulation results are shown in **Figure 17(a-d)**.

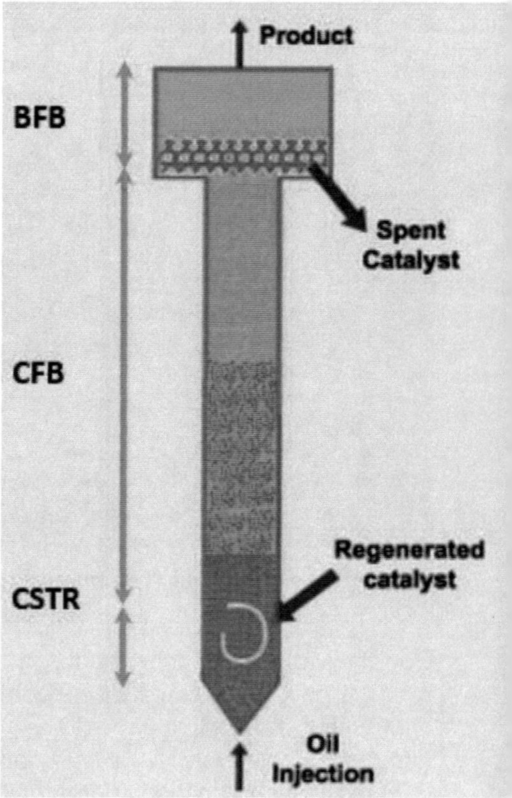

Figure 15: 3-zone simulation of an FCC performance

Reaction system: FCC 5-Lump reactions with GASOIL feed

GASOIL	GASOLINE	Rxn 1
GASOIL	DRYGAS	Rxn 2
GASOIL	LPG	Rxn 3
GASOIL	COKE	Rxn 4
GASOLINE	DRYGAS	Rxn 5
GASOLINE	LPG	Rxn 6
GASOLINE	COKE	Rxn 7
LPG	DRYGAS	Rxn 8

Figure 16: FCC 5-lump reactions with GASOIL feed

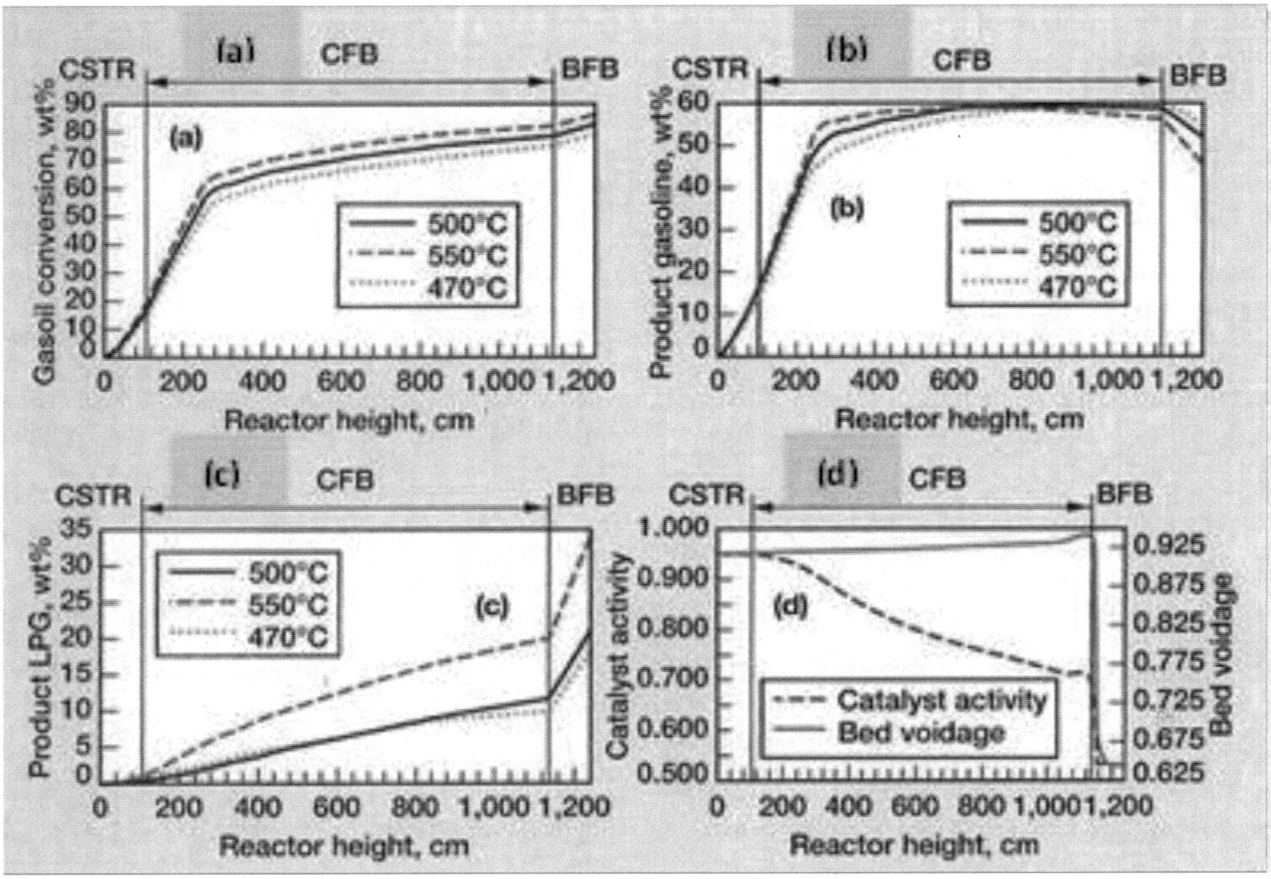

Figure 17: FCC simulation using a 5-lump reaction scheme

Emerging Technologies and Particulate Systems

Is ***Methanol*** the fuel of the future? There is a worldwide push to capture CO_2 from atmosphere and power plant exhausts and convert this and H_2O into methanol. One of the emerging technologies is ***PTM*** [(solar) Power to Methanol] that may lead to zero carbon footprint – as solar energy seems to be becoming increasingly more competitive and economical. The particulate technology can play a vital role in the PTM process, where the CO_2 could be captured and concentrated ***economically*** by using a regenerable particulate sorbent in a fluidized bed process. The other is the conventional route of ***syngas-to-methanol*** through syngas production from coal or biomass – the ***renewable energy route***. In both of these technologies, the particulate technology plays a vital role. Syngas production by gasification of coal or biomass in a fluidized bed reactor is one of the most attractive means for the second route. Due to a rather unique and perhaps the most diverse expertise and experience in both fluidization and reaction engineering, and also in coal combustion and gasification technology[28-31], DIFREX can play an instrumental role in implementing a rapid and successful move to ***the fuel of the future***. An integrated CTSM/ BTSM (Coal-to-Syngas-Methanol/ Biomass-to-Syngas-Methanol) plant can be designed and built promptly with help from DIFREX. This will be further facilitated by DIFREX already having an advanced and proprietary syngas to methanol (S2M) reactor design available.

References

1. Dutta, S. and R. Gualy, "Build Robust Reactor Models", Chemical Engg. Progress, 96(10), October 2000, p. 37.
2. Dutta, S. and R. Gualy, "Super-modeling Reactors for Profitability", Chem. Engg., 107(6), 2000, p. 72.
3. Dutta, S. and R. Gualy, "Modernize Process Reactors", Hydrocarbon Processing, November, 1999, p.91.
4. Dutta, S. and R. Gualy, "Overhaul Process Reactors", Hydrocarbon Processing, September 1999, p. 43.
5. Dutta, S. and R. Gualy, "General Reactor Model Improves HPI Applications", Hydrocarbon Processing, July 1999, p. 45.
6. Dutta, S., "Circulating Versus Bubbling Fluidized Bed for Hot Gas Desulfurization by Zinc Titanate", Proc. 5th Int'l. Circulating Fluidized Bed Conf., Beijing, PRC, May, 1996, p. pr-7.
7. Dutta, S., "Reactor Model: Usefulness of a Lesser Known Tool in Process Plants", paper presented at SABIC Tech. Meeting, May, 1996.
8. Dutta, S., "Fluidized and Moving Bed Desulfurization by Zinc Titanate", AIChE Symp. Ser., 301(90), 1994, p. 157.
9. Dutta, S. and B. Jazayeri, "Alternate Reactor Concepts for Oxidative Coupling of Methane", Proc. 7th Int'l. Fluidization Conf., ed. Potter and Nicklin, Brisbane, Australia, May, 1992, p. 445.
10. Dutta, S. and G. D. Suciu., "An Experimental Investigation on the Effectiveness of Baffles in Breaking Bubbles in Fluid Beds", J. Chem. Engg. Japan, 25(3), 1992, p. 345.
11. Dutta, S., "A Novel CFB Application: A Chemical Compressor", Proc. 3rd Int'l. Circulating Fluidized Bed Conf., ed. Basu, Horio and Hasatani, Nagoya, Japan, Oct., 1990, p. 465.
12. Dutta, S., "A Novel Circulating Bed Metal Hydride Compressor", paper presented at the 8th World Hydrogen Conf., Honolulu, Hawaii, 1989.
13.Dutta, S., B. R. Christian and M. F. Ratterman, "Circulating Fluid Bed Reactor Model Developed for FCC/ART Catalyst Regenerator", Proc. 6th Int'l. Fluidization Conf., ed. Grace, Shemilt and Bergougnou, Banff, Canada, May, 1989, p. 9.
14. Dutta, S. and G. D. Suciu, "Unified Model Applied to the Scale-up of Catalytic Fluid Bed Reactors of Commercial Importance", Proc. 6th Int'l. Fluidization Conf., ed. Grace, Shemilt and Bergougnou, Banff,Canada,May,1989, p.311.
15. Dutta, S., "A Unified Approach to the Modeling of Catalytic Fluidized Bed Reactors", Proc. World Cong. III Chem. Reaction Engg., Tokyo, 1986, p. 287.
16. Dutta, S., S. C. Arnold, G. D. Suciu and L. Verde, "Scale-Up of a Commercial Catalytic Fluid Bed Reactor Involving Complex Kinetics", Inst. Chem. Engg. Symp. Ser. 87, 1984, p. 517.
17. Dutta, S., "Measurement of Bubble Characteristics by Optical Probe", paper presented at the AIChE Annual Meeting, Chicago, Nov., 1980.
18. Dutta, S., "A Simple Probe for Fluidized Bed Measurements", Cana. J. Chem. Engg., 57, 1979, p. 115.
19. Dutta, S., "Dead-zone Heights near the Grids of Fluidized Beds", Fluidization, Cambridge University Press, 1978, p. 32.
20. Wen, C. Y. and S. Dutta, "Research Needs for Analysis, Design and Scale-up of Fluidized Bed Reactors", AIChE Symp. Ser., 73(161), 1977, p.1.
21. Fluidization Engineering, 2nd edition, by Daizo Kunii and Octave Levenspiel, Butterworth-Heinemann, 1991.
22. Chemical Reaction Engineering, 3rd edition, by Octave Levenspiel, John Wiley & Sons, 1999.
23. Werther, J., Chem. Engg. Sci., 35, 1980, p. 372.
24. Kato, K., and C. Y. Wen, Chem. Engg. Sci., 24, 1969, p. 1351.
25. Hatano, H., T. Okuma and M. Ishida, J. Chem. Engg. Japan, 17, 1984, p. 491.
26. Kwauk, M. N., N. Wang, Y. Li, B. Chen and Z. Shen, Circulating Fluidized Bed Technology, ed. P. Basu, Pergamon Press, 1986, p. 33.
27. Yang, W. C. and D. L. Keairns, AIChE Symp. Ser., 141(70), 1974, p.27.
28 Wen, C.Y. and S. Dutta, Book Chapter "Kinetics of Coal Pyrolysis and Gasification Reactions", Ch.2 Coal Conversion Technology, Addison-Wesley Publishing Company, 1979.
29. Wen, C.Y. and S. Dutta, "Solid-Gas Reactions in Coal Conversion Processes", Coal Processing Technology, CEP Tech. Manual, IV, 1978, p. 40.
30. Wen, C.Y. and S. Dutta, "Reactivity of Coal and Char: Part I – In Carbon Dioxide Atmosphere", I&EC Process Design & Development, 16, 1977, p.20.
31. Wen, C.Y. and S. Dutta, "Reactivity of Coal and Char: Part II – In Oxygen-Nitrogen Atmosphere", I&EC Process Design & Development, 16, 1977, p.31.
32. www.difrex.com
33. http://difrex.com/pdf/BrochureContents.pdf.
34 Dutta, Subhash, Stephen Arnold, Aashish Gaurav and Jim Brenner, "DIFREX Nexgen Expands and Advances Reactor & Technology Solutions", paper presented at AIChE Spring Meeting, Orlando, April, 2018.

508152 Erosion and Mobilisation of Highly ACTIVE Simulant Suspensions with Impinging Vertical Jets

Abstract:

Based on the nuclear industry in the U.K., storage of nuclear waste poses a particularly complex risk. Part of the waste during storage before being disposed of begins to form colloidal particles. These particles begin to settle out, and can compromise the integrity of their storage container. Using an Impinging vertical liquid jet is the means to prevent this. However the operational effectiveness of these jets is unknown due to the nature of the containers. Reduced scale modelling is required to assess their efficacy, and understand the dimensionless results as they allow systems of scale to be compared. This is of particular interest due to the wide reaching applications to all industries that may require utilisation of impinging jets, or require mass transfer, or even heat transfer.

An experimental rig of 0.5 m^3 has been used to model the central jet of the storage tank system. A Honite simulant (SiO_2, glass spherical particles) of varying size distributions has been used for the initial non-cohesive particle testing, and modelling. Each experiment uses 10 different jet diameters. Data is collected with an Ultrasonic Velocity Profiler, which scans the simulant bed in real time under impingement utilising an automated traverse. Acoustic backscatter data was recorded to yield a dynamic scour depth, and dynamic crater radius to allow dimensionless analysis of the erosion for each size distribution. Dimensionless analysis takes the form of the Erosion Parameter, E_c, which allows comparison for systems of scale and a means of correlating multiple simulants against one another, so patterns are easily identifiable. Part of the dimensionless analysis was also to deduce the Shieldâ€™s Parameter from vane viscometry to allow for a comparison of erosion behaviours taking into account particle properties, flow properties and the overall influence these have on the erosion profile.

505818 REAL WORLD Examples of Influences of Particles and Cake Formation on SOLID-Liquid Separation Technology Operation

Abstract:

The paper begins with a general discussion of the bench-top laboratory tests that are conducted for solid-liquid separation technology problem analysis, process development, technology selection and scale-up. The tests include pressure, filter media, filter aids, and similar process parameters.

Solid-liquid separation technologies are explained for continuous pressure filtration and batch pressure filtration. This provides an overview for the testing and troubleshooting.

Three different case histories are presented illustrating different scenarios of the impact of particles, particle size distribution (PSD) and cake formation on the separation technology operations. The first is a specialty chemical application for a continuous rotary pressure filter where the particles changed from lab scale to production scale. The second case is a clarification application where the system was running successfully for over one-year but inexplicitly, the performance changed drastically and the filter media began plugging very quickly during the cycle. The final application looks at lab testing and pilot scale testing where the cake formation changed resulting is a unique operating technique.

Sherlock Holmes and Dr. John Watson are fictional characters of Sir Arthur Conan Doyle. Process engineers who live in the real-world can learn many things from the two of them for solving process filtration problems. One example that Holmes proves time and again is that there is no benefit to â€œjumping to conclusions.â€ As discussed in the illustrations, there are many choices of filtration technologies to achieve the necessary level of quality with a cost-effective and a reliable process. Also, in terms of troubleshooting, there are many causes of filtration symptoms. With careful analysis and taking a creative and "non-conventional approach" by looking at what is behind the process data and the operating parameters, the optimum solution will be realized.

508441 Performance in Multi-Stage Fluidized Bed Reactors: From Theory to Industrial Application

Abstract:

The absence of back-mixing combined with the increasing solids residence time in multi-stage fluidized beds creates the opportunity to perform different reaction conditions in one reactor. Back-mixing of gas and solids can be prevented effectively by baffles, resulting from the phase separation, i.e. the formation of â€˜gas cushionâ€™ across baffles. In present work, the equilibrium Mach number (Mae) is introduced here as the indication of dense and dilute phase in fluidization. Then, gas solids flow across a baffle is treated as equilibrium flow of gas solids mixture through a nozzle. And the principle of gas solids phase separation through a nozzle can be detected by stability analysis. Furthermore, various industrial applications with multi-stage fluidized beds are discussed in light of our model.

507683 Spray Granulation in Continuously Operated Horizontal Fluidized Beds: Investigation of the Particle Transport Behaviour and the Product Properties in a Multi-Staged System

Abstract:

Fluidized bed spray granulation is an important process to produce granules with a desired particle size and structure. The particle properties can be influenced by several process parameters like gas inlet temperature, liquid feed rate as well as droplet size of the atomized liquid feed.

For continuous treatment of powders progressively horizontal fluidized beds are used. These horizontally structured apparatuses are characterized by a rectangular cross section. An advantage of a horizontal fluidized bed is the flexible partitioning of the process chamber into several stages due to weirs to apply different processes simultaneously, like granulation, drying or cooling in one single apparatus. The separation into several compartments provides a targeted impact on the transport behavior and the residence time of the particles, which are feed into the process chamber.

In this contribution, we present an overview on the process, investigating the dependency of the resulting product properties on the process conditions and highlighting possibilities of influencing the transport behavior in a horizontal fluidized bed.

This work focuses on the particle properties of sodium benzoate granules, which are produced by fluidized bed spray granulation during different process conditions. The drying process has a decisive influence on the kinetics of the growth rate and thus also on the properties of the particles, like the morphological structure, particle moisture and porosity.

Based on the granulation process, particles, which reach a certain product grain size, should be discharged as fast as possible in order to prevent further growth. To achieve this objective the residence time of the particles in the multi-staged process chamber has been investigated experimentally for different weir configurations (over, under and lateral flow). For this purpose Tracer particles with a magnetic coating have been used to analyze the transport behavior across the stages under different weir configurations, process conditions and particle size ranges.

Additionally, the investigation according to the particle transport and the fluid mechanics in the horizontal fluidized bed was performed by simulations based on Discrete Particle Modelling (DPM) and verified by experiments in a pilot plant. Using DPM, it is possible to examine the particle dynamics on a microscopic level in order to analyze particle exchange rates between the individual stages of the fluidized bed while utilizing different weir configurations and process conditions.

The financial support of DFG (Deutsche Forschungsgemeinschaft) within the priority program SPP 1679 "Dynamic simulation of interconnected solids processes DYNSIM-FP" is gratefully acknowledged.

Direct Numerical Simulation of Reactive Particulate Flows using an Immersed Boundary Method

Jiangtao Lu*, Michael D. Tan, Elias A.J.F. Peters, Johannes A.M. Kuipers
Multiphase Reactors Group, Department of Chemical Engineering and Chemistry, Eindhoven University of Technology, P.O. Box 513, 5600 MB Eindhoven, The Netherlands
*E-mail: J.Lu1@tue.nl

Abstract

In this paper, we report the extension of an earlier developed Direct Numerical Simulation (DNS) model to study coupled heat and mass transfer problems in particulate flows. The DNS model builds on an efficient ghost-cell based Immersed Boundary Method (IBM) which implicitly incorporates the boundary conditions into the discretized momentum, thermal and species conservation equations of the fluid phase. On the particles an exothermic surface reaction takes place. The heat and mass transport is coupled through the particle temperature, which offers a dynamic boundary condition for the fluid phase thermal energy equation.

The present simulations are performed for four fluid-solid systems. Following the case of the unsteady mass and heat diffusion in a large pool of quiescent fluid, we consider a stationary sphere under forced convection. In both cases variable reaction rates are imposed at the particle surface, and the particle temperatures obtained from DNS show a good agreement with analytical/empirical solutions. After that, we apply our DNS model to the three-bead reactor and finally, a dense particle array consisting of hundreds of particles distributed in a random fashion is studied. The concentration and temperature profiles are compared with a 1D heterogeneous reactor model and the heterogeneity inside the array is discussed.

Keywords: Direct numerical simulation, Immersed boundary method, Gas-solid flow, Damköhler number, Coupled heat and mass transfer.

Introduction

Particulate flows are frequently encountered in a wide range of industrial processes, such as synthesis of chemicals by heterogeneous catalysis, combustion of pulverized coal and coating on particle surfaces. Simultaneously with the mass transfer these processes are often accompanied with significant heat effects, which introduces additional complexity to the system. Accurate predictions of the gas-solid interactions are of great help to improve performance and facilitate optimal design of process equipment. Therefore it is important to understand the heat and mass transport processes in a coupled manner in such complex heterogeneous systems.

In the part decades extensive experimental investigations have been conducted for particulate flows, from which various correlations have been proposed for heat and mass transfer [1-6]. These correlations are very helpful for a quick estimation of the average heat and mass transfer rates for engineering purposes, however detailed information such as local values, spatial distribution and temporal evolution cannot be easily quantified. 3D transient simulation of complex multiphase flows has attracted considerable interest due to the fast development of computational capabilities, through which detailed quantitative information can be produced instead of an average value in heat and mass transfer processes. As the most detailed level of the multiscale modelling approach [7], DNS is a powerful tool to resolve all the details at the smallest relevant length scales to gain fundamental insight in fluid-solid interactions and quantitatively derive closures for application in more coarse-grained models. In recent years the Immersed Boundary method (IBM), as a branch of DNS, has received a lot of attention. Taking the advantages of easy grid generation and efficient CPU utilization, IBM has been applied in various studies including moving particles, complex geometries and deformable immersed objects [8-13]. Following the fluid flow equations, additional equations for species and thermal energy transport can be added using the same methodology.

For the IBM fluid-solid coupling schemes can be roughly categorized into two types: continuous forcing method (CFM) and discrete forcing method (DFM). In CFM a Cartesian grid is used for fluid flow simulation whereas the immersed boundary is represented by Lagrangian marker points [14-18]. The interaction between the immersed object and the fluid is accounted for by explicitly introducing a forcing term in the governing equations, and a regularized Dirac delta function is used to distribute this singular force over the Eulerian cells surrounding each Lagrangian point. In DFM the immersed boundary is treated as a sharp interface [19-23]. Rather than a forcing term determined by the feedback mechanism, the predefined boundary condition on the immersed object surface is enforced by extrapolating the surrounding fluid variables to the ghost cells (cells inside the immersed object but possessing at least one fluid cell neighbor). The virtual ghost value is then directly incorporated into the fluid phase equations at the discrete level. DFM is also known as the ghost cell approach, and it is the one utilized in our DNS model.

In contrast to numerous studies of momentum transfer in particulate flows, much less IBM computational results are available in the field of mass and heat transfer [24-29]. Only a few studies on reactive systems involving coupled mass and heat transfer processes, have been reported. Dierich et al. [30] applied 2D modeling to study the behavior of ice particles melting in the water, where the interfacial heat transfer as well as the phase change phenomena play a significant role. The Dirichlet boundary condition is applied on the ice surface. Kedia et al. [31] introduced a second order method for simulations of low-Mach number chemically reacting flow around heat-conducting immersed solid objects in combustors. A buffer zone is used to account for the fluid-solid interactions, where the conjugate heat transfer and the zero species penetration are applied. Deen and Kuipers [32] applied the directional quadratic interpolation scheme to study the coupled heat and mass transfer in a dense particle array, where they assume an infinitely fast chemical reaction proceeding at the particle surface. Also in the combustion field, Abdelsamie et al. [33] studied low-Mach number turbulent reacting two-phase flows in arbitrary geometries. In their work, CFM and DFM are implemented for moving particles and static boundaries respectively, and the source terms for species and thermal energy are directly incorporated in the governing equations.

Building on our previous study [34], the DNS model which is based on an efficient ghost-cell based immersed boundary method is extended to the simulation of reactive particulate flows. The coupling of heat and mass transfer arises as a consequence of an exothermic chemical reaction proceeding at the exterior surface of the particles that the particle is heated up by the liberated reaction heat and subsequently transfers the thermal energy to the fluid phase. As the Robin boundary condition is realized at the exact fluid-solid interface through a second order quadratic interpolation scheme in our IBM, a surface reaction rate is incorporated to describe the interplay between chemical reaction and mass transfer processes. For heat transfer processes, the Robin boundary condition simply switches into the Dirichlet boundary condition with the solid temperature determined by the particle thermal energy equation, which serves as a dynamic boundary condition for the fluid phase thermal energy equation.

The organization of this paper is as follows. First, the description of the model is given, including governing equations, numerical solutions and fluid-solid coupling. After that four reactive fluid-particle systems with increasing complexity are studied: unsteady diffusion around a single sphere, a single sphere under forced convection, the three-bead reactor and a dense particle array. Finally, the conclusions are presented.

Model description

In this part, we describe the governing equations that need to be solved in DNS, the numerical details of the solution methods and the fluid-solid coupling. For the model presented in this paper, the following main assumptions are applied:

- Both fluid and solid phase have constant physical properties.
- The fluid phase is incompressible and Newtonian.
- Diffusion is Fickian.
- The solid phase consists of spherical particles, on which external surface a chemical reaction proceeds.
- The intra-particle temperature gradients are negligible, and no heat effect on the reaction rate is considered.

Governing equations

The transport phenomena in the fluid phase are governed by the conservation equations for mass, momentum, species and thermal energy, respectively defined as:

$$\nabla \cdot \boldsymbol{u} = 0 \tag{1}$$

$$\frac{\partial \rho_f \boldsymbol{u}}{\partial t} + \nabla \cdot \left(\rho_f \boldsymbol{u}\boldsymbol{u}\right) = -\nabla p + \mu_f \nabla^2 \boldsymbol{u} + \rho_f \boldsymbol{g} \tag{2}$$

$$\frac{\partial c_f}{\partial t} + \nabla \cdot \left(c_f \boldsymbol{u}\right) = D_f \nabla^2 c_f \tag{3}$$

$$\rho_f C_{p,f}\left[\frac{\partial T_f}{\partial t} + \nabla \cdot \left(T_f \boldsymbol{u}\right)\right] = \lambda_f \nabla^2 T_f \tag{4}$$

In above equations, ρ_f is the fluid density, μ_f is the fluid viscosity, D_f is the species mass diffusivity in the fluid and $C_{p,f}$ and λ_f are the heat capacity and thermal conductivity of the fluid phase respectively.

The particle temperature is governed by the following equation assuming of a uniform particle temperature:

$$V_s C_{p,s} \frac{dT_s}{dt} = \Phi_{th,f\to s} + \Phi_{sp,f\to s}\left(-\Delta H_r\right) \tag{5}$$

In this equation, V_s is the particle volume and $C_{p,s}$ is the volumetric heat capacity of the solid phase. The first term on the right hand side is the fluid-solid heat transfer rate while the second term represents the rate of reaction heat liberated from a chemical reaction. The heat transfer rate and mass transfer rate, with the normal pointing outward of the solid, are calculated by the following two equations respectively:

$$\Phi_{th,f\to s} = -\iint_{S_s} \left(-\lambda_f \nabla T_f \cdot \boldsymbol{n}\right) dS \tag{6}$$

$$\Phi_{sp,f\to s} = -\iint_{S_s} \left(-D_f \nabla c_f \cdot \boldsymbol{n}\right) dS \tag{7}$$

Considering an exothermal chemical reaction proceeding at the exterior surface of the particles, the heat liberation is assumed to be rapidly transported to the interior of the particle with a negligible intra-particle temperature gradient. The coupling between the fluid thermal energy equation and the fluid species conservation equation is fulfilled through the particle thermal energy equation. In other words, the particle temperature of individual particle offers a dynamic boundary condition for the thermal energy equation of the fluid phase.

Numerical solution method

The governing equations are solved by a finite difference scheme implemented for a 3D staggered Cartesian grid with a uniform grid spacing in all three directions. Following our previous work [34], the numerical solution of the fluid phase governing equations is acquired by embedding second order discretization schemes as well as compact computational stencils. The momentum, species and thermal energy conservation equations are discretized temporally by applying the Adams-Bashforth scheme for the convective transport and the Euler backward scheme for the diffusive transport:

$$\rho_f \boldsymbol{u}^{n+1} = \rho_f \boldsymbol{u}^n + \Delta t \left[-\nabla p^{n+1} - \left(\frac{3}{2} \boldsymbol{C}_m^n - \frac{1}{2} \boldsymbol{C}_m^{n-1} \right) + \boldsymbol{D}_m^{n+1} + \rho_f \boldsymbol{g} \right] \tag{8}$$

$$c_f^{n+1} = c_f^n + \Delta t \left[-\left(\frac{3}{2} C_{sp}^n - \frac{1}{2} C_{sp}^{n-1} \right) + D_{sp}^{n+1} \right] \tag{9}$$

$$T_f^{n+1} = T_f^n + \frac{\Delta t}{\rho_f C_{p,f}} \left[-\left(\frac{3}{2} C_{th}^n - \frac{1}{2} C_{th}^{n-1} \right) + D_h^{n+1} \right] \tag{10}$$

In these equations, n is the time step index. The convective momentum $\boldsymbol{C}_m$, molar flux C_{sp} and heat flux C_{th} are given by:

$$\boldsymbol{C}_m = \rho_f (\nabla \cdot \boldsymbol{uu}) \tag{11}$$

$$C_{sp} = \nabla \cdot (c_f \boldsymbol{u}) \tag{12}$$

$$C_{th} = \rho_f C_{p,f} \nabla \cdot (T_f \boldsymbol{u}) \tag{13}$$

and the diffusive momentum fluxes $\boldsymbol{D}_m$, molar flux D_{sp} and heat flux D_{th} are computed as:

$$\boldsymbol{D}_m = \mu_f \nabla^2 \boldsymbol{u} \tag{14}$$

$$D_{sp} = D_f \nabla^2 c_f \tag{15}$$

$$D_{th} = \lambda_f \nabla^2 T_f \tag{16}$$

The convection and diffusion terms are spatially discretized by a second-order total variation diminishing scheme and a standard second-order central differencing scheme, respectively. The solution of Equation (8) is achieved by using a two-step projection method [12, 24] where a tentative velocity field $\bar{u}^{**}$ is first computed by using the old pressure gradient. As the second step, the velocity field at the new time step $n+1$ is obtained based on the new pressure gradient calculated from the Poisson equation.

The solid phase equation is solved after the fluid phase equations, and the trapezoidal rule is applied for the time integration:

$$T_s^{n+1} = T_s^n + \frac{\Delta t}{V_s C_{p,s}} \left[\frac{\Phi_{th,f\to s}^n + \Phi_{th,f\to s}^{n+1}}{2} + (-\Delta H_r) \frac{\Phi_{sp,f\to s}^n + \Phi_{sp,f\to s}^{n+1}}{2} \right] \tag{17}$$

Fluid-solid coupling

The fluid-solid coupling is the key element of our DNS model, which enforces the Robin boundary condition exactly at the immersed object surface and implicitly at the level of the discretized equations. The method is described in more detail in our early publication [34], and we only outline the essential elements in this paper.

After discretization, the differential equations are transformed into algebraic equations providing the relationship between the fluid variable at any position of the simulation domain and its six neighbors. Ghost points are first defined as points inside the solid phase but possessing at least one neighbor in the fluid phase. During the computation, each fluid point is checked if any of its six neighbors represents a ghost point. If this is the case, the IBM is invoked which incorporates the predefined boundary condition. A second order quadratic interpolation scheme is used in the reconstruction procedures, where a generic variable ϕ in the vicinity of the immersed object surface can be approximated in terms of a second-order polynomial:

$$\phi = \sum_{i=0}^{2} \sum_{j=0}^{2} \sum_{k=0}^{2} c_{ijk} x^i y^j z^k, i+j+k \le 2 \tag{18}$$

where x, y and z are relative coordinates with respect to the origin located at the boundary point. This equation is in fact the approximation of ϕ using the Taylor expansion near the boundary point:

$$\phi(x,y,z) = \phi_B + \frac{\partial \phi_B}{\partial x} x + \frac{\partial \phi_B}{\partial y} y + \frac{\partial \phi_B}{\partial z} z + \frac{1}{2} \frac{\partial^2 \phi_B}{\partial x^2} x^2 + \frac{1}{2} \frac{\partial^2 \phi_B}{\partial y^2} y^2 + \frac{1}{2} \frac{\partial^2 \phi_B}{\partial z^2} z^2 + \cdots \tag{19}$$

The first four coefficients:

$$c_{000} = \phi\big|_B,\ c_{100} = \frac{\partial \phi}{\partial x}\bigg|_B,\ c_{010} = \frac{\partial \phi}{\partial y}\bigg|_B \text{ and } c_{001} = \frac{\partial \phi}{\partial z}\bigg|_B \tag{20}$$

are required for the enforcement of the Robin boundary condition:

$$\alpha\phi_B + \beta\frac{\partial\phi_B}{\partial n} = f \tag{21}$$

which are obtained from nine neighboring fluid points and one image point as a function of relative coordinates and ϕ values. After that the image point value can be back evaluated by satisfying the boundary condition, from which the ghost point value can be finally computed. With that, the coefficients in the algebraic equations are updated. Altered coefficients within the original stencil are incorporated in the implicit scheme, while neighbors outside the original stencil are accounted for in an explicit way.

As indicated in the introduction, the Robin boundary condition switches to the Dirichlet boundary condition straightforwardly for the fluid phase thermal energy calculation, whereas the surface reaction rate is incorporated into the Robin boundary condition for the mass transfer calculation. Considering a first order irreversible chemical reaction proceeding on the external surface of spherical particles, at steady state the diffusion flux to the sphere surface equals the rate of consumption due to reaction:

$$D_f\frac{\partial c_{f,s}}{\partial \boldsymbol{n}} = kc_{f,s} \tag{22}$$

where D_f is the species mass diffusivity, $\boldsymbol{n}$ is the normal direction pointing outward of the sphere, and k is the surface reaction rate coefficient. This equation is in fact of a functional form of the Robin boundary condition, from which the Damköhler number quantifying the ratio of the reaction rate to the diffusion rate is obtained by nondimensionalization:

$$Da = \frac{kr_s}{D_f} \tag{23}$$

where r_s is the particle radius.

Results and discussions

In this part, four cases with increasing complexity are presented for coupled heat and mass transfer in particulate flows, with an exothermic first order irreversible chemical reaction proceeding at the particle surface. Following the comparison between DNS results and analytical (empirical) solutions for two single particle systems, our DNS model is applied to systems with multiple particles.

Unsteady heat and mass transport

Here we consider the unsteady diffusion of a certain species to a sphere, followed by the unsteady conduction of thermal energy to the surrounding fluid. The sphere is positioned in the center of a large pool of quiescent fluid. The governing equations for unsteady mass and heat diffusion in the fluid phase are described by Equation (3) and (4) respectively, with $\boldsymbol{u}$ set as zero. The boundary conditions are:

$$c_f\Big|_{wall,t} = c_{f,0} \tag{24}$$

$$T_f\Big|_{wall,t} = T_{f,0} \tag{25}$$

at the boundaries of the simulation domain, and

$$D_f \frac{\partial c_f}{\partial r}\bigg|_{r_s,t} = kc_f\Big|_{r_s,t} \tag{26}$$

$$T_f\Big|_{r_s,t} = T_s \tag{27}$$

at the sphere surface. $c_{f,0}$ and $T_{f,0}$ are the initial conditions for species and thermal energy equation respectively, and Equation (24) is valid as long as the diffusion fronts have not reached the confining walls. The particle temperature T_s in Equation (26) serves as a dynamic boundary condition for the fluid phase thermal energy equation, and governed by the particle thermal energy equation (Equation (5)).

For the DNS the particle is located in the center of a cubic box with a length of 0.04 m. Five reaction rates varying from slow reaction to fast reaction are imposed at the sphere surface, which correspond to the Damköhler number of 0.01, 0.1, 1, 10 and 100. The data used for the numerical simulation are given in Table 1.

Table 1: Data used for the simulation of unsteady heat and mass transport.

Parameter	Value
Time step [s]	5×10^{-5}
Grid size [m]	2.5×10^{-4}
Sphere diameter [m]	0.005
Fluid density [kg/m^3]	1.0
Fluid thermal conductivity [W/m/K]	0.025
Fluid heat capacity [J/kg/K]	1000
Species diffusivity [m^2/s]	2×10^{-5}
Particle volumetric heat capacity [J/m^3/K]	1000
Reaction enthalpy [J/mol]	-10^{-5}
Fluid initial concentration [mol/m^3]	1.0
Fluid initial temperature [K]	293
Particle initial temperature [K]	293

An "exact" solution for the particle temperature is obtained by solving the spherically symmetric model using a standard second order finite difference technique. In this case, the governing equations for unsteady mass and heat diffusion in the fluid phase are only r dependence, and respectively described as:

$$\frac{\partial c_f}{\partial t} = \frac{D_f}{r^2} \frac{\partial}{\partial r}\left(r^2 \frac{\partial c_f}{\partial r} \right) \tag{28}$$

$$\rho_f C_{p,f} \frac{\partial T_f}{\partial t} = \frac{\lambda_f}{r^2} \frac{\partial}{\partial r}\left(r^2 \frac{\partial T_f}{\partial r} \right) \tag{29}$$

It should be noted that a very large number of grid points in the radial direction was used to obtain this highly accurate numerical solution.

Table 2: Comparison between "exact" solutions and DNS results for the temperature difference (in K) between the particle and the bulk fluid far from the particle at 3 s.

Da	0.01	0.1	1	10	100
"exact"	0.66	6.19	36.89	72.55	80.21
DNS	0.67	6.25	37.75	73.78	81.16

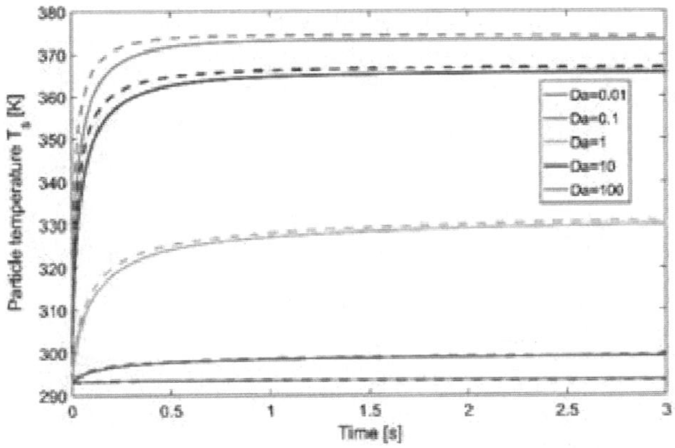

Figure 1: Comparison of particle temperature evolution profiles between the "exact" solutions and the DNS results, which are indicated by the solid lines and the dashed lines respectively.

The comparison of the temperature difference between the particle and the bulk fluid for the "exact" solutions and the DNS results at 3 s are listed in Table 2. From the table, a good agreement is observed for all reaction rates. In Figure 1, the particle temperature evolution profiles are plotted against time, to compare the DNS results and the "exact" solutions. As clearly demonstrated in Figure 1, these two solutions are in good agreement. For all reaction rates, the species flux is relatively high initially, so that the heat liberated from the exothermic reaction rapidly heats the particle up from the initial temperature. After that, a temperature difference between the particle and the surrounding fluid is established, so that the heat is transferred from the particle to the fluid through unsteady heat conduction. At the final stage, the heat removal rate is approaching the heat liberation rate, and therefore a low reaction rate results in a low particle temperature. The final stage is achieved faster with higher reaction rates, due to the dominating role of the Damköhler number. In other words, the solid temperature evolution is fully controlled by the unsteady diffusion, namely the Sherwood number, for the case of $Da = 0.01$.

Single sphere under forced convection

Building on the last case, we now consider a single stationary sphere under forced convection. The sphere is located at the center of the domain laterally while it is positioned at a distance of two times of the sphere size from the inlet in the flow direction. Fluid flows into the system with constant inlet concentration of 1 mol/m^3 and constant inlet temperature of 293 K. For this system, we consider 3 reaction rates corresponding to the Damköhler number of 0.1, 1 and 10. Besides the same data used in the last case, fluid viscosity is also required for the current numerical simulation which is specified to be 2×10^{-5} kg/m/s. The simulations are computed in a 160×160×160 grid with uniform grid spacing in all directions. At the inlet, uniform fluid velocity is imposed, which is varied to give the particle Reynolds number of 20, 40, 60, 100, 200 and 300. Free slip boundary conditions are applied at the transversal domain boundaries for velocity calculation, whereas the standard atmospheric pressure is set at the outlet. Homogeneous Neumann boundary condition is used for both concentration and thermal energy equations at all domain boundaries except the inlet.

According to Equation (5), the particle temperature at steady state is described by:

$$\Phi_{sp,f\to s}\left(-\Delta H_r\right) = -\Phi_{th,f\to s} \tag{30}$$

The particle temperature can be calculated from the mass transfer coefficient k_m and heat transfer coefficient α_h.

$$S_s k_m \left(c_{f,0} - c_{f,s}\right)\left(-\Delta H_r\right) = -S_s \alpha_h \left(T_{f,0} - T_s\right) \tag{31}$$

where S_s is the particle surface area. Equation (31) can be rearranged to obtain the following expression:

$$T_s = T_{f,0} + \frac{2Da}{\left(2Da + Sh_s\right)} \frac{Sh_s}{Nu_s} \frac{c_{f,0}\left(-\Delta H_r\right)}{\rho_f C_{p,f} Le} \tag{32}$$

with the external mass and heat transport coefficient predicted by the well-known empirical Frössling and Ranz-Marshall correlations:

$$Sh_s = \frac{k_m d_s}{D_f} = 2.0 + 0.6\left(Re_s\right)^{\frac{1}{2}}\left(Sc\right)^{\frac{1}{3}} \tag{33}$$

$$Nu_s = \frac{\alpha_h d_s}{\lambda_f} = 2.0 + 0.6\left(Re_s\right)^{\frac{1}{2}}\left(Pr\right)^{\frac{1}{3}} \tag{34}$$

where Re_s is the particle Reynolds number, Sc is the Schmidt number and Pr is the Prandtl number, respectively defined as:

$$Re_s = \frac{\rho_f u_0 d_s}{\mu_f} \tag{35}$$

$$Sc = \frac{\mu_f}{\rho_f D_f} \tag{36}$$

$$Pr = \frac{\mu_f C_{p,f}}{\lambda_f} \tag{37}$$

In Equation (32), Le is the Lewis number defined as the ratio of the thermal diffusivity to the mass diffusivity:

$$Le = \frac{\frac{\lambda_f}{\rho_f C_{p,f}}}{D_f} = \frac{Sc}{Pr} \tag{38}$$

In the current work, $Sc = 1$ (and $Pr = 0.8$) is used, which means that the momentum and mass (temperature) boundary layers have a similar thickness.

Table 3: Comparison of the particle temperature at steady state between DNS results and empirical values (indicated as EMP).

	$Da = 0.1$		$Da = 1$		$Da = 10$	
	DNS	EMP	DNS	EMP	DNS	EMP
$Re_s = 20$	296.7	296.4	319.7	318.0	363.6	360.6
$Re_s = 40$	296.0	295.8	315.7	314.5	360.6	358.1
$Re_s = 60$	295.6	295.5	313.4	312.5	358.3	356.2
$Re_s = 100$	295.1	295.1	310.5	309.9	354.7	353.4
$Re_s = 200$	294.6	294.6	306.9	306.6	348.8	348.7
$Re_s = 300$	294.4	294.4	305.0	304.8	344.9	345.6

In Table 3, the comparisons between the simulation results obtained from our DNS model and the empirical values calculated from Equation (32) are shown. As observed from the table, the particle temperature increases with higher reaction rates and decreases with larger Reynolds numbers. The former behavior is understood as the result of more heat liberated from the chemical reaction, whereas the latter behavior is well explained by the stronger convective heat transfer from the particle to the fluid. All results are in good agreement. It is also verified that in the final steady state, the calculated ratio of the heat and mass transfer rates equals the reaction enthalpy (-100kJ/mol).

Three-bead reactor

In this section, we consider an in-line array of three spheres, the so-called three-bead reactor. The spheres are positioned in a cuboidal domain with the length of 0.10 m in the flow direction and 0.01 m in the lateral direction. At the domain boundaries, velocity obeys the free slip condition and the system is isolated and adiabatic for species and the thermal energy calculation. At the outlet, the standard atmospheric pressure is set, and zero gradient boundary condition is set for both concentration and temperature. The first sphere is located at a distance of two times of the sphere size from the inlet in the flow direction, and the other two spheres are located in such a way that the mutual distance between all sphere centers is one and half times of the sphere diameter. In this simulation, the reaction rate at the sphere surface is maintained at $Da = 1$ indicating the equivalent time scale for reaction and diffusion, whereas 3 particle based Reynolds numbers $Re_s = 60$, 100 and 200 are applied to the system. All other simulation parameters are the same as those in the previous case.

The particle temperature evolution profiles are presented in Figure 2. There are two contributors to the rise of the particle temperature: liberated reaction heat and convective heat transfer. From the figure, it can be inferred that a thermal energy wave is moving through the in-line array of three spheres. For all Reynolds numbers, the particle temperature increases from the 1st sphere to the 3rd sphere. All spheres rapidly heat up due to the exothermic chemical reaction proceeding at the surface. Due to a temperature difference between the particle and the surrounding fluid, the thermal energy is transferred from the solid phase to the fluid phase and further transported downstream by the fluid flow. For the 1st sphere, the particle temperature soon reaches a constant value as the removed heat equals the generated reaction heat. For the 2nd and 3rd spheres, the unconverted reactant is partly converted on their surface and the thermal energy in the fluid phase gives additional heating to these spheres. Due to this reason, the last sphere takes the longest time to reach the steady state. From the figure, it is also clear that with higher Reynolds numbers the final solid temperature is lower and achieved faster for all particles.

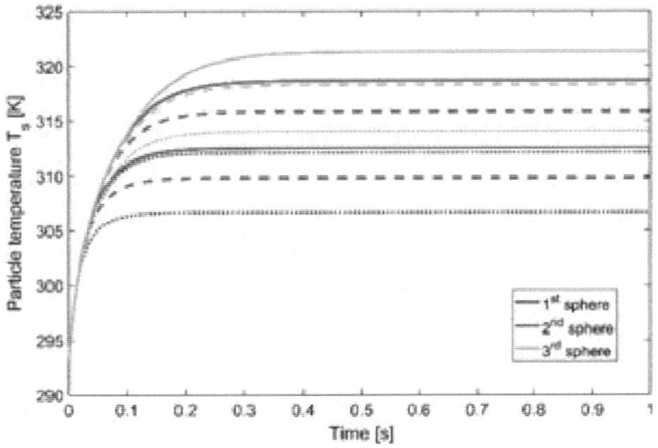

Figure 2: Evolution profiles of the particle temperature for the three-bead reactor. Particle Reynolds number of 60, 100 and 200 are presented by the solid lines, dashed lines and dotted lines respectively.

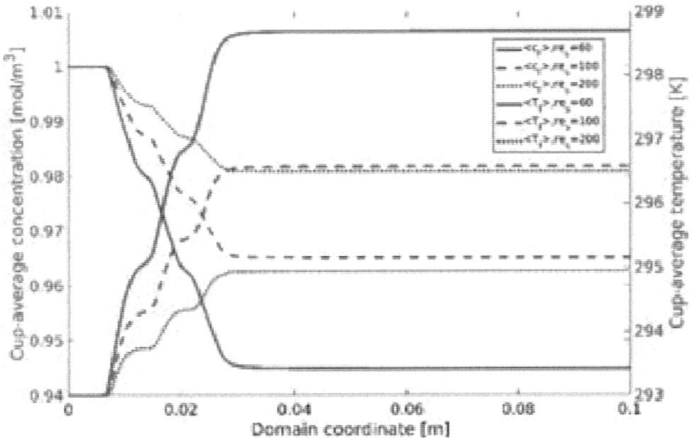

Figure 3: Fluid phase cup-average concentration and temperature profiles along the flow direction. Three simulation cases with the particle Reynolds number 60, 100 and 200 are presented by the solid lines, dashed lines and dotted lines respectively, whereas the blue and red colour indicate the concentration and temperature respectively.

The axial profiles of cup-average concentration and temperature in the fluid phase are shown in Figure 3, which demonstrates the relative contribution of the individual spheres to the overall reactant conversion and temperature rise. The cup-average value ϕ (namely concentration or temperature in the current case) is defined by:

$$\langle \phi_f \rangle = \frac{\iint_{S_f} u(x,y,z)\phi_f(x,y,z)\,dydz}{\iint_{S_f} u(x,y,z)\,dydz} \tag{39}$$

In this equation, the integration is performed over a cross section S_f perpendicular to the flow direction x, and $u(x,y,z)$ is the x component of the fluid velocity at this certain point (x,y,z). It should be noted that,

S_f only accounts for the part occupied by the fluid phase while the points inside the solid phase in the current cross section are not taken into consideration in the calculation. From the figure, as expected, it can be seen that three spheres have the same contribution to the overall species conversion and temperature rise. This is due to the low reaction rate specified at the sphere surface and hence all three spheres possess almost equivalent reaction consumption. Higher Reynolds number will decrease the species conversion at the sphere surface, and consequently lead to lower fluid temperature in the system. This corresponds well with the lower particle temperature in Figure 2.

From the simulation, the overall conversion of the reactant in the three-bead reactor is obtained, which is 0.055, 0.035 and 0.019 for the particle Reynolds number of 60, 100 and 200 respectively. The theoretical fluid outlet temperature can be computed from the adiabatic temperature rise:

$$T_{f,out} = T_{f,in} + \frac{(-\Delta H_r)\left(\langle c_{f,in}\rangle - \langle c_{f,out}\rangle\right)}{\rho_f C_{p,f}} \tag{40}$$

The fluid outlet temperatures given by the simulations are 298.69 K, 296.59 K and 294.95 K for increasing Reynolds numbers, which are in good agreement with the theoretically calculated values: 298.51 K, 296.48 K and 294.91 K.

Dense particle array

For the last case, our proposed DNS model is applied to a dense stationary array consisting of a relatively large number of particles with the same size. The particle array is created by the hard-sphere Monte-Carlo method and distributed in a random configuration over the computational domain with a predefined solid phase packing density of 0.4. The simulations are computed on a 3D domain with the packing height of 0.15 m in the flow direction and a length of 0.025 m in the lateral direction. In total, 573 spheres are positioned inside the simulation domain with a periodic configuration in the cross-sectional directions. For the simulation three prescribed uniform fluid velocities are imposed at the inlet, leading to the particle Reynolds number of 60, 100 and 200 respectively. A reaction rate equaling the diffusion rate, namely $Da = 1$, is specified at the particle surface. Other parameters used in the simulation are the same as those in the second case. The two parameters Schmidt number and Prandtl number, describing the relative thickness of the momentum boundary layer to the mass and thermal boundary layer, are 1.0 and 0.8 respectively. The periodic boundary condition is applied at the lateral domain boundaries for all velocity, concentration and temperature calculations. The pressure at the outlet is set as the standard atmospheric pressure, and a zero slope boundary condition is set there for species and thermal energy equations.

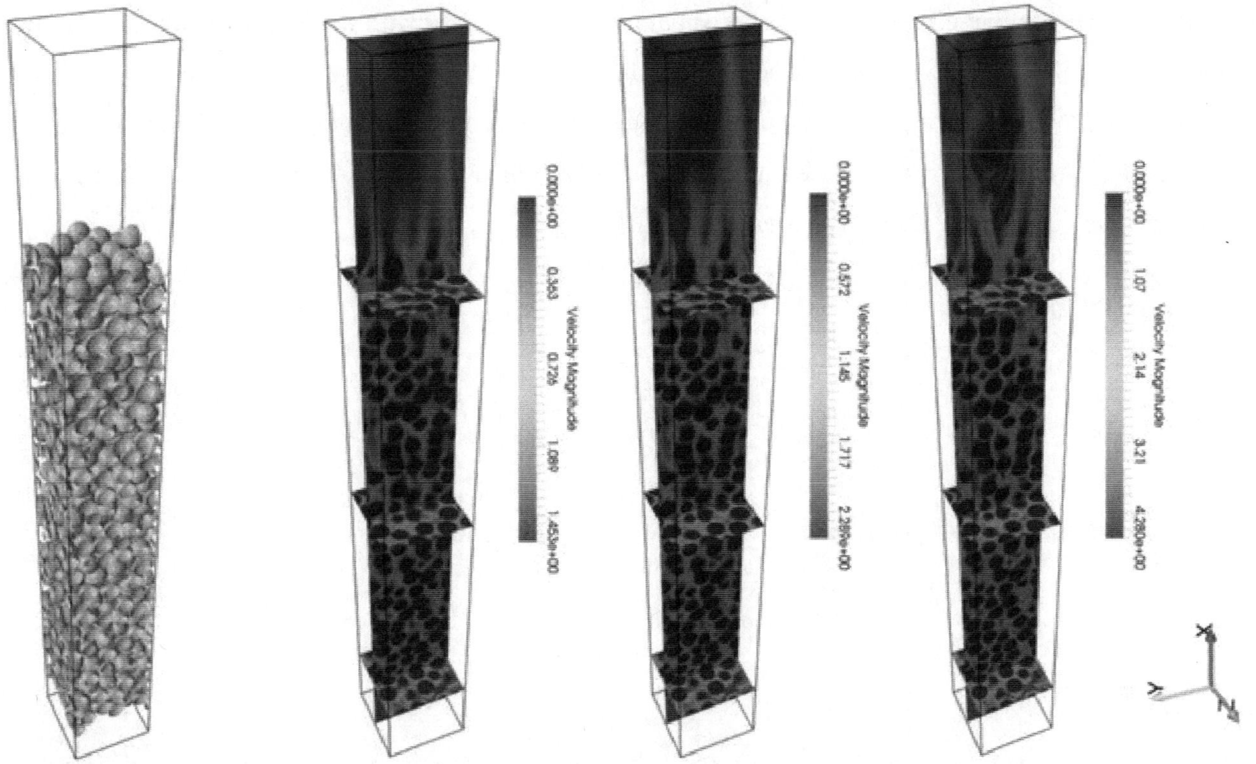

Figure 4: The particle configuration for the dense array simulation (left), and the computed velocity field (right) for the central plane and three lateral cross sections with the particle Reynolds number of 60, 100 and 200 (from left to right).

The computational domain and particle arrangement are shown in Figure 4. It should be noted that, two empty regions are reserved for the inlet and outlet, with a length of 0.015 m and 0.05 m respectively. The incorporation of these two regions is essential to avoid problems for inflow development and outflow recirculation. For our simulations the mesh resolution N, defined as the ratio of the particle diameter to the grid size, is 20. This value is selected by performing a mesh convergence test in a small sub-array, following the methodology published in our previous paper [34]. The deviation of the total heat transfer rate in the current work using $N = 20$ is 4.14%, which is slightly higher than the one in our earlier work [34] and can be explained by the increased solid phase packing density. The computed velocity field for all three Reynolds numbers are plotted in the longitudinal and transversal cross sections. The longitudinal cross section is the central plane of the particle array, whereas the three transversal cross sections are located at the packing height of 0.005 m, 0.075 m and 0.145 m respectively. In velocity maps, the periodic flow field in the lateral directions and the preferred flow pathways in the array are clearly observed. Note that with our DNS model, the velocity field is obtained for the entire computational domain, i.e. also inside the particles, however it is zero there due to the particular enforcement of the no-slip boundary condition at the particle surface. The complexity of the flow structure is depicted by the transversal cross sections along the packing that the local increase of the fluid velocity significantly varies with the local porosity (caused by the variation of the local particle structure). With increased Reynolds numbers, besides a more heterogeneous flow field inside the array, a more unsteady wake is observed at the rear of the particle array which develops from straight streamlines into vortex. The periodically distributed particles, which are 'cut' by the domain boundaries, are clearly visualized in Figure 4.

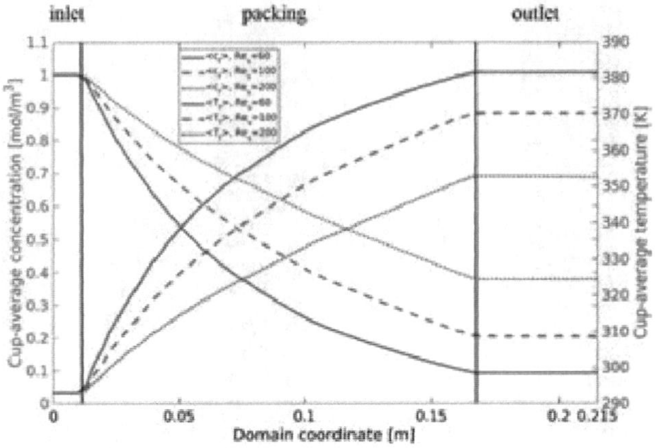

Figure 5: Cup-average concentration and temperature profiles obtained from the dense particle array simulations. The case with the particle Reynolds number 60, 100 and 200 are respectively indicated by the solid, dashed and dotted lines, whereas the concentration and temperature profiles are indicated by the blue and red colour respectively.

The evolution of the cup-average concentration and temperature along the particle array is of high interest for industrial applications, as it may indicate an approximate reactor length for the design purpose or an appropriate feedstock amount for the production process. The calculations of the cup-average concentration and temperature are given in the preceding section (Equation (39)). In Figure 5, the cup-average concentration and temperature profiles of the fluid phase are shown as a function of the domain coordinate in the flow direction. The empty inlet and outlet regions are clearly visible as here the profiles are constant. Within the packing range, the species concentration decreases due to the chemical reaction proceeding at the external surface of the particles and the fluid temperature increases due to the particles that are heated up by the heat liberated from the exothermic reaction. As expected, higher Reynolds number will decrease the decay rate of the concentration and consequently lead to lower species conversion, which will accordingly reduce the temperature rise of the system. Similarly to the three-bead reactor, the computed fluid temperature at the outlet, 381.4 K, 370.2 K and 352.8 K for the particle Reynolds number of 60, 100 and 200 respectively, can be compared with the value predicted by Equation (40). For increasing Reynolds numbers, the overall conversion obtained from the simulations are 0.908, 0.796 and 0.620, and the sequential calculation gives the values of 383.8 K, 372.6 K and 355.0 K respectively. The concentration and temperature distribution are shown in Figure 6 for the central plane of the particle array. It should be noted that, not only the information of the fluid phase but also the information of the solid phase is visualized in this figure. The concentration inside the particles is zero due to the assumption of the reactive external surface, whereas the temperature inside individual particles is its real solid temperature computed by our DNS model. This is a very important point, as it may help us to find out where local high temperatures are located. Although the cup-averaged values in Figure 5 show that the fluid temperature increases along the flow direction, from Figure 6 it can be clearly observed that both fluid and particle temperature vary in the transversal cross section. High solid temperature occurs at regions with local dense particle configuration, especially for particles suffering from blockage effect. The increased solid temperature consequently heats up the surrounding fluid and results in the nonuniform temperature distribution in the fluid phase. This heterogeneity is intensified at higher Reynolds numbers. In Figure 6, it is clearly observed that the system's thermal energy continuously accumulates along the flow direction. In other words, the heat produced by the exothermic chemical reaction is stored in the particles and propagates to downstream particles by the convective heat transfer of the fluid.

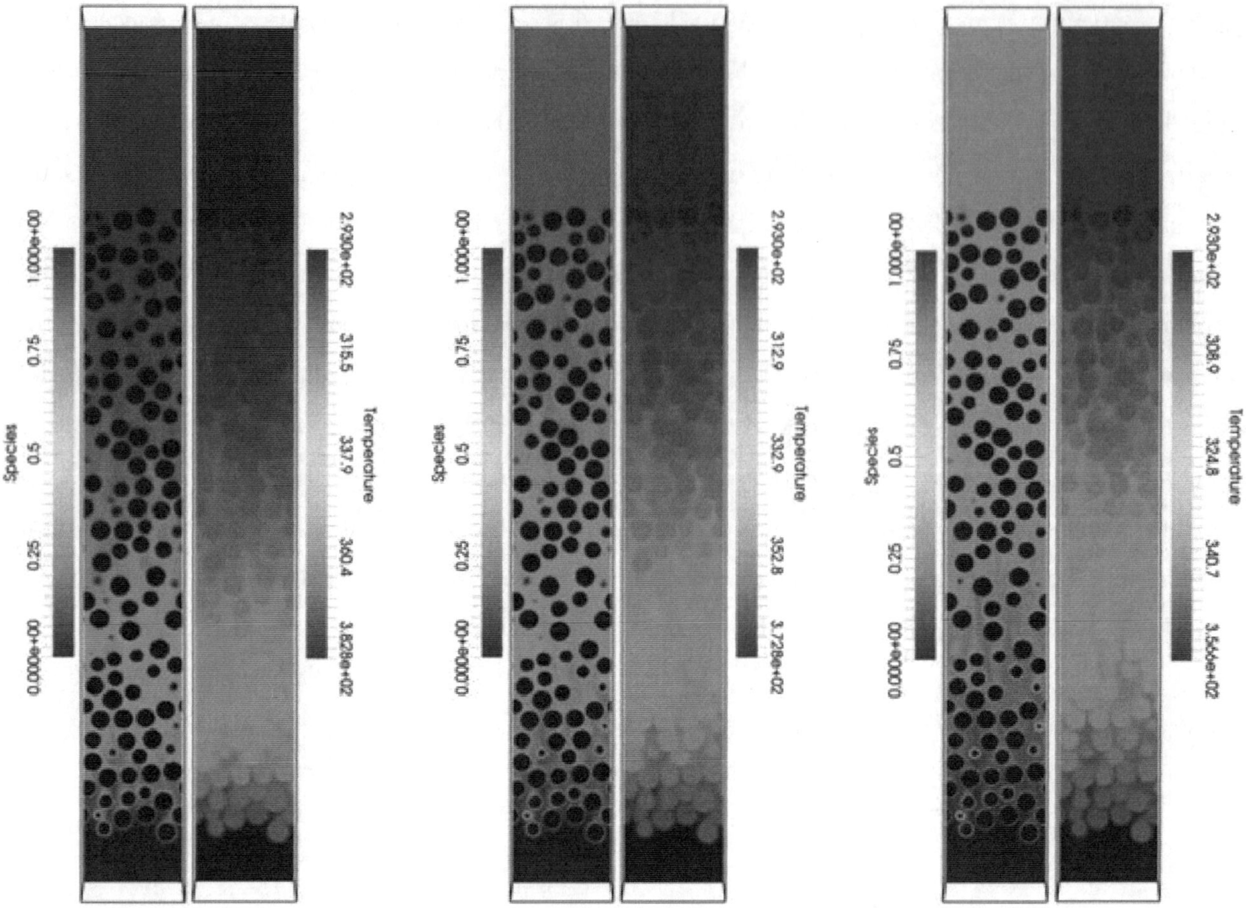

Figure 6: Concentration and temperature distribution for the central plane of the particle array, with the particle Reynolds number of 60, 100 and 200 (from left to right). Note that the particle temperature is also indicated in the figure.

In more empirical approaches the transport behavior of packed beds is commonly described by the classic one dimensional heterogeneous model. By incorporating the chemical reaction rate, the following equations are obtained for the fluid phase at final steady state:

$$u_{in}\frac{\partial c_f}{\partial x} = D_{ax}\varepsilon\frac{\partial^2 c_f}{\partial x^2} - 2a_s\frac{D_f}{d_s}\frac{DaSh_s}{2Da+Sh_s}c_f \tag{41}$$

$$u_{in}\rho_f C_{p,f}\frac{\partial T_f}{\partial x} = \lambda_{ax}\varepsilon\frac{\partial^2 T_f}{\partial x^2} + 2a_s\frac{D_f}{d_s}\frac{DaSh_s}{2Da+Sh_s}c_f(-\Delta H_r) \tag{42}$$

where ε is the void fraction of the particle array and a_s is the specific fluid-particle contact surface area given by:

$$a_s = \frac{6}{d_s}(1-\varepsilon) \tag{43}$$

D_{ax} and λ_{ax} are the axial dispersion coefficient and the axial conductivity respectively. The Damköhler number is a predefined value in our simulations, whereas the Sherwood number is computed from the empirical Gunn correlation [6]:

$$Sh_s = \left(7 - 10\varepsilon + 5\varepsilon^2\right)\left(1 + 0.7 Re_s^{\frac{1}{5}} Sc^{\frac{1}{3}}\right) + \left(1.33 - 2.40\varepsilon + 1.20\varepsilon^2\right) Re_s^{\frac{7}{10}} Sc^{\frac{1}{3}} \quad (44)$$

In this equation, Re_s is the particle Reynolds number calculated with the inlet fluid superficial velocity u_{in}. For more details of this one dimensional heterogeneous model, we refer interested readers to the Appendix. In Figure 7, the comparisons of the cup average concentration and temperature profiles between the DNS results and the 1D model are shown. As presented by the figure, good agreements are reached for the overall features of the profiles. The important discrepancy is that, the DNS model gives a faster and higher species consumption and consequently a similar trend for thermal energy than the ones predicted by the 1D model. This discrepancy further increases at higher Reynolds numbers, and the reason is thought to be the inhomogeneous flow pattern inside the particle array. In the figure, we also plot the 1D profiles without the axial dispersion effect (for both species and thermal energy transport). As expected, the axial dispersion plays a more pronounced role at low Reynolds numbers and it is negligible for the range of the Reynolds number considered in the current work due to the dominating role of the convective transport.

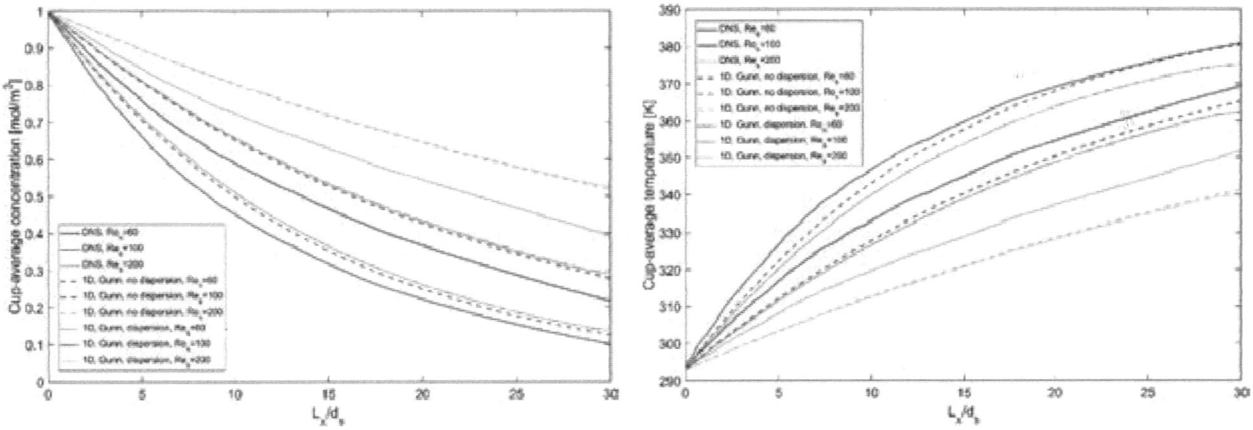

Figure 7: Comparisons of the fluid cup average concentration (left) and temperature (right) profiles among the DNS results, the 1D model using the Sherwood number obtained from the Gunn correlation and the 1D model without the axial dispersion effect using the same Sherwood number.

The average Sherwood number obtained from the slice-based computation is as well used in the 1D model. It is calculated by the following expression in planes perpendicular to the flow direction:

$$Sh_{slice} = \frac{\overline{\nabla c_f \cdot \boldsymbol{n}}}{\langle c_f \rangle - \overline{c_{f,s}}} d_s \tag{45}$$

where the numerator and denominator account for the average concentration gradient and the average driving force respectively, for all parts of the particle surface within the current plane. As shown in Figure 8, the slice-based Sherwood number describes the local variation of the mass transfer performance in the streamwise direction, which oscillates in the full packing region due to the varying fluid-particle interface area in each transversal plane. From the figure it can be concluded that the Sherwood number is statistically homogeneous because it is independent of the streamwise coordinate except for a very small slab at the end of the array. The Sherwood numbers within this slab significantly deviate from the homogeneous value, and hence is excluded in the average value calculation. In Table 4, the average Sherwood number is compared with the value obtained from the Gunn correlation. As predicted by the correlation, the Sherwood number increases with higher Reynolds numbers, however accompanied by larger deviations. Similar behavior has been reported by other studies [27, 29, 35], and the values obtained from their refitted correlations are also listed in the table. The utilization of the average value does not help to improve the prediction of the 1D model, as depicted by Figure 9. Due to the lower values applied in the 1D models, the concentration profiles are somewhat higher than the ones computed using the Gunn correlation and the temperatures lower.

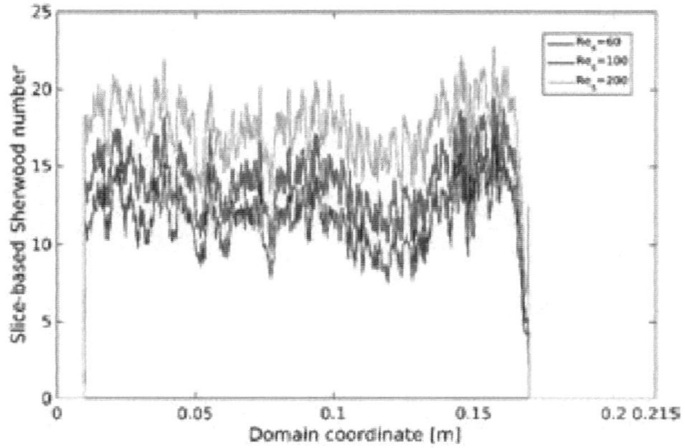

Figure 8: Slice-based Sherwood number along the flow direction for all three Reynolds numbers.

Table 4: Comparison of the average Sherwood number between the Gunn correlation and the DNS results.

Re_s	Gunn	Tavassoli [27]	Sun [29]	Slice-based	Particle-based
60	12.90	10.99	10.30	11.65	14.47
100	15.81	14.30	13.03	14.13	15.73
200	21.59	-	-	17.69	18.60

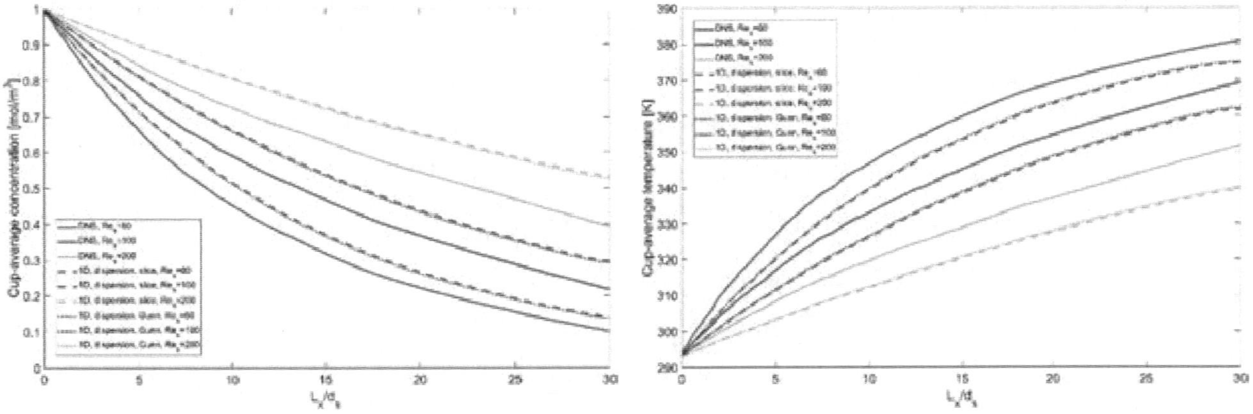

Figure 9: Comparisons of the fluid cup average concentration (left) and temperature (right) profiles among the DNS results, the 1D model using the Sherwood number obtained from the slice-based computation and the 1D model using the Sherwood number obtained from the Gunn correlation.

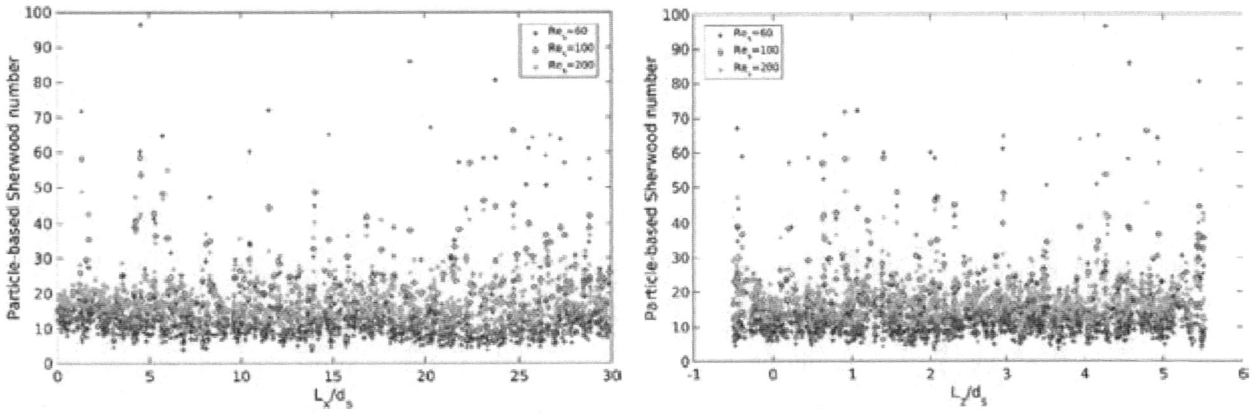

Figure 10: Distribution of the Sherwood number of individual particles in the flow (left) and lateral (right) directions.

The previous results indicate that the variable flow field inside the array, which results from the absence of a homogeneous packing structure, has considerable influence on the local transport performance of individual particles. Monitoring the variation from particle to particle may help us to gain deeper understanding of the heterogeneity. For this purpose, the Sherwood number of individual particle is computed, which is defined as:

$$Sh_{particle,i} = \frac{\left(\overline{\nabla c_f \cdot \boldsymbol{n}}\right)_i}{\left\langle c_{f,i} \right\rangle - \overline{c_{f,s,i}}} d_{s,i} \tag{46}$$

The numerator is the average concentration gradient, whereas the denominator is the local driving force for individual particle which is calculated as the difference between the cup-average concentration at the center position of the particle and the average surface concentration of the particle. There are two advantages by applying this definition. On one hand, it is closely related to the slice-based calculation, as the average over particles around a certain position should be close to the Sh_{slice} there. On the other hand, rather than the overall value obtained from the slice-based calculation, this particle-based quantity is commonly used in the coarser scale models to characterize the interfacial transfer. In Figure 10, the values of the computed Sherwood number of individual particles are plotted as a function of the streamwise (x) and spanwise (z is shown here) directions. As wall effects are eliminated by applying the periodic boundary condition at the spanwise boundaries, no entrance region is observed for $Sh_{particle,i}$ in the flow direction (left panel) and no predominant high values are observed at the lateral boundaries (right panel). The average value over all particles are listed in Table 4 as well, and the standard deviation decreases with increasing Reynolds number which is 12.14, 9.21 and 8.13 respectively. From the figure, it seems that the Sherwood number of individual particles have a quite uniform and homogeneous distribution in all directions. However, some details are worth being discussed. First, decrease of the Sherwood number at high Reynolds number. Although most particles behave in the same way as the overall Sherwood number, namely that the value increases with higher Reynolds numbers, some particles have a completely opposite behavior, namely that the maximal value occurs at $Re_s = 60$. This indicates the preferred flow path inside the array that the fluid channels somewhere and leads to stagnation in other places, especially at high Reynolds number. Second, variation of the Sherwood number in the same plane perpendicular to the flow direction. The local heterogeneity can be more pronouncedly revealed by looking at the particles at the same position along the flow direction, which can even differ up to 100%. Since $Sh_{particle,i}$ is defined as the ratio of the concentration gradient and the concentration driving force, it can be concluded that at least one of them is not constant at the same cross section perpendicular to the flow. This is in contrast to the assumptions employed in the 1D model that both gradient and driving force mainly change in the flow direction, and this may explain the deviation in our previous results. In Figure 11, the concentration gradient and the concentration driving force are plotted at the particle Reynolds number of 200 in the streamwise direction. As expected, both gradient and driving force decrease along the particle array. In the figure, it is clearly demonstrated that the particles in the same transversal plane experience some differences for the concentration gradient, however differences for the concentration driving force are more significant. To confirm our finding, the gradient and the driving force for thermal energy transfer is also plotted in Figure 11, which show the same behavior. This well explains the variations in the previous results, and may indicate the improper usage of the cup-average concentration for the driving force calculation. In other words, particles in the same spanwise plane may not feel the same driving force, instead, they are more likely affected by a local driving force which is based on the local microstructure around each particle. However, this may trigger further questions on how to derive an overall value from individual values, which can be then applied in phenomenological models. This raises a challenge for future researches to analyze really local and detailed information, which is realizable by using DNS.

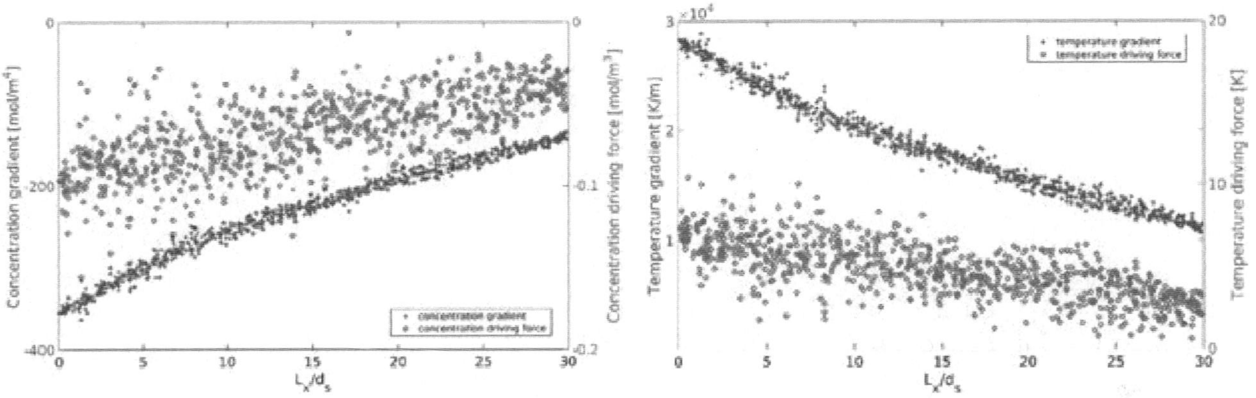

Figure 11: Distribution of the mass (left) and thermal energy (right) gradient and the driving force of individual particles along the flow direction, at the particle Reynolds number of 200.

Conclusion and outlook

In this paper a DNS model based on a ghost-cell based immersed boundary method is extended for simulations of reactive particulate flows, namely the coupled heat and mass transfer processes. An exothermic first order irreversible chemical reaction proceeds the exterior surface of the particle, whose reaction rate is incorporated into the Robin boundary condition which is further incorporated into the fluid species equation at the discrete level implicitly. The liberated reaction heat heats up the particle, whose temperature offers a dynamic boundary condition for the fluid thermal energy equation.

Using our extended DNS model, four fluid-solid systems are studied. The particle temperature obtained from DNS agrees with the value obtained from the well-established solutions for a single sphere under both unsteady and convective situation. In the three-bead reactor, the particle temperature increases from the 1st sphere to the 3rd sphere and the species conversion decreases at higher Reynolds numbers. The adiabatic temperature rise obtained from DNS is in a good agreement with the theoretical value. For the dense particle array, the fluid concentration and temperature profiles are obtained and compared with the 1D heterogeneous model. The qualitative agreement is good, however the quantitative discrepancy is thought to be caused by the heterogeneity inside the array which is further amplified at higher Reynolds numbers. This point is visualized in both velocity and temperature distributions. Further analysis of the Sherwood number of individual particles reveals that the particles in the same transversal plane experience large variations which mainly come from the widely scattered driving force.

In this paper, our DNS model reveals its strong power in modeling of reactive particulate flows for engineering applications. Future work can be expanded in two aspects. First, the current DNS model needs to be extended towards more realistic systems, which considers the temperature dependent reaction rate and incorporates the realistic reaction kinetics. Second, the local transport behavior of individual particles needs to be studied, from which a better estimate of the transfer performance is expected by quantitatively understanding the effect of local parameters.

Nomenclature

a_s	Specific fluid-particle contact surface area, [1/m].
c_f	Species molar concentration, [mol/m^3].
$c_{f,0}$	Initial species molar concentration, [mol/m^3].
$c_{f,in}$	Inlet species molar concentration, [mol/m^3].
$c_{f,s}$	Molar concentration on particle surface, [mol/m^3].
$\langle c_f \rangle$	Cup-average species molar concentration, [mol/m^3].
c_{ijk}	Coefficients in second-order polynomial.
$C_{p,f}$	Fluid heat capacity, [J/kg/K].
$C_{p,s}$	Solid volumetric heat capacity, [J/m^3/K].
C_h	Convective heat transport per unit of volume, [J/m^3/s].
C_{sp}	Convective species transport per unit of volume, [mol/m^3/s].
d_s	Sphere diameter, [m].
D_{ax}	Axial dispersion coefficient, [m^2/s].
D_f	Mass diffusivity, [m^2/s].
D_h	Diffusive heat transport per unit of volume, [J/m^3/s].
D_{sp}	Diffusive species transport per unit of volume, [mol/m^3/s].
f	Coefficient of Robin boundary condition.
k	Surface reaction rate coefficient, [m/s].
k_m	External mass transfer coefficient, [m/s].
n	Time step index.
p	Pressure, [Pa].
r	Spherical coordinate, [m].

r_s	Sphere radius, [m].
S_s	Particle surface area, [m^2].
t	Time, [s].
T_f	Fluid temperature, [K].
$T_{f,0}$	Initial fluid temperature, [K].
$T_{f,in}$	Inlet fluid temperature, [K].
$T_{f,out}$	Outlet fluid temperature, [K].
$\langle T_f \rangle$	Cup-average fluid temperature, [K].
T_s	Solid temperature, [K].
u_{in}	Fluid superficial velocity at inlet, [m/s].
V_s	Particle volume, [m^3].
x, y, z	Relative coordinate directions, [m].

Greek letters

α, β	Coefficients of Robin boundary condition.
α_h	Heat transfer coefficient, [W/m^2/K].
ε	Void fraction, [1].
λ_f	Fluid thermal conductivity, [W/m/K].
λ_{ax}	Axial conductivity, [W/m/K].
μ_f	Fluid dynamic viscosity, [kg/m/s].
ρ_f	Fluid density, [kg/m^3].
ϕ	General fluid variable.
ΔH_r	Reaction enthalpy, [J/mol].
Δt	Time step, [s].
ΔT_a	Adiabatic temperature rise, [K].
$\Phi_{th,f\rightarrow s}$	Heat transfer rate from fluid to solid, [J/s].
$\Phi_{sp,f\rightarrow s}$	Molar transfer rate from fluid to solid, [mol/s].

Vectors

$\boldsymbol{C}_m$	Convective momentum flux, [N/m^3].
$\boldsymbol{D}_m$	Diffusive momentum flux, [N/m^3].
$\boldsymbol{g}$	Gravitational acceleration, [m/s^2].
$\boldsymbol{n}$	Unit normal vector, [1].

$\boldsymbol{u}$ Velocity, [m/s].

Subscripts and superscripts

B Boundary point.

f Fluid phase.

s Solid phase.

Operators

$\frac{\partial}{\partial t}$ Partial time derivative, [1/s].

∇ Gradient operator, [1/m].

$\nabla \cdot$ Divergence operator, [1/m].

∇^2 Laplace operator, [1/m^2].

Acknowledgements

This work was supported by the Netherlands Center for Multiscale Catalytic Energy Conversion (MCEC), an NWO Gravitation programme funded by the Ministry of Education, Culture and Science of the government of the Netherlands.

Appendix: One dimensional heterogeneous model

In this appendix, we present the one dimensional heterogeneous model applied for the comparison in the dense particle array case. The governing equations are the species and thermal energy conservation equations for the fluid phase:

$$(1-\eta)\frac{\partial c_f}{\partial t}+u_{in}\frac{\partial c_f}{\partial x}=D_{ax}(1-\eta)\frac{\partial^2 c_f}{\partial x^2}-k_m a_s\left(c_f-c_{f,s}\right) \tag{A1}$$

$$(1-\eta)\rho_f C_{p,f}\frac{\partial T_f}{\partial t}+u_{in}\rho_f C_{p,f}\frac{\partial T_f}{\partial x}=\lambda_{ax}(1-\eta)\frac{\partial^2 T_f}{\partial x^2}-\alpha_h a_s(T_f-T_s) \tag{A2}$$

and for the solid phase:

$$\eta\frac{\partial c_{f,s}}{\partial t}=-k_m a_s\left(c_{f,s}-c_f\right)-a_s k c_{f,s} \tag{A3}$$

$$\eta C_{p,s}\frac{\partial T_s}{\partial t}=-\alpha_h a_s(T_s-T_f)+a_s k c_{f,s}(-\Delta H_r) \tag{A4}$$

In above equations, η is the solid phase packing density, u_{in} is the fluid superficial velocity at inlet, ρ_f is the fluid density, $C_{p,f}$ is the fluid heat capacity, $C_{p,s}$ is the solid volumetric heat capacity, ΔH_r is the reaction enthalpy, a_s is the specific fluid-particle contact surface area

$$a_s = \frac{6\eta}{d_s} \tag{A5}$$

k_m is the external mass transfer coefficient

$$k_m = \frac{Sh_s D_f}{d_s} \tag{A6}$$

α_h is the heat transfer coefficient

$$\alpha_h = \frac{Nu_s \lambda_f}{d_s} \tag{A7}$$

k is the surface reaction rate coefficient

$$k = \frac{DaD_f}{r_s} \tag{A8}$$

D_{ax} is the species axial dispersion coefficient

$$D_{ax} = D_f \left(\frac{1}{\sqrt{2}} + \frac{Re_s Sc}{2(1-\eta)} \right) \tag{A9}$$

and λ_{ax} is the thermal axial conductivity [36] (for details of ω, A and Γ we refer to the original work, as they are beyond the scope of this paper)

$$\lambda_{ax} = \lambda_{ax,stat} + \lambda_{ax,dyn} \tag{A10}$$

$$\frac{\lambda_{ax,stat}}{\lambda_f} = \left(1-\sqrt{\eta}\right) + \sqrt{\eta}\left[\omega A + (1-\omega)\Gamma\right] \tag{A11}$$

$$\frac{\lambda_{ax,dyn}}{\lambda_f} = \frac{Re_s Pr}{2(1-\eta)} \tag{A12}$$

In first order irreversible surface reaction, the external mass flux to the sphere surface is equal to the consumption rate on the surface:

$$k_m \left(c_f - c_{f,s}\right) = kc_{f,s} \tag{A13}$$

With above procedures, Equation (A1) to (A4) are simplified and sequentially computed to produce the concentration and temperature profiles in both fluid and solid phases.

Reference

[1] N. Frössling, Uber die Verdunstung Fallender Tropfen, Gerlands Beitr. Geophys., 52 (1938) 107-216.

[2] W.E. Ranz, W.R. Marshall, Evaporation from Drops, Chem. Eng. Prog., 48 (1952) 141-146 and 173-180.

[3] D. Thoenes, H. Kramers, Mass transfer from spheres in various regular packings to a flowing fluid, Chem. Eng. Sci., 8 (1958) 271-283.

[4] A.S. Gupta, G. Thodos, Direct analogy between mass and heat transfer to beds of spheres, AlChE J., 9 (1963) 751-754.

[5] K. Daizo, S. Motoyuki, Particle-to-fluid heat and mass transfer in packed beds of fine particles, Int. J. Heat Mass Transfer, 10 (1967) 845-852.

[6] D.J. Gunn, Transfer of heat or mass to particles in fixed and fluidised beds, Int. J. Heat Mass Transfer, 21 (1978) 467-476.

[7] M.A. van der Hoef, M. van Sint Annaland, N.G. Deen, J.A.M. Kuipers, Numerical Simulation of Dense Gas-Solid Fluidized Beds: A Multiscale Modeling Strategy, Annu. Rev. Fluid Mech., 40 (2008) 47-70.

[8] E.A. Fadlun, R. Verzicco, P. Orlandi, J. Mohd-Yusof, Combined Immersed-Boundary Finite-Difference Methods for Three-Dimensional Complex Flow Simulations, J. Comput. Phys., 161 (2000) 35-60.

[9] H.S. Udaykumar, R. Mittal, P. Rampunggoon, A. Khanna, A Sharp Interface Cartesian Grid Method for Simulating Flows with Complex Moving Boundaries, J. Comput. Phys., 174 (2001) 345-380.

[10] Y.-H. Tseng, J.H. Ferziger, A ghost-cell immersed boundary method for flow in complex geometry, J. Comput. Phys., 192 (2003) 593-623.

[11] J. Kim, D. Kim, H. Choi, An Immersed-Boundary Finite-Volume Method for Simulations of Flow in Complex Geometries, J. Comput. Phys., 171 (2001) 132-150.

[12] S. Das, N.G. Deen, J.A.M. Kuipers, Immersed boundary method (IBM) based direct numerical simulation of open-cell solid foams: Hydrodynamics, AlChE J., 63 (2017) 1152-1173.

[13] A. Gilmanov, S. Acharya, A computational strategy for simulating heat transfer and flow past deformable objects, Int. J. Heat Mass Transfer, 51 (2008) 4415-4426.

[14] C.S. Peskin, Numerical analysis of blood flow in the heart, J. Comput. Phys., 25 (1977) 220-252.

[15] D. Goldstein, R. Handler, L. Sirovich, Modeling a No-Slip Flow Boundary with an External Force Field, J. Comput. Phys., 105 (1993) 354-366.

[16] E.M. Saiki, S. Biringen, Numerical Simulation of a Cylinder in Uniform Flow: Application of a Virtual Boundary Method, J. Comput. Phys., 123 (1996) 450-465.

[17] M. Uhlmann, An immersed boundary method with direct forcing for the simulation of particulate flows, J. Comput. Phys., 209 (2005) 448-476.

[18] S. Xu, Z.J. Wang, An immersed interface method for simulating the interaction of a fluid with moving boundaries, J. Comput. Phys., 216 (2006) 454-493.

[19] J. Mohd-Yusof, Combined Immersed-Boundary and B-spline methods for simulations of flow in complex geometries, Cen. Turb. Res., Annual Research Brief (1997) 317-327.

[20] S. Marella, S. Krishnan, H. Liu, H.S. Udaykumar, Sharp interface Cartesian grid method I: An easily implemented technique for 3D moving boundary computations, J. Comput. Phys., 210 (2005) 1-31.

[21] R. Ghias, R. Mittal, H. Dong, A sharp interface immersed boundary method for compressible viscous flows, J. Comput. Phys., 225 (2007) 528-553.

[22] N. Zhang, Z.C. Zheng, An improved direct-forcing immersed-boundary method for finite difference applications, J. Comput. Phys., 221 (2007) 250-268.

[23] N.G. Deen, S.H.L. Kriebitzsch, M.A. van der Hoef, J.A.M. Kuipers, Direct numerical simulation of flow and heat transfer in dense fluid–particle systems, Chem. Eng. Sci., 81 (2012) 329-344.

[24] N.G. Deen, J.A.M. Kuipers, Direct Numerical Simulation of Fluid Flow and Mass Transfer in Dense Fluid–Particle Systems, Ind. Eng. Chem. Res., 52 (2013) 11266-11274.

[25] Z.-G. Feng, S.G. Musong, Direct numerical simulation of heat and mass transfer of spheres in a fluidized bed, Powder Technol., 262 (2014) 62-70.

[26] N. Zhang, Z.C. Zheng, S. Eckels, Study of heat-transfer on the surface of a circular cylinder in flow using an immersed-boundary method, Int. J. Heat Fluid Flow, 29 (2008) 1558-1566.

[27] H. Tavassoli, S.H.L. Kriebitzsch, M.A. van der Hoef, E.A.J.F. Peters, J.A.M. Kuipers, Direct numerical simulation of particulate flow with heat transfer, Int. J. Multiphase Flow, 57 (2013) 29-37.

[28] J. Xia, K. Luo, J. Fan, A ghost-cell based high-order immersed boundary method for inter-phase heat transfer simulation, Int. J. Heat Mass Transfer, 75 (2014) 302-312.

[29] B. Sun, S. Tenneti, S. Subramaniam, Modeling average gas–solid heat transfer using particle-resolved direct numerical simulation, Int. J. Heat Mass Transfer, 86 (2015) 898-913.

[30] F. Dierich, P.A. Nikrityuk, S. Ananiev, 2D modeling of moving particles with phase-change effect, Chem. Eng. Sci., 66 (2011) 5459-5473.

[31] K.S. Kedia, C. Safta, J. Ray, H.N. Najm, A.F. Ghoniem, A second-order coupled immersed boundary-SAMR construction for chemically reacting flow over a heat-conducting Cartesian grid-conforming solid, J. Comput. Phys., 272 (2014) 408-428.

[32] N.G. Deen, J.A.M. Kuipers, Direct numerical simulation of fluid flow accompanied by coupled mass and heat transfer in dense fluid–particle systems, Chem. Eng. Sci., 116 (2014) 645-656.

[33] A. Abdelsamie, G. Fru, T. Oster, F. Dietzsch, G. Janiga, D. Thévenin, Towards direct numerical simulations of low-Mach number turbulent reacting and two-phase flows using immersed boundaries, Comput. Fluids, 131 (2016) 123-141.

[34] J. Lu, S. Das, E.A.J.F. Peters, J.A.M. Kuipers, Direct numerical simulation of fluid flow and mass transfer in dense fluid-particle systems with surface reactions, Chem. Eng. Sci., 176 (2018) 1-18.
[35] B. Kravets, H. Kruggel-Emden, Investigation of local heat transfer in random particle packings by a fully resolved LBM-approach, Powder Technol., 318 (2017) 293-305.
[36] P. Zehner, E.U. Schlünder, Wärmeleitfähigkeit von Schüttungen bei mäßigen Temperaturen, Chem. Ing. Tech., 42 (1970) 933-941.

FLUIDIZATION CHARACTERISITCS OF COHESIVE POWDER IN A PILOT PLANT SCALE VIBRATED FLUIDIZED BED DRYER AT LOW VIBRATION FREQUENCIES

Sören E. Lehmann, Hamburg University of Technology, Hamburg, Germany
Ernst-Ulrich Hartge, Hamburg University of Technology, Hamburg, Germany
Alfred Jongsma, Tetra Pak CPS, Heerenveen, The Netherlands
Ilse-Marte deLeeuw, Tetra Pak CPS, Heerenveen, The Netherlands
Fredrik Innings, Tetra Pak Processing Systems, Lund, Sweden
Stefan Heinrich, Hamburg University of Technology, Hamburg, Germany

Abstract

Mechanical vibration is often applied in industrial scale fluidized bed dryers in food and pharmaceutical processes to overcome operational problems due to cohesiveness. However, the understanding of the process is still insufficient regarding detailed modelling, apparatus design, up-scaling and process optimization.

Almost all of the experimental research in the field of vibrating fluidized beds is conducted on a lab-scale. Within the current project a pilot plant scale unit with a cross section of 250 x 500 mm² and a total height of 3 m was built. The influence of process parameters, such as gas velocity, vibration intensity, powder moisture content and bed mass, on the fluidization characteristics of whole milk powder was investigated. In order to characterize the lower limit of fluidization from an operational point of view, the velocity of full fluidization (u_{ff}) was introduced. It is defined as the lowest superficial gas velocity at which the entire bed is fully fluidized.

It could be observed that an increase in powder moisture content results in significant increase of u_{ff}. Introduction of mechanical vibration into the bed resulted in the reduction of u_{ff}, expansion of the bed and reduction of bubble volume fraction. This expansion of the suspension phase explains reported increase in heat and mass transfer in vibrated fluidized beds.

Introduction

Drying in fluidized beds is a suitable process for temperature sensitive materials. Due to very good particle mixing and high heat and mass transfer rates, homogenous product properties are achieved and the drying can take place at relatively low temperatures. Hence, fluidized bed drying is often applied in the pharmaceutical and food industry [1]. Many of the products are fine powders (Geldart classifications A and C) [2] or show cohesive behavior due to their chemical composition (i.e. amorphous lactose) [3]. It is well known, that fluidization of cohesive powders is often hindered by the occurrence of channeling (rat holes) and formation of agglomerates in the bed. Various methods are investigated to improve fluidization, such as pulsed air flow, introduction of acoustic or mechanical vibration, magnetic or electric fields, modification of internals (i.e. baffles, agitators) or combinations of those. Aim of all these approaches is the introduction of additional forces into the system which counteract the cohesive forces. Thereby reduction of agglomeration and channeling is achieved. Improved fluidization or in some cases enabling fluidization is the intended result [4].

The only option that is applied today on the industrial scale is mechanical vibration of the fluidized bed. Hence, the focus of this study lies on mechanical vibration of fluidized beds at low frequencies (≤ 10 Hz). The vibration is commonly characterized by the dimensionless vibration number or vibration intensity Λ. It is defined as the ratio of acceleration induced by vibration to gravitational acceleration g. Acceleration due to vibration depends on the frequency f and the amplitude of vibration A_f.

$$\Lambda = \frac{(2\pi f)^2 \cdot A_f}{g} \qquad (1)$$

Vibro-fluidization or vibrated fluidized beds (VFB) have been studied extensively in the past. However, almost all experimental investigations have been conducted in small lab-scale apparatuses and mostly at frequencies between 10 to 50 Hz. Consistently the main effects of vibration on the fluidization of cohesive particles have been reported to reduce agglomerate size and channeling, reduce minimum fluidization velocity and cause an increase in bed expansion [2, 5].

For experiments with cohesive powders in small setups it is very likely that wall effects influence the results significantly. Thus, a pilot-plant-scale VFB dryer was designed and constructed to reduce the impact of wall effects. Another design factor was the capability to operate the test plant at process parameters closely oriented to industrial conditions. Thereby, the gap between lab-scale experiments and industrial application shall be closed. In this work the focus lies on the VFB drying of milk powder. Cruz at al. [6] investigated the drying of whole milk powder in a lab-scale fluidized bed dryer at a very shallow fixed bed height of 2 cm and at a vibration frequency of 60 Hz. Furthermore, their focus lay on drying kinetics and left questions regarding fluidization characteristics unanswered [6]. Before analysis of drying kinetics in VFB dryers, the influence of process parameters (i.e. vibration) on fluidization characteristics of cohesive powders needs to be investigated and understood. Therefore, parameter studies regarding the fluidization characteristics of whole milk powder are conducted and discussed in this paper.

Fluidized beds are distinguished in a bubble phase and a suspension phase. In the suspension phase heat and mass transfer takes place between the particles and the suspension gas. The rest of the fluidization gas bypasses the particles in form of bubbles and has no direct interaction with the solids [7]. In analogy to widely applied correlations used in the modeling of fluidized beds, fluidization characteristics are evaluated by analysis of bubble volume ratio, bed expansion and minimum fluidization velocity (u_{mf}) [8, 9]. However, the determination of u_{mf} for cohesive powders can be challenging. The lower limit of fluidization will be characterized by the lowest gas flow rate that allows complete fluidization.

Lower limit of fluidization

The minimum fluidization velocity u_{mf} is a rather theoretical parameter and can vary significantly from superficial gas velocities that lead to complete fluidization, especially for beds with broad particle size distributions. Only in the idealized case with extremely narrow PSD and spherical particles u_{mf} represents a superficial gas velocity at which the bed is fully fluidized. The cohesiveness of the investigated powder poses an additional challenge in the determination of u_{mf}.

One aim of this work is to identify the influence of process parameters on the lower limit of fluidization from an operational point of view. Thus, reduction of superficial gas velocity from the fully fluidized state until the fixed bed region is investigated. Observations during these experiments show that partial fluidization of the bed can occur below a certain value of superficial gas velocity. Thereby, some parts of the bed are fluidized, some show channeling and others show fixed bed behavior. This characteristic gas velocity marks the lower limit of full fluidization or short: full fluidization velocity u_{ff}. At even lower gas velocities, fluidization is no longer occurring and stable channels form in the fixed bed.

The full fluidization velocity u_{ff} can be determined by plotting the measured pressure drop across the bed against the superficial gas velocity. The point, where the pressure drop begins to deviate from the constant level in the fluidized bed range with decreasing gas velocity is the full fluidization velocity. Figure 1 shows the comparison of determination of u_{mf} and u_{ff} for exemplary cases with idealized, realistic and cohesive bed material. In the case of idealized bed material (same-sized, spherical particles) u_{mf} and u_{ff} are equal. For the realistic case (with broad PSD and non-spherical particles) and the cohesive case (with channeling or partial fluidization in the fixed bed range) u_{ff} can be significantly larger than u_{mf}. The operational parameter u_{ff} will be used in the following for identification of influence of process parameters on the lower limit of fluidization.

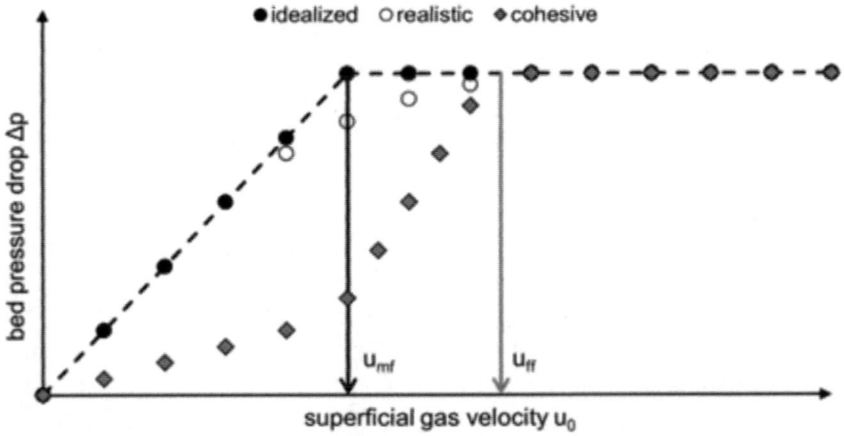

Figure 1: Determination and comparison of minimum fluidization velocity u_{mf} and full fluidization velocity u_{ff} for exemplary cases of idealized bed (same-sized, spherical particles), realistic bed (wider size distribution) and cohesive bed material (with channeling in the fixed bed range).

Materials and Methods

Commercially available whole milk powder is used within the scope of this work. For characterization of the milk powder particle density, size distribution and moisture content are analyzed. Apparent particle density ρ_{app} is measured by dispersion of the sample in a defined volume of Glycerol (≥99.5 %, p.a., anhydrous) and noting change in mass and volume. Untapped bulk density is measured filling a cup of known volume. Mass readings are taken with a balance. Powder moisture content is determined by thermo-gravimetric investigation at 105 °C for 10 minutes in a Precisa EM 120-HR.

A simplified flow chart of the pilot-plant-scale VFB dryer is shown in figure 2. A rotary blower delivers ambient air as fluidization gas. The flow rate is measured with an orifice plate (according to DIN EN ISO 5167). The air is then conditioned to desired temperature and relative humidity in an air conditioning unit consisting of a pre heater, a humidifier and an after heater. After leaving the VFB, the air passes a filter before being released to the environment. In the filter entrained particles are recovered for further investigation.

Comprehensive measurement equipment is installed in the system. Relative humidity and temperature of the process air are measured before the air conditioning, at the entrance and exit of the fluidized bed. Pressure readings are taken at the orifice plate and at different positions in the dryer with relative pressure sensors HXCM from Sensor Technics, Germany. Relative pressure drop is measured across the distributor as well as five height intervals in the fluidized bed and the free board. Furthermore, bed temperature is measured with PT100 at the same height intervals inside the dryer. Data acquisition is conducted with LabView and a PC.

The rectangular drying chamber measures 250 by 500 mm and 2 m in height (from the distributor). The front wall of the fluidization zone is made of transparent glass to allow visual observation, reading of bed heights with a ruler attached to dryer as well as investigations of bubble characteristics via high speed camera (100 fps; with IDT-NX 4-S2 from Imaging Solutions GmbH, Germany). Temperature and pressure probes are located in the back wall. An expansion zone is located on top. During operation material can be fed through an air lock in the expansion zone or samples can be taken for analysis. The vibrating parts (plenum, distributor and drying chamber) are decoupled from the expansion zone by a rubber gaiter. The vibration is brought into the system by two synchronized eccentric motors. The vibration frequency is controlled continuously by a frequency converter. The vibrating parts are suspended by a set of springs which allow free movement in all directions. The direction of vibration is controlled by the orientation of the vibrators (mounted to the bottom of the plenum chamber). With the orientation applied in the scope of this work, strictly vertical vibration is induced. Vibration frequency and amplitude is measured with a pen recorder mounted to the VFB dryer.

Influence of process parameters on the fluidization characteristics of whole milk powder is investigated. Thereby, superficial gas velocity u_0, vibration intensity Λ, bed height (bed mass) and powder moisture content x are varied. Powder moisture content is set by fluidizing the bed with defined relative air humidity until the powder reached equilibrium moisture content. All experiments are conducted at air temperatures around 20 °C to ensure reproducibility.

Characterization of fluidization characteristics is done by investigating the effect varying process parameters on bubble volume fraction, bed expansion respectively bed porosity ε and lower limit of fluidization u_{ff}. The lower limit of fluidization is investigated by measuring the pressure drop across the bed for varying superficial gas velocities. By segmentation of grey scale high speed camera images with ImageJ the bubble volume fraction can be calculated (black pixels represent bubbles, white pixels the suspension phase). As a simplifying approach it is assumed that the glass wall represents a cut through the isometric fluidized bed. Bed expansion is evaluated via the bed porosity ε, calculated based on the bed height h, mass of bed material m, cross sectional area of the bed A and apparent particle density ρ_{app}.

$$\varepsilon = 1 - \frac{m}{A \cdot h \cdot \rho_{app}} \qquad (2)$$

It has to be mentioned that sticking of bed material to the apparatus walls (including the glass wall) occurs. Especially for elevated moisture contents sticking increases to a level, where exact determination of bed height as well as generation of evaluable images of the bubble behavior through the glass wall become impossible. Hence, only limited data for elevated moisture contents for bed expansion and bubble volume fraction is available.

Table 1: Summary of investigated vibration parameters

Frequency [Hz]	f	0	4	6	8	10
Amplitude [mm]	A_f	0	5	4	3.5	3
Intensity [-]	Λ	0	0.32	0.58	0.9	1.21

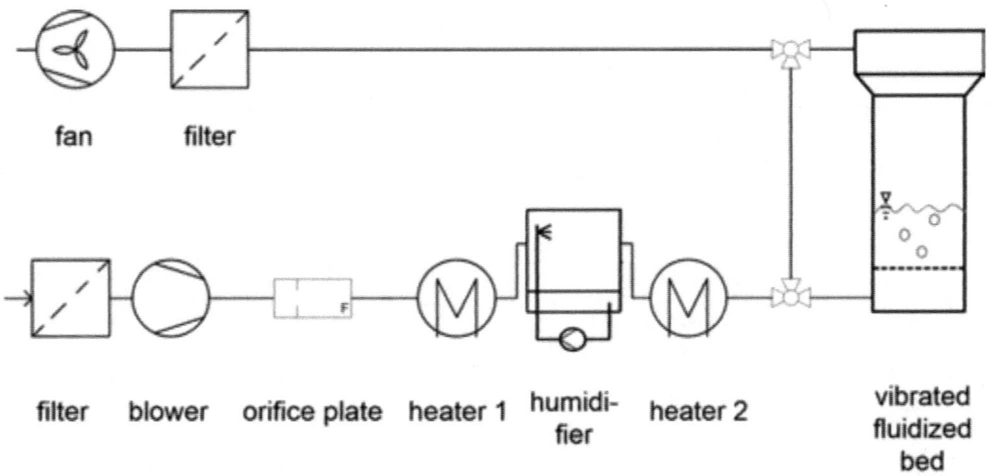

Figure 2: Simplified flow chart of the VFB drying test plant.

Results and Discussion

In total five different vibration intensities are investigated regarding their effect on fluidization characteristics. Parameter combinations are shown in table 1. Furthermore, powder moisture content (4, 6 and 9 wt.-%) as well as initial bed mass (12 and 25 kg, corresponding to a fixed bed height of ~19 resp. ~38 cm) are varied. Some parameter settings were conducted repeatedly (a, b) to investigate reproducibility. In order to maintain constant mass of material in the bed, the superficial gas velocity is varied from fixed bed range up into the fully fluidized state, but below levels with significant entrainment.

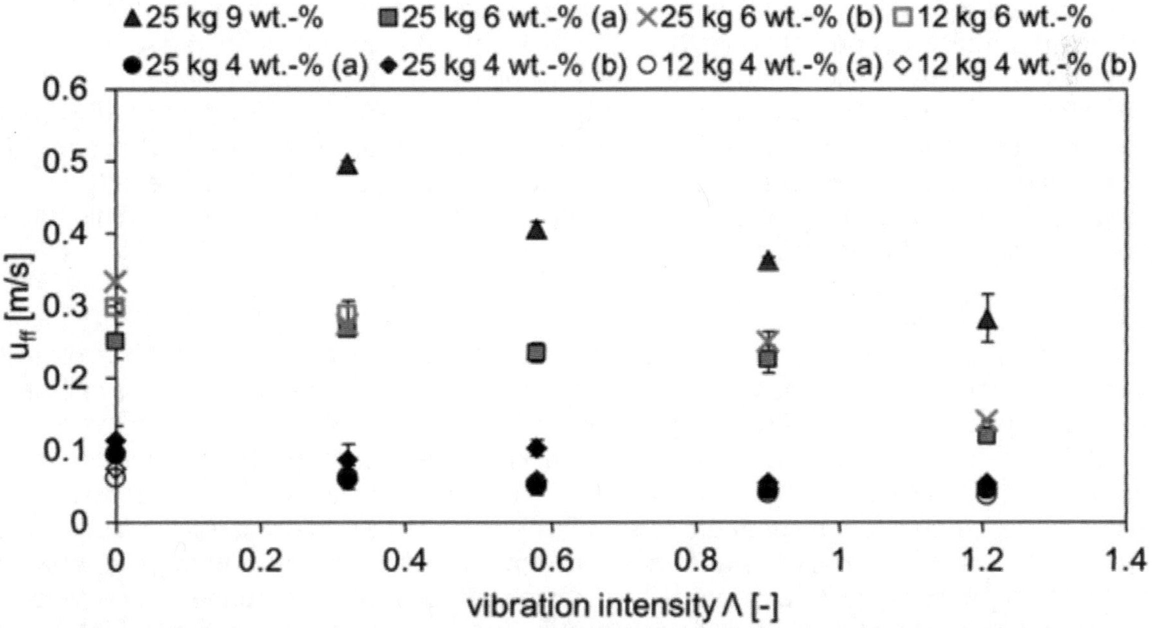

Figure 3: Full fluidization velocity u_{ff} of whole milk powder vs. vibration intensity in dependence of bed mass and powder moisture content as well repetition of measurements at constant process parameters (a and b).

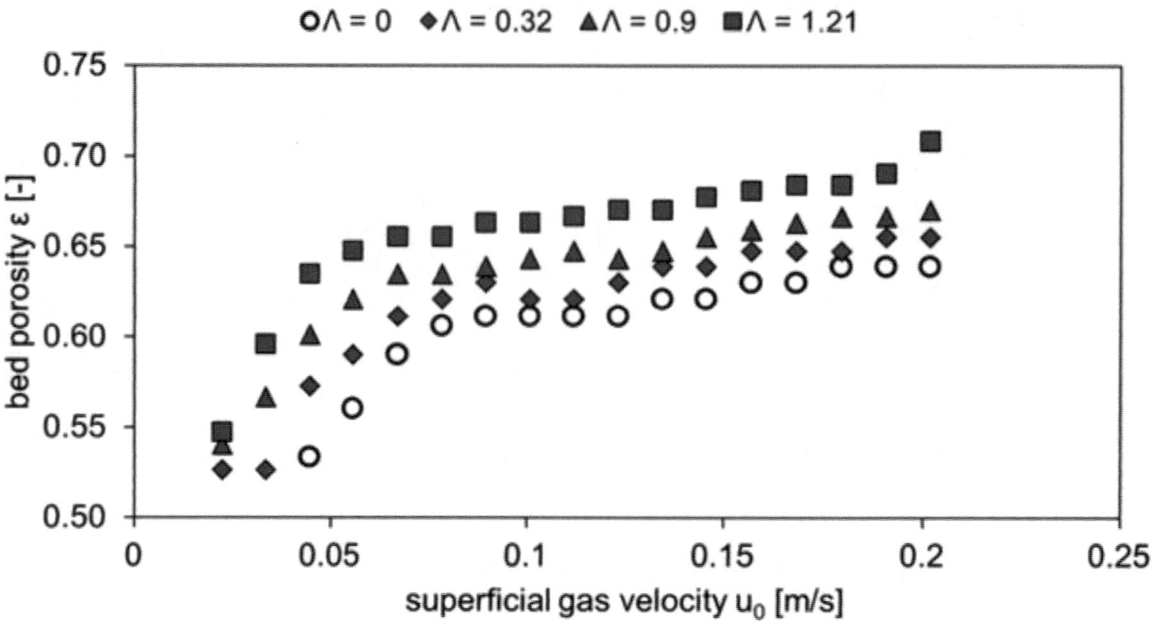

Figure 4: Bed porosity of 25 kg whole milk powder (x = 4 wt.-%) vs. superficial gas velocity in dependence of vibration intensity.

Influence on lower limit of fluidization

Determined values of u_{ff} are plotted in figure 3 against the vibration intensity for varying initial bed mass and powder moisture content as well as repeated experiments under the same conditions. The data from repeated experiments shows very similar results. It can thus be concluded that the results are reproducible. Furthermore, it can be observed that variation of the bed mass within the investigated range has no impact on the lower limit of fluidization.

Powder moisture content however has significant influence on u_{ff}. Increase of moisture content from 4 to 6 wt.-% leads to an increase in u_{ff} by roughly a factor of three. Increase from 4 to 9 wt.-% causes an increase by a factor of six. This holds for all investigated vibration intensities. Additionally, it is shown in figure 3 that introduction of vibration results in decrease of u_{ff}. For higher vibration intensities, u_{ff} is decreasing stronger. Vibration intensity of 1.21 leads to a reduction of u_{ff} by a factor of two, compared to the non-vibrated case. This can be observed for all investigated moisture levels.

Influence on bed expansion

Figure 4 shows the bed porosity plotted against superficial gas velocity for different vibration intensities for initial bed mass of 25 kg and moisture content of 4 wt.-%. In all cases the fixed bed porosity is 0.53. Focusing on the non-vibrated case, the transition range from fixed to fluidized bed is characterized by a strong increase in porosity with increasing superficial gas velocity (from 0.04 to 0.09 m/s). After full fluidization is reached, the bed porosity reaches a constant level of 0.61 (from 0.09 to 0.13 m/s). State of full fluidization was confirmed with visual observations and measurement of u_{ff} via pressure drop (see figure 3). Steady and slight increase of bed porosity is observed for further increase in gas velocity. This qualitative course of bed porosity can be observed for all applied vibration intensities and coincides with fluidization characteristics of Geldart A powders. Furthermore, increasing vibration intensity shows two simultaneous effects. Firstly, a strong influence in bed porosity and thereby bed expansion can be observed. Porosity was increased from 0.61 to 0.66 at the fully fluidized state comparing the non-vibrated case and a vibration intensity of 1.21.

Secondly, the onset and end of the transition from fixed to fully fluidized bed shifts to lower superficial gas velocities. This means a reduction in u_{ff} due to vibration and coincides with the findings described in the previous chapter.

The same trends regarding decrease of lower limit of fluidization and increase in bed expansion due to vibration have been reported by Mawatari et al. [2] and other researchers for other materials than milk powder [5]. Thus, the generally reported impact of vibration on the fluidization of cohesive powders could be confirmed and quantified for whole milk powder in a pilot-plant-scale VFB dryer at low vibration frequencies.

Influence on bubble volume fraction

Influence of vibration on the bubble volume fraction for different superficial gas velocities is shown in figure 5. Bubble volume fraction is determined as the mean number of black pixels from all images of a series taken at constant process parameters. Error bars indicate the standard deviation through each image series. A clear influence on bubble volume fraction due to superficial gas velocity cannot be identified for the investigated range. However, a slight decrease in bubble volume fraction with increasing vibration intensity can be observed. Up- and downwards movement of the bed material due to vibration apparently reduces the bubble gas flow.

This is particularly interesting, regarding the finding that vibration leads to significant increase in bed expansion. In combination this leads to the conclusion that vibration results in significant expansion in the suspension phase. Considering the widely accepted assumption that heat and mass transfer between particle and gas only take place in the suspension phase, this would explain the reported enhancement of drying kinetics in vibrated fluidized beds.

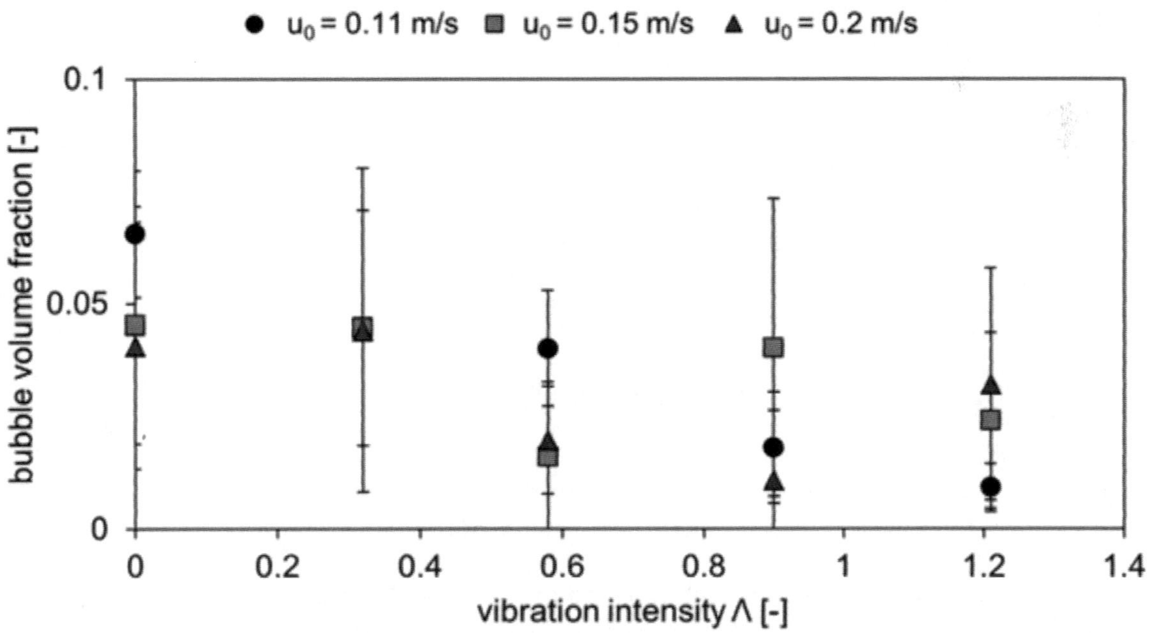

Figure 5: Bubble volume fraction of 12 kg whole milk powder (x = 4 wt.-%) vs. vibration intensity in dependence of superficial gas velocity.

Conclusion

A pilot-plant-scale vibrated fluidized bed dryer was designed and constructed to conduct parameter studies on fluidization behavior and drying of cohesive powders (i.e. milk powder) at low vibration frequencies. Process parameters such as superficial gas velocity, vibration frequency and amplitude, powder moisture content and bed height were investigated with regard to their influence on the fluidization characteristics of whole milk powder. For characterization of the lower limit of fluidization from an operational point of view, the full fluidization velocity u_{ff} was introduced. It was shown that increasing powder moisture content leads to strong increase in u_{ff}. Increase in powder moisture content from 4 to 9 wt.-% resulted in an increase of u_{ff} by a factor of six. Mechanical vibration was found to reduce u_{ff}. A vibration intensity of 1.21 resulted in a reduction of u_{ff} by 50 % compared to non-vibrated fluidization. Strong bed expansion due to vibration could be observed, while bubble volume fraction decreases slightly with increasing vibration intensity. Consequently, expansion takes place in the suspension phase and would explain improved heat and mass transfer in vibrated fluidized beds.

List of Symbols

A	cross sectional area of fluidized bed
A_f	amplitude of vibration
ϵ	bed porosity
u_0	superficial gas velocity
u_{ff}	full fluidization velocity
u_{mf}	minimum fluidization velocity
f	frequency of vibration
g	gravitational acceleration
h	bed height
Λ	vibration intensity
ρ_{app}	apparent particle density

References

1 Daud, W. R. W. (2008), "Fluidized Bed Dryers — Recent Advances," *Advanced Powder Technology*, 19, pp. 403–418.

2 Mawatari, Y.; Hamada, Y.; Yamamura, M. and Kage, H. (2015), "Flow Pattern Transition of Fine Cohesive Powders in a Gas-Solid Fluidized Bed under Mechanical Vibrating Conditions," *Procedia Engineering*, 102, pp. 945–951.

3 Palzer, S. (2005), "The effect of glass transition on the desired and undesired agglomeration of amorphous food powders," *Chemical Engineering Science*, 60, pp. 3959–3968.

4 van Ommen, J. R. (2009), "Reshaping the Structure of Fluidized Beds," *CEP*, pp. 49–57.

5 Xu, C. and Zhu, J. (2006), "Parametric study of fine particle fluidization under mechanical vibration," *Powder Technology*, 161, pp. 135–144.

6 Cruz, M. A. A.; Passos, M. L. and Ferreira, W. R. (2005), "Final Drying of Whole Milk Powder in Vibrated-Fluidized Beds," *Drying Technology*, 23, pp. 2021–2037.

7 Burgschweiger, J. and Tsotsas, E. (2002), "Experimental investigation and modelling of continuous fluidized bed drying under steady-state and dynamic conditions," *Chemical Engineering Science*, 57, pp. 5021–5038.

8 Hilligardt, K. and Werther, J. (1986), "Local bubble gas hold-up and expansion of gas-solid fluidized beds," *German Chemical Engineering*, 9, pp. 215–221.

9 Hilligardt, K. and Werther, J. (1987), "The influence of temperature and solids properties on the size and growth of bubbles in gas fluidized beds," *Chemical Engineering & Technology*, 10, pp. 272–280.

508019 Scale up Practices for Today's Breakthrough Technologies

Abstract:

Fluidized bed reactors have distinct advantages over other reactor unit operations such as superior heat transfer and solids mobility. While the superior heat transfer has sparked breakthrough technologies in processes such as gasification, acrylonitrile and oxychlorination, it is the solids mobility benefits that are being exploited today. This mobility allows solids or sets of solids to move from reduction to oxidation zones in a continuous fashion thereby reducing operational complexity and capital costs. As a result, processes once considered impossible are being implemented where particulate feeds stocks and media are combined and separated in fluidized bed unit operations to provide superior yields unattainable on a commercial-scale with other reactor designs.

Encina Chemicals has capitalized on this solids mobility benefit for a coal upgrading process. Along with coal upgrading, the process provides the continuous regeneration of heat transfer media together with the separation of ash byproduct. The effect of complex economic drivers and resulting design criteria on required fluidization hydrodynamics are managed by combining cold-flow hydrodynamic testing with CFD modeling insights. This enables rapid hypothesis testing and scale-up practices with faster time to commercialization of an optimum design. This paper discusses the approach used to combine cold-flow hydrodynamics research with CFD simulation for rapid process commercialization considering relevant economic and technological drivers.

509649 Study of FCC Riser Hydrodynamics Using CFD Tools

Abstract:

Past experiences of commercial FCC units indicate that hydrodynamics have a pronounced effect on the riser reactor performance. A study of FCC riser hydrodynamics was undertaken to understand the phenomena observed in the field. It encompassed computational fluid dynamics (CFD), cold flow modeling, radioactive tracer studies, kinetic modeling and yield estimation. The ultimate goal of the study was to isolate the effects of riser diameter, lift steam velocity, and solids flux on the hydrodynamic behavior and yield performance of FCC risers.

Results of the cold flow studies were used to select and validate the best CFD modeling approach. Predictions of several models developed within the Eulerian-Eulerian and Eulerian-Lagrangian frameworks were compared with the experimental solids flux profiles in the cold riser operating at commercially-relevant velocities. Both model classes showed good agreement with experiment in the "normal" riser operation. The Eulerian-Lagrangian approach showed better predictions of experiment when the riser was "pushed" in terms of velocity and solids flux.

The general trends of riser hydrodynamics, per CFD predictions, were identified as follows. Lower vapor velocity, larger riser diameter and higher solids flux through the riser all contribute to hydrodynamic non-idealities, such as catalyst downflow at the wall and vapor channeling through the core of the riser. In practice, these "non-idealities" lead to undesirable shifts in conversion (under-conversion in the core) and selectivity (overcracking in the annulus) of the riser.

508584 Investigation of Particle Sheeting of Monodisperse Particles with an Euler-Euler Two-Fluid Model Incorporating Electrostatic Charging Effects

Abstract:

Electrostatic charging of particles is a disruptive phenomenon commonly encountered in gas-solid fluidization. In commercial applications such as catalytic polymerization of polyethylene, such events lead to particle aggregation as well as "sheeting" on the reactor wall. The sheets lead to halting the production process for reactor clean up and in turn results in large revenue losses.

Computational Fluid Dynamic (CFD) modeling has received interest over the past few years to predict the occurrence and magnitude of charging in gas-solid fluidized beds. Rokkam et al. (2013) had carried out CFD simulations using fixed particle charge densities to demonstrate the effect of the electrostatic forces in influencing the hydrodynamics of fluidization and wall sheeting. The model used three types of particles, namely wall, dropped and fines. The size and charge density of each of these types was based on the average values of these variables as observed in the fluidization experiments carried out by Sowinski (2012). It was shown that at the end of fluidization, a small layer of highly charged particles sticks to the wall in a similar way to that observed in experiments. The goal of the present work is the extension of Rokkam's work to include a model for charge generation and transport. To this purpose, two phenomena need to be described by the model: charging due to particle-wall collisions, and charging due to particle-particle collisions. The model for particle-wall charging hypothesized by Matsusaka et al. (2000) and Matsusaka and Masuda (2003) was used to obtain a boundary condition describing particle charging due to contact with the wall. The methodology based on the kinetic theory of the granular flow described by Johnson and Jackson (1987) was used to derive this boundary condition.

The charging model of Matsusaka et al. (2000) was used to describe charging due to particle-particle collisions by determining the rate of change of charge due to collisions through the solution of the appropriate collision integral, as defined in accordance to the kinetic theory for monodisperse granular spheres (Jenkins and Savage, 1983). Consequently, expressions for the charge transport coefficient due to particle-particle collisions and charge flux due to particle-wall collisions are found. The model was implemented into an Euler-Euler two-fluid model with kinetic theory closures (Jenkins and Savage, 1983) to describe the properties of the particulate flow. The resulting model was applied to the simulation of gas-solid fluidized beds.

The domain size and initial conditions used to perform the numerical simulations reproduce the experimental conditions of Sowinski (2012). The fluidized bed was 0.1 m in diameter and the particles had a diameter of 300 μm. The fluidization velocity was varied between 1.5 and 4 times the minimum fluidization velocity, encompassing both bubbling and slug flow regimes. For the electrostatic properties, work function values of 4.4 eV and 5.6 eV (Masuda et al., 2006) were used for the reactor wall (carbon steel) and polyethylene particles respectively. A saturation charge density of 117 μC/kg, based on experimental observations was used. The bed was fluidized until a steady state of electrification was observed. The gas flow was then interrupted, and particles were allowed to settle at the bottom of the bed. Results of the simulations showed that the profile of charge density had its highest value near the wall, and decreased towards the center of the fluidized bed. A force balance was used to determine regions where particles would stick to the wall if the distributor plate was removed and the particles be allowed to fall. The radial component of the electrostatic force predicted by the model was significantly larger than its axial counterpart, and its magnitude also declined rapidly with distance from the wall. It was determined that the particles within a 2 mm thickness layer on the wall would not drop, in agreement with experimental observation in Mehrani's group. The average charge density within this layer was 47 μC/kg while it was 70 μC/kg for a thickness layer of 0.5 mm on the wall. The results are in agreement with the experimental observations obtained by Sowinski (2012).

References

Jenkins, J.T., Savage, S.B., 1983. A theory for the rapid flow of identical, smooth, nearly elastic, spherical particles. Journal of Fluid Mechanics 130, 187. doi:10.1017/S0022112083001044

Johnson, P.C., Jackson, R., 1987. Frictional–collisional constitutive relations for granular materials, with application to plane shearing. Journal of Fluid Mechanics 176, 67. doi:10.1017/S0022112087000570

Masuda, H., Higashitani, K., Yoshida, H. (Eds.), 2006. Powder technology handbook, 3rd ed. Taylor & Francis

Matsusaka, S., Ghadiri, M., Masuda, H., 2000. Electrification of an elastic sphere by repeated impacts on a metal plate. Journal of Physics D: Applied Physics 33, 2311–2319

Matsusaka, S., Masuda, H., 2003. Electrostatics of particles. Advanced Powder Technology 14, 143–166. doi:10.1163/156855203763593958

Rokkam, R.G., Sowinski, A., Fox, R.O., Mehrani, P., Muhle, M.E., 2013. Computational and experimental study of electrostatics in gas–solid polymerization fluidized beds. Chemical Engineering Science 92, 146–156. doi:10.1016/j.ces.2013.01.023

Sowinski, A., 2012. The study of reactor wall fouling in gas-solid fluidized beds caused by electrostatic charge generation (Thesis dissertation). University of Ottawa

506383 Large-Scale DEM-CFD Method for an Industrial Fluidized Bed

Abstract:

Numerical simulations of fluidized bed are often performed for the better understanding of complex phenomena. In these numerical simulations, Eulerian-Lagrangian approach [1] is often employed, where solid and fluid phases are modeled by discrete element method (DEM) and computational fluid dynamics (CFD), respectively. Indeed, the DEM-CFD method is applied to various solid-fluid coupling problems. Adequacy of the DEM-CFD method is proven not only in a gas-solid flow system such as a fluidized bed [2,3] but also a wet ball mill [4,5]. Although the DEM-CFD method seems to be established, there is a couple of critical problems such as substantial limit of number of calculated particles and difficulty for stable computation for a gas-solid flow in a complex-shaped boundary. In order to solve these problems, we develop a new methodology that is referred to as DEM-CGM-CDF-SDF-IBM method. In the DEM-CGM-CFD-SDF-IBM method, signed distance function (SDF) [6] and immersed boundary model [7] are used for boundary wall of solid and fluid phases. The coarse grain model (CGM) [8] of the DEM is employed as the scaling law model, where a model particle represents a group of original particles. To show adequacy of the DEM-CGM-CDF-SDF-IBM method, a validation test is performed in a fluidized bed. In the simulation, 200,000 spherical particles whose density, original particle diameter and coarse grain ratio are, respectively, 2500 kg/m^3, 0.1 mm and 5. Experiments are performed under fair conditions of the numerical simulations. It is shown that the DEM-CGM-CDF-SDF-IBM method can simulate the bed height and pressure drop obtained from the oroginal system accurately through the validation test. Consequently, we conclude that the DEM-CGM-CDF-SDF-IBM method can simulate industrial scale fluidized bed, and hope that our method becomes standard in numerical simulation method in industrial systems.

References

[1] Y. Tsuji, T. Kawaguchi, T. Tanaka, Powder Technol., 77 79â€“87 (1993)

[2] M. Sakai, H. Takahashi, C.C. Pain, J-P Latham, J. Xiang, Adv. Powder Technol., 23, 673-681 (2012)

[3] M. Sakai, M. Abe, Y. Shigeto, S. Mizutani, H. Takahashi, A. Vire, J.R. Percival, J. Xiang, C.C. Pain, Chem. Eng. J., 244, 33-43 (2014)

[4] X. Sun, M. Sakai, Y. Yamada, J. Comput. Phys., 248, 147-176 (2013)

[5] X. Sun, M. Sakai, M-T. Sakai, Y. Yamada, Chem. Eng. J., 246, 122-141 (2014)

[6] Y. Shigeto, M. Sakai, Chem. Eng. J., 231, 464-476 (2013)

[7] X. Sun, M. Sakai, Chem. Eng. Sci., 139, 221-240 (2016)

[8] M. Sakai and S. Koshizuka, Chem. Eng. Sci., 64, 533-539 (2009)

513870 Predicting Performance of Industrial FCC Regenerator Using Computational Fluid Dynamics

Abstract:

In the Fluid Catalytic Cracking (FCC) process, uniform regeneration of spent catalyst is key to the successful operation of the FCC units. The degree and uniformity of catalyst regeneration has a substantial impact on the product yields. The regenerator combustion efficiency strongly depends on good distribution of both air and spent catalyst, and air catalyst mixing and flow patterns within the regenerator are greatly influenced by both process conditions as well as regenerator hardware configuration. Therefore, understanding the regenerator hydrodynamics and kinetics, and the impact of hardware modifications on regenerator performance is essential for efficient regenerator design, optimization and troubleshooting.

In this work, a CFD based kinetic tool has been developed to simultaneously model both hydrodynamics and coke combustion characteristics in the FCC regenerator. The tool enables us to evaluate the influence of the fluidized bed hydrodynamic on the catalyst regeneration in an industrial FCC regenerator, and serves as a useful tool for future technology development. TechnipFMC employed Barracuda CFD software and coupled the code with proprietary coke combustion kinetic equations and rate constants, developed experimentally at IFP Energyies nouvelles (IFPEN). TechnipFMC and IFPEN have tested and compared the model against several commercial regenerator designs. This study also provides an example of how the CFD based kinetic tool is used in the modeling of an industrial FCC regenerator. The results provide detailed qualitative mapping of the regenerator behavior in terms of flue gas composition, temperature distribution including afterburning, and regenerator bed density in both axial and radial planes. The results proved to be instrumental to recommend hardware and operational modifications.

508514 CFD Analysis of CO_2 Removal Using an Amine-Based Solid Sorbent

Abstract:

Amine sorbents have been used for scrubbing of carbon dioxide (CO_2) from a high-volume flue gas stream. In addition, reasonably high capacity of amine-based solid sorbents also has been shown experimentally by National Energy Technology Lab (NETL) using an integrated adsorption and regeneration system of fluidized beds. In this study, we use amine-based solid sorbents to perform three and two dimensional CO_2 capture CFD simulations using a bubbling bed and the riser section of a circulation fluidized bed systems. We also compare our simulation results with NETL experiments. Simulations are performed at three different temperatures, to consider the relation of rate of reaction and CO_2 capture capacity of sorbents with temperature variation.

The Eulerian-Eulerian two fluid model using kinetic theory approach is used to perform our numerical simulations. For dense flow of solids, frictional stress and viscosity are also considered. In addition, the spices, momentum and energy conservation equations are solved to calculate the rate and extend of CO_2 sorption in the system.

507726 Direct Numerical Simulation of Flow and Mass Transfer in Complex Geometries Using a Coupled Volume of Fluid and Immersed Boundary Method

Abstract:

Currently innovative solutions are being sought to enhance the perception of specific tastes by delivering the markers more effectively, without affecting the taste intensity and the consumer preference, for example by creating regions with different concentrations of the tastant in a product. However, solutions for the implementation of a heterogeneous taste distribution are still lacking today, especially in beverage products. Dissolution and distribution of tastants in liquids poses a complex multiphase flow problem. With an exponential increase in the computational power over last few decades, and through a combination of different simulation techniques, Computational Fluid Dynamics (CFD) has been emerging as an effective tool for obtaining in-depth knowledge of the involved scale and phase effects in multiphase flows. The goal of present work is to use a multiscale modeling approach to investigate flow and mixing dynamics of beverages in a cylindrical cup.

In this work, a fully resolved direct numerical simulation of flow and mass transfer are coupled for three-dimensional complex geometries on a non-body conformal Cartesian computational domain. A direct forcing implicit, second-order accurate immersed boundary method (IBM) (Deen et al., 2012) is used to resolve the coupling between fluid phase and solid wall. The fluid-fluid interface is tracked by a mass conservative sharp interface volume of fluid (VOF) method (Baltussen et al., 2017). Moreover, the fluid-fluid interfaces in contact with solid boundaries leads to a three-phase contact line, in addition to the interface motion. An apparent contact angle is imposed as a boundary condition at the three-phase contact line (Patel et al., 2017). A second-order implicit (semi-implicit) method is employed to incorporate the species conservation equation of the fluid in a structured grid. Neumann boundary conditions are incorporated at the fluid-wall interface for mass transfer, while Dirichlet boundary conditions are applied for the coupling between fluid-particle interfaces.

The proposed approach is first applied to simulate the mass transfer of static flow in a complex geometry. The simulation results are validated against corresponding analytical solutions. Furthermore, present model is used to investigate the influence of the tilting motion of the geometry on the resulting mass transfer in the fluid.

Keywords: Direct Numerical Simulation (DNS), Volume of Fluid (VOF), Immersed Boundary Method (IBM), contact line dynamics, mass transfer, tilting

References:

1. Deen, N.G., Kriebitzsch, S.H., van der Hoef, M.A. and Kuipers, J.A.M., Direct numerical simulation of flow and heat transfer in dense fluidâ€"particle systems. Chemical Engineering Science, 81 (2012): 329-344.

2. Baltussen, M. W., Segers, Q. I. E., Kuipers, J. A. M., and Deen, N. G., Cutting bubbles with a single wire. Chemical Engineering Science 157 (2017): 138-146.

3. Patel, H.V., Das, S., Kuipers, J.A.M., Padding, J.T. and Peters, E.A.J.F., A coupled Volume of Fluid and Immersed Boundary Method for simulating 3D multiphase flows with contact line dynamics in complex geometries. Chemical Engineering Science, 166 (2017): 28-41.

508602 Computational Study on Biomass Fast Pyrolysis: Hydrodynamic Effects in a Laboratory-Scale Fluidized Bed

Abstract:

Fast pyrolysis is a leading candidate process for converting biomass to liquid fuels and high value chemicals. During fast pyrolysis in bubbling bed or circulating bed reactors, biomass particles are rapidly heated through contacting with hot gases and solids, and their constituent components decompose into volatiles, ash, and char. The product vapor/gas composition, which determines the yield of fuel and/or chemical feedstock-compatible molecules, is highly dependent on the bubbling intensity which promotes heat and mass transfer and also controls the particle residence time. As with other energy-intensive processes, accurate characterization of biomass pyrolysis reactors at smaller lab and pilot scales is a vital first step in developing successful commercial versions.

In this study, we simulate a lab-scale bubbling fluidized bed biomass fast pyrolysis reactor to explore the effects of hydrodynamics on biomass conversion as the gas flow is increased through the bubbling-to-slugging transition, with all the other operating variables held constant. We employ a 3D implementation of MFiX, which is an open-source software package supported by DOE that utilizes a continuum (two-fluid) approach for modeling the fluidzed reactor hydrodynamics. Bubbling intensity and dynamics and the resulting impact on pressure measurements have been reported for a similar reactor set up previously[http://dx.doi.org/10.1016/j.cej.2016.08.113].

We compare mixing and elutriation predictions from our computational model to experimental studies in the literature, as well as experimental pyrolysis yield measurements when reaction chemistry is coupled to the hydrodynamics. We also discuss implications from the results of this study for future numerical and experimental studies of similar reactors.

509429 Kinetics Study and Its Implementation in a First Principles Model for the Riser of a Resid Fluid Catalytic Cracking Unit

Abstract:

Fluid catalytic cracking is one of the most important conversion processes in modern refining operations, converting heavy petroleum feedstock into valuable lighter products like LPG and cracked naphtha, and is a major contributor to the commercial gasoline market. However, coke is also deposited on the catalysts as a considerable by-product of cracking reactions, causing rapid catalyst deactivation. The spent catalysts are transported to a regenerating section, where the activity is restored, and the hot catalysts transported back to the reactor section, thereby also supplying the heat requirement to the otherwise endothermic cracking process. When a high CCR (Conradson Carbon Residue) atmospheric residue is introduced as feed to the catalytic cracking unit, high amounts of coke are deposited on the spent catalyst. In such a case, the catalyst regeneration by coke combustion needs to be conducted in a 2-stage regenerator to avoid hydrothermal deactivation of the catalyst in the regenerator.

Since the RFCC (Resid Fluid Catalytic Cracking) unit is the heart of the refinery, optimization of the operating conditions to maximize the production of valuable products can generate significant additional revenue for the refinery. This necessitates availability of robust predictive models, developed from first principles for the two risers and the two regenerators, and effectively integrating them.

Several experimental runs using an equilibrated commercial catalyst were conducted in a bench-scale, bubbling fluidized bed, multi receiver ACE unit, and kinetics of the RFCC process describing the cracking reactions and coke formation derived using 10 product lumps. Experiments were designed to accurately estimate the kinetics of the intermediate steps. To achieve this, intermediate liquid products, were also injected as feed in some experiments. The two-phase fluidization model is composed of a two-phase bed section, with a dense phase modelled as a CSTR and a bubble phase modelled as a PFR, and a freeboard section modelled as a PFR. The extracted kinetics has been further utilized for developing a detailed reactor model for the riser of the RFCC unit, which was modeled as a fast fluidized bed reactor. It considers the heat and mass transfer between the gas and solids, and additional hydrodynamic parameters (e.g. slip velocity) have been captured in detail.

The hydrodynamics considerations and reactor model details for both the ACE unit and the commercial riser will be presented in detail in the presentation. The models have been executed using an Equation-Oriented modeling platform gPROMS, with properties of the different lumps estimated simultaneously during model solution using Multiflash property estimation package which enhances the fidelity of the model solutions. Parametric runs were executed for varying riser temperature, cat-to-oil ratio, catalyst circulation rate, and the metal and additive contents in the equilibrium catalyst, to examine their effects on selectivities to the valuable products and coke formation on the catalyst. This reactor model, currently developed using design data, will be further tuned with plant operating data, and will be utilized to account for the pressure balance and heat balance of the unit, which are key for proper operation of the unit, as well as for operational optimization of the RFCC unit.

507243 The Influence of Meso-Scale Flow Structure on Transport Phenomena in a Circulating Fluidized Bed

Abstract:

A circulating fluidized bed (CFB) has been applied widely in the industrial process due to homogeneous mass and heat transfer. However, due to lacking of clear image of understanding the influence of meso-scale flow structure, up to now, directly scaling up this process based on the theoretical model is still a challenge for chemical engineering. In the present study, the heterogeneous flow structure in circulating fluidized bed is firstly divided into three homogeneous theoretical structures, for example, the dispersed phase, the cluster phase and the inter-phase phase. And thus, in these homogeneous structures, momentum transfer, mass transfer and heat transfer coefficient can be evaluated by the equations from the experimental or theoretical results based on homogeneous assumptions.

Firstlyï¼Œ the parameters characterizing heterogeneous flow structure for the dispersed phase, the cluster phase and the inter-phase phase should be solved based on the models, for example, EMMS and some semi-empirical equations from fitting experimental data in literatures. And thus, the momentum transfer coefficient between gas and solid for TFM model can be evaluated by solving the force balance equations among three homogeneous phases. Mass and heat transfer coefficient between gas and solid can be obtained by respectively solving the mass and energy conservation equations in the dispersed phase and the cluster phase and then summing up it. As a result, the relationship between meso-scale flow structure and transport coefficient in CFB can be obtained.

Two flow model (TFM) has been widely used to simulate the fluidized bed. However, due to the limitation of computerâ€™s ability, coarse calculating grid has to be used to simulate the industrial fluidized bed process. And thus, there are still heterogeneous flow structures in each calculating grid with employing it to simulate CFB. Therefore, in order to obtain the reasonably predicted results through TFM theory, which should be improved by using transport coefficients considering the influence of meso-scale flow structure in each calculating grid.

In our study, TFM model improved by considering the relationship between heterogeneous flow structure and transport coefficient is used to simulate circulating fluidized bed. Momentum and heat transfer for TFM has been verified by comparing for the axial or radial distribution of viodage and temperature between theoretical results and experimental data in literature. Regarding to mass transfer, the relative experiments were performed in a circulating fluidized bed with ID 50mm and Length 4m by using the catalytic oxidization of carbon monoxide as a model reaction. The experimental results are analyzed by the Muti Scale Mass Transfer (MSMT). And there are good agreements between experimental data and simulating results for the axial distribution of carbon monoxide.

Keywords: Circulating fluidized bed, Transport coefficient, Meso-scale, Cluster phase.

AUTHOR INDEX

AUTHOR INDEX

AUTHOR INDEX

AUTHOR INDEX

AUTHOR INDEX

AIChE
120 Wall Street, FL 23
New York, NY 10005-4020

ISBN 978-1-5108-6979-0

CURRAN ASSOCIATES INC.
proceedings.com